# Science Networks. Historical Studies

Birkhäuser

Science Networks. Historical Studies
Founded by Erwin Hiebert and Hans Wußing
Volume 53

---

Edited by Eberhard Knobloch, Helge Kragh and Volker Remmert

Editorial Board:

K. Andersen, Amsterdam
H.J.M. Bos, Amsterdam
U. Bottazzini, Roma
J.Z. Buchwald, Pasadena
K. Chemla, Paris
S.S. Demidov, Moskva
M. Folkerts, München
P. Galison, Cambridge, Mass.
J. Gray, Milton Keynes
R. Halleux, Liége
S. Hildebrandt, Bonn
D. Kormos Buchwald, Pasadena
Ch. Meinel, Regensburg
J. Peiffer, Paris
W. Purkert, Bonn
D. Rowe, Mainz
Ch. Sasaki, Tokyo
R.H. Stuewer, Minneapol
V.P. Vizgin, Moskva

More information about this series at
http://www.birkhauser-science.com/series/4883

Kathleen Clark

# Jost Bürgi's Aritmetische und Geometrische Progreß Tabulen (1620)

## Edition and Commentary

Birkhäuser

Kathleen Clark
School of Teacher Education
The Florida State University
Tallahassee, FL, USA

Videos can also be accessed at http://link.springer.com/book/10.1007/978-1-4939-3161-3

ISSN 1421-6329 ISSN 2296-6080 (electronic)
Science Networks. Historical Studies
ISBN 978-1-4939-7991-2 ISBN 978-1-4939-3161-3 (eBook)
DOI 10.1007/978-1-4939-3161-3

Mathematics Subject Classification (2010): 01-02, 01-08, 01A45, 03-03

Springer New York Heidelberg Dordrecht London
© Springer Science+Business Media New York 2015
Softcover re-print of the Hardcover 1st edition 2015
This work is subject to copyright. All rights are reserved by the Publisher, whether the whole or part of the material is concerned, specifically the rights of translation, reprinting, reuse of illustrations, recitation, broadcasting, reproduction on microfilms or in any other physical way, and transmission or information storage and retrieval, electronic adaptation, computer software, or by similar or dissimilar methodology now known or hereafter developed.
The use of general descriptive names, registered names, trademarks, service marks, etc. in this publication does not imply, even in the absence of a specific statement, that such names are exempt from the relevant protective laws and regulations and therefore free for general use.
The publisher, the authors and the editors are safe to assume that the advice and information in this book are believed to be true and accurate at the date of publication. Neither the publisher nor the authors or the editors give a warranty, express or implied, with respect to the material contained herein or for any errors or omissions that may have been made.

Printed on acid-free paper

Springer Science+Business Media LLC New York is part of Springer Science+Business Media (www.springer.com)

# Preface

The primary aim for writing this book was simple: to provide an edition and English translation of Jost Bürgi's *Aritmetische vnd Geometrische Progreß Tabulen/sambt gründlichem unterricht/wie solche nützlich in allerley Rechnungen zu gebrauchen/ vnd verstanden werden sol*[1] (1620). To clarify, when I refer to the *Aritmetische und Geometrische Progreß Tabulen* (the abbreviated title will be used hereafter), I mean the manuscript that contains both Bürgi's tables, which were printed with title page, and 23 pages of handwritten text (a 2-page foreword and 21 pages of "instruction" for how to use the tables). There are precious few copies of the *Aritmetische und Geometrische Progreß Tabulen* (and even fewer that contain the handwritten foreword and "instruction"), and the copy that was used to write this book is held in the Department of Special Collections of the Library of the Karl-Franzens-University Graz, in Graz, Austria.

This book is organized into the following chapters. Chapter 1 contains biographical and contextual content to familiarize readers for whom Bürgi is relatively unknown. Several biographies of Bürgi exist, which range from quite brief (e.g., the entry by Nový that appears in the *Dictionary of Scientific Biography*) to book length (e.g., Staudacher's recent book (in its second edition, with a third edition planned), in German and published in 2014). In Chapter 1, I provide enough detail about Bürgi's life and mathematical contributions in order to introduce the reader to a broader story than is typically provided in survey of history of mathematics textbooks. Thus, a secondary aim of this book is to offer readers the opportunity to examine Bürgi's role in the development of what John Wallis identified as one of "two developments that had greatly eased the labour of calculation" (Wallis 1685, pp. 22–23)[2] and to highlight an accurate telling of Bürgi's mathematical prowess that has not previously appeared in English.

---

[1] *Arithmetic and Geometric Progression Tables/together with detailed instruction/how to use these in all sorts of useful calculations/and how they should be understood.*

[2] Wallis identified the two developments as the introduction of decimal fractions by Simon Stevin in 1585 and the invention of logarithms by John Napier in 1614 (Stedall 2008, p. 34).

By way of "full disclosure"—and with his permission—I have heavily drawn upon Fritz Staudacher's lovely book, *Jost Bürgi, Kepler und der Kaiser* (2014), for the purpose of providing a fluid timeline of Bürgi's life. Also, I chose to rely more on Staudacher's text than that of Ludwig Oechslin (*Jost Bürgi*, 2001; also only in German), since Oechslin concentrated more on Bürgi's mechanics, astronomy, and horology.

Chapter 2 provides brief descriptions for the known copies of the *Aritmetische und Geometrische Progreß Tabulen*, e.g., those that are printed (tables only) and those that include the "Kurzer Bericht" (printed tables and handwritten instructions), as well as a detailed description of the copy that is the focus of this book and which is located in the Department of Special Collections of the Library of the Karl-Franzens-University Graz, in Graz, Austria.

Chapter 3 begins with an orientation to the chapter and a few comments for reading the transcription and translation. Then, the complete facsimile of Bürgi's *Aritmetische und Geometrische Progreß Tabulen* (i.e., its title page and the text of the foreword and instruction for use of the tables) is given.[3] This facsimile is also available for download from www.springer.com/us/book/9781493931606. Next, I provide a corresponding transcription, as it was written, in order to preserve the original text (including errors and idiosyncrasies), as well as Bürgi's tone and style. Alongside this transcription, I also include a transcription of the Gdańsk (Poland)[4] manuscript, which is the copy used by Hermann Gieswald in his 1856 edition, so that readers may conveniently and closely examine the subtle and not-so-subtle differences between the two manuscripts. Finally, the translation and commentary is divided into seven subsections, according to the purpose of the text and the type of examples discussed. Heinz Theo Lutstorf published a similar work in 2005 (in German, with no accompanying English translation), in which he analyzed the copy of Bürgi's *Aritmetische und Geometrische Progreß Tabulen* that is held in Gdańsk, Poland. When appropriate, I have included references to Lutstorf's commentary to emphasize important points.

Chapter 4 summarizes my perspective on two questions that have been asked numerous times: Who is the copyist of the Graz manuscript of the *Aritmetische und Geometrische Progreß Tabulen*? And, what is the relationship between the Graz and Gdańsk manuscripts?

Although I have received much assistance from very competent writers, mathematics historians, and scholars while working on this project, I am not a traditionally trained historian. Consequently, if you have found your way to this book, I ask that you read it with the two stated aims in mind, as opposed to imposing a critical edition structure on what follows. Finally, I hope that this book provides an important addition to the known scholarship on Jost Bürgi.

Tallahassee, FL Kathleen Clark

---

[3] However, the facsimile of the 58 pages of tables is given in Appendix C and can be downloaded from www.springer.com/us/book/9781493931606.

[4] Formerly Danzig, Prussia/Germany.

# Acknowledgments

My interest in the history of logarithms dates back to my dissertation research that involved working with high school teachers on ways to teach students about logarithms and logarithmic functions using a historical perspective. I was greatly influenced by the writing of historians of mathematics dedicated to exploring the role of history of mathematics in teaching, particularly the work of the late John Fauvel. In his introduction to *Revisiting the History of Logarithms*, Fauvel (1995) quoted and shared the following:

> *My father was* l'ingegné *(the engineer), with his pockets always bulging with books and known to all the pork butchers because he checked with his logarithmic ruler the multiplication for the prosciutto purchase.* [Primo Levi, 12, p. 19]
>
> The subject of logarithms, like the notorious "asses' bridge" in Euclid (*Elements* I,5) for an earlier generation, seems to mark an intellectual rite of passage: before going over there is a sense of unfathomable mystery, even danger, ahead; afterwards there is still some wonder and perplexity at just what it is one has learned. Some stumble at the hurdle and feel forever excluded, like the lame boy of Hamelin; others press on and on and still do not come to the end of what is undeniably a paradigm of the rich complexity of mathematical concerns.
>
> All this remains true, even now that a traditional calculational justification for studying logarithms has passed into history.... (Fauvel 1995, p. 39)

This passage very much set the tone for my dissertation research, as I always believed "something more" could be cultivated (mathematically, culturally, historically) in the teaching of logarithms.

As part of my dissertation research, the classroom materials that I constructed and used with teachers (and, for subsequent use with their students) were informed by the work of John Napier, Henry Briggs, William Oughtred, and Leonhard Euler. Jost Bürgi was mentioned only briefly when I worked with the teachers, and this was primarily because of how the resources I used at the time treated his role in the development of the logarithmic relation. Even then, the brief references struck me as afterthoughts, as found in a short paragraph in Cajori (1915):

> The only possible rival of John Napier in the invention of logarithms was the Swiss Joost Bürgi or Justus Byrgius (1552–1632). ... Bürgi published a crude table of logarithms six years after the appearance of Napier's *Descriptio*, but it seems that he conceived the idea

> and constructed that table as early, if not earlier, than Napier did his. However, he neglected to have the results published until after Napier's logarithms...were known and admired throughout Europe. (pp. 166–167)[1]

In 2009 I was awarded a research fellowship at the University of Canterbury (Christchurch, New Zealand) to conduct research in history of mathematics. The opportunity at Canterbury was the result of an effort by Clemency Montelle (and supported by the School of Mathematics and Statistics) to increase the production of research in history of mathematics. The first task of my fellowship was to respond to Clemency's request to describe possible connections for our research collaboration, and I immediately responded that I was keen to pursue what I felt was "the rest of the story" regarding the development of the logarithmic relation. Consequently, I felt Jost Bürgi's contribution would provide the missing piece to an incomplete story about the independent invention of logarithms.

I located the Graz copy of *Aritmetische und Geometrische Progreß Tabulen* (1620)[2] in January 2009, and with the assistance of Michaela Scheibl at the Department of Special Collections of the Library of the Karl-Franzens-University Graz, Clemency and I received a digital scan of the complete copy held there. Although we have presented papers and published articles about the parallel insights of John Napier and Bürgi in the early years of the seventeenth century, our initial research developed into something more from my perspective, and this book represents my desire to provide access to the life and one mathematical contribution of Jost Bürgi to non-German language readers.

This book would not have been possible without the encouragement and assistance of several individuals and institutions. I am indebted to Clemency Montelle for introducing me to many tools that made this scholarly "labor of love" a reality, not the least of which is having the confidence to live and work in multiple academic environments (mathematics, mathematics education, and mathematics history). I am fortunate to have been awarded the time and resources to live and work in Christchurch, New Zealand, and I will be forever grateful to the School of Mathematics and Statistics at the University of Canterbury.

I am also grateful to Michaela Scheibl at the Department of Special Collections of the Library of the Karl-Franzens-University Graz (Graz, Austria) for her assistance in providing me with the digital copies of the texts I needed for my research. She also assisted in reaching a publication agreement from her department and the library so that I and Birkhäuser Mathematics are able to provide others access to the manuscript that is the subject of this book.

I was fortunate that Fritz Staudacher contacted me in the autumn of 2013, inquiring about the insights we might discuss with each other concerning our shared interest in Jost Bürgi. His initial email led to an eventful trip to Zürich in March 2014

---

[1] Almost 100 years later it is still often easier to find references to the development of logarithms that omit mention of Bürgi, including this example from Pesic (2010): "...long before John Napier, Stifel seems to have invented logarithms independently" (p. 506).

[2] The complete title is: *Aritmetische und Geometrische Progreß Tabulen/sambt gründlichem unterricht/wie solche nützlich in allerley Rechnungen zu gebrauchen/vnd verstanden werden sol.*

where I met Fritz, Jörg Waldvogel (Professor Emeritus, ETH-Zürich), and Christelle Wick (Toggenburger Museum, Lichtensteig, Switzerland). The assistance, encouragement, and discussions that I shared with each of these new friends (including Irene Waldvogel, Jörg's lovely wife) are what made the completion of this project possible, and I publicly offer them my sincerest thanks.

Ewa Lichnerowicz of the Library of the Polish Academy of Sciences (Gdańsk, Poland) provided much needed assistance at the end of my revision work, and I am very thankful for her kindness, patience, and ability to communicate with me in English.

The many hours, weeks, and years, as well as the financial commitment to see this book to fruition, were lovingly and consistently supported by my partner in life, Todd Clark. Without him, and the sacrifices he made, this book would not be possible.

Finally, I dedicate this book to my parents, John Edward McGarvey, who passed away suddenly at the age of 74 on 29 March 2014, and Mary Regina McGarvey, who lost her brave battle with cancer at the age of 72 on 3 October 2015. I miss you, Mom and Dad.

# Contents

**Figure 1** Frontispiece of Benjamin Bramer's *Bericht zu M. Jobsten Burgi seligen Geometrischen Triangular Instruments mit schönen Kupfferstücken hierzu geschnitten* (1648; image courtesy of Toggenburger Museum, Lichtensteig, Switzerland)

# Chapter 1
# A Brief Biography of Jost Bürgi (1552–1632)

## Introduction

Several German- and French-language resources contain brief biographies of Jost Bürgi (e.g., Cantor 1900; Lutstorf 2005; Montucla 1758; Naux 1966; Wolf 1858). No substantial personal information on Jost Bürgi[1] exists in the English language, other than the short (just over one page) account by Nový (1970) in the *Dictionary of Scientific Biography*. We can, however, construct a decent timeline of Bürgi's life from German-language resources (see Appendix A), particularly when it is situated with respect to Bürgi's contemporaries who were engaged in or aided in the development of scientific work dependent upon the logarithmic relationship. Staudacher (2014) published (in German) a quite extensive account of Bürgi's life, which included content on his mathematical and scientific achievements and contributions, as well as accompanying obstacles, family relationships, and other personal attributes. Using translations of Staudacher's text, as well as more traditional sources of biographical information on Bürgi, the major aspects of Bürgi's professional life are highlighted in the brief biography presented here.

## Lichtensteig and Surrounds: Bürgi's Early Life and Work (1552–1579)

Bürgi was born 28 February 1552, in Lichtensteig in the Toggenburg, a 70 km long alpine highland valley along the Thur River and southwest of Mount Säntis in the Canton of St. Gallen, Switzerland. Jost and his parents were Protestant, which was

[1] Bürgi's given name is sometimes given as Joost, Jobst, or Justus (when used with the Latinized version of his surname, Byrgius).

© Springer Science+Business Media New York 2015 
K. Clark, *Jost Bürgi's Aritmetische und Geometrische Progreß Tabulen (1620)*,
Science Networks. Historical Studies 53, DOI 10.1007/978-1-4939-3161-3_1

representative of the majority of Roman Catholic and Protestant families living in this small village of approximately 400 inhabitants (Figure 1.1). We do not know anything of substance about Bürgi's early learning (Waldvogel 2012, p. 3), except that he probably received an almost complete 6-year formal education that was typical of boys in Bürgi's time and until the beginning of the twentieth century (Staudacher 2014). In 1564, Bürgi finalized his formal education, but due to religious battles as a result of the Counter-Reformation in Switzerland, Lichtensteig was often left without a teacher. Consequently, Jost and his classmates may have lost 1 year of the 6-year formal education. Although the majority of people of the Toggenburg Valley supported and followed the Protestant teachings of Ulrich Zwingli (1483–1531), the citizens were almost always overruled by the duke-abbot of the St. Gallen monastery. According to Staudacher (2014), the lessons in public schools were composed of up to 50 % choral singing lessons, with the remainder in computing, reading, and writing. Bürgi did not know Latin (and certainly did not write or publish in Latin) and regarding his knowledge of scientific languages, Bürgi stated:

> Weil mir auß mangel der sprachen die thür zu den authoribus nit alzeitt offen gestanden, wie andern, hab jch etwas mehr, als etwa die glehrte vnd belesene meinen eigenen gedanckhen nachhengen vnd newe wege suechen müessen. (List and Bialas 1973, p. 7)[2]

After his early and brief education and beginning in 1565 (Staudacher 2014), Bürgi began training in various trades that later contributed to the craftsmanship necessary for instrument making by working with his father, who was a locksmith.[3] Bürgi possibly trained as a goldsmith between 1565 and 1567 with David Widiz (~1535–1596), when Widiz relocated to Lichtensteig from Augsburg (Staudacher, p. 52).

Bürgi most likely apprenticed with someone with experience in making technical instruments, such as clock- and watch-making. Faustmann (1997) and Naux (1966) noted that Bürgi possibly worked as a traveling apprentice in Straßburg, where he may have come in contact with the teachings of Conradus Dasypodius[4] (~1531 to ~1601). According to Sesiano (in the *Historical Dictionary of Switzerland*, 1986), Dasypodius was a mathematics professor at the Academy of Straßburg from 1562, where he also took care of Swiss fellows studying there. Dasypodius also continued the design and construction of the second version of the astronomical clock for the Straßburg Cathedral (built during 1570–1574), and Bürgi may have participated in the construction of this clock (Waldvogel 2014). Some experts still believe this hypothesis, put forth by Rudolf Wolf (1858), made sense at the time due to Bürgi's potential training trajectory.

---

[2] *Because I did not know other languages, the doors to the well-known scientists were not always open for me. So, opposite to the well-educated scholars, I had to think a little bit more by myself and find my own ways.*

[3] In the sixteenth century, the professions of locksmithing and making clocks were closely connected.

[4] Dasypodius' German surname was "Rauchfuss." Rauchfuss followed the practice of his time and grecianized his name to "Dasypodius."

**Figure 1.1** Part of a stained glass coat of arms that mentions the grandparents of Bürgi (photo courtesy of Toggenburger Museum, Lichtensteig, Switzerland)

The construction of the second version of Straßburg Cathedral's clock was carried out by the well-known clockmakers Isaac and Josias Habrecht (of the Canton of Thurgau, Switzerland). This version of the astronomical clock, which operated well into the eighteenth century, was well known for its complexity because of its numerous devices, including indicators for planets and eclipses, calendar dials, and the astrolabe. Wolf's speculation that Bürgi apprenticed under the Habrechts during the construction of the cathedral's clock in Straßburg has persisted for more than 150 years, but today it is denied by experts such as Roegel, Oechslin, and Oestermann. Waldvogel (2014) and Staudacher (2014) speculated that Bürgi might have acquired his skills in Schaffhausen, Switzerland, which is closer to Lichtensteig in eastern Switzerland and where the Habrecht family built clocks until at least 1572 before moving to Straßburg. The Habrechts designed and constructed the Bern, Solothurn, and Schaffhausen astronomical clocks, as well as clocks in many cities of southern Germany, including Heilbronn, Donaueschingen, Ulm, and Altdorf near Nürnberg (Staudacher 2014, pp. 55–56).

In 1570 or 1571, Bürgi most probably completed his professional trades training, and from about 1571 he worked as a clockmaker in various locations, possibly in Augsburg due to the many connections he held with people from there (e.g., Widiz), and later in Nürnberg. In 1576, Christoph Heiden (1526–1576), a famous mathematician and celestial-terrestrial globe inventor, died in Nürnberg, and Bürgi, who was in Nürnberg as well, finalized a celestial-terrestrial globe that was under construction in Heiden's workshop.[5] Heiden received orders directly from Emperor Maximilian II and also served as first president of Altdorf University in Nürnberg.

Also in 1576, Maximilian II died, and his son Rudolf II von Habsburg (1552–1612) was named successor and emperor of the Holy Roman Empire. Rudolf II was deeply interested in the arts and sciences, including alchemy. Since he was not as engaged in the political, ceremonial, and daily managerial duties of his position, he moved the seat of the Habsburg Empire from Vienna to Prague in 1583, to serve as better protection against the Ottoman Turks. In 1592 and upon the recommendation of Vice Chancellor Jacob Curtius (1554–1594), Rudolf II selected Nicolaus Reimers Baer, or Nicolaus Reimers Ursus[6] (1551–1600), as imperial mathematician. Then, in 1599 and after recommendation of his Imperial Physician Thaddäus Hagecius (1525–1601), he named Tycho Brahe (1546–1601) of Denmark as imperial astronomer to his court in Prague. Eventually, in 1601, Rudolf selected Johannes Kepler (1571–1630) as Brahe's successor, and, by following his own interest in goldsmithing and clockmaking, Rudolf selected Jost Bürgi as his imperial clockmaker in 1604.

However, before Bürgi worked in Nürnberg, close connections developed between Duke Wilhelm IV (1532–1592) and Georg Joachim Camerarius (1534–1598), as well as between Heiden, Camerarius, and Bürgi. In 1579, the duke invited Bürgi to court in Kassel to work as a clockmaker and also as a craftsman in his observatory (Staudacher 2014). To receive such an invitation from the duke would have meant that Bürgi was already established with most of the skills and knowledge to deserve such a prestigious appointment in the observatory in Kassel.

## Connections in Kassel: 1579–1603

After arriving in Kassel in 1579, Bürgi was engaged in clock and instrument making, and later in astronomy and mathematics, as well. In 1580 he built his first Kassel celestial sphere, worked with astronomical instruments, and developed various metal sextants in brass, steel, and copper. In 1583, Bürgi invented his own type of proportional compass, and in 1584, he created the world's first clock precise to the second and which indicated seconds both visually and auditorily. As a prerequisite to this revolutionary observatory clock, Bürgi had to invent new methods and mechanical systems for smoothly and steadily distributing the initial forces of a weight or of a spring, which was realized by his inventions of the cross-beating escapement and of the rewound weight. Notably, both of these Bürgi inventions were in place 70 years before Huygens' and Newton's pendulum clocks and 120

[5] This is a newly discovered fact taken from the inventory list of Emperor Rudolf II's Kunstkammer (i.e., a "collector's cabinet," which contains a collection of curiosities and treasures) in Prague (Staudacher 2014, p. 76).

[6] Several variations exist for Reimers' name, some of which include Reimarus Ursus, Raimarus Ursus, and Nicolaus Reymers Baer. In this chapter, I will use Reimers.

years before John Harrison's chronometer (Staudacher 2014). It is not surprising then that in a letter to Brahe in 1586, Wilhelm IV said: "…unsers Uhrmachers M. Just [Bürgi], *qui quasi indagine alter Archimedes* ist."[7]

Most importantly for the time period 1584/1585, Bürgi, Christoph Rothmann (1551–1600), and Wilhelm IV—all as astronomers in Kassel—began a new measurement program of the stars in order to obtain better data for navigation, astronomy, and astrology. Two years after beginning their work, the *Grand Hessiae Register of Stars* (in the original German: *Grosses Hessisches Sternverzeichnis*) was completed and included 383 newly measured stars (Staudacher 2014, p. 134).

In 1584, Paul Wittich (~1546–1586) arrived in Kassel and stayed several months, and during the same time period, Bürgi began a search for ways in which to improve methods and formulae for prosthaphaeresis.[8] As a result of his extraordinary mathematical and technical talent and from his experience in calculating and formulating gearings, Bürgi was well positioned to contribute to innovations necessary to improve upon astronomical calculations. And, in order to improve upon such work at the time, Bürgi would have needed to be knowledgeable of the notion of prosthaphaeresis and computation involving sines.

Prosthaphaeresis, a process that converts more complicated multiplication (or division) into simpler addition (subtraction), was probably well known to Islamic scientists from at least the eleventh or twelfth century. Prosthaphaeretic formulas, in modern trigonometric notation, include the identities

$$\cos(a+b)=\cos(a)\cos(b)-\sin(a)\sin(b)$$

and

$$\cos(a-b)=\cos(a)\cos(b)+\sin(a)\sin(b)$$

To observe the "product to sum" transformation, we first subtract the second formula from the first

$$\cos(a-b)-\cos(a+b)=2\sin(a)\sin(b);$$

and isolating the product term yields

$$\sin(a)\sin(b)=\frac{1}{2}\left[\cos(a-b)-\cos(a+b)\right].$$

Thus, when two angle measures are known, an easier calculation is made when subtracting the cosine of their sum from the cosine of their difference and then dividing the result by 2, as opposed to multiplying two sine values.

---

[7] "*…our clockmaker Jost Bürgi,* who is almost on the way of another [a second] Archimedes" (Roegel 2010a, p. 5).

[8] *Prosthaphaeresis*, from the Greek *prosthesis* (addition) and *aphaeiresis* (subtraction).

There has been much speculation about Bürgi's contribution to the improvement of prosthaphaeresis, as well as his construction of a table of sines. For example, Thoren (1988) discussed Bürgi's role in the evolution and publication of the trigonometric formulas that reduce a more complicated operation (multiplication) into a simpler one (subtraction), as in the formula above. In his account, Thoren traced the first publication of the method of prosthaphaeresis to Reimers, who first mentioned Bürgi's calculations in 1588. Attributing this "first" to Reimers is questionable, according to Thoren, and he discussed the potential contribution of Tycho Brahe, Paul Wittich, and Jost Bürgi to the use, publication, and geometrical proof of prosthaphaeretic formulas (e.g., for computing $\sin(a)\sin(b)$). Moreover, Thoren stated that:

> Ursus…issued a disclaimer in 1597…. According to him, Wittich…brought the *method* to the astronomical observatory of the Landgrave [Landgraf] of Hessen-Cassel in 1584; but what he brought was only one prosthaphaeretic equation (for $\sin A \sin B$), and no *proof* for it! It had been the Landgrave's [Duke's] clock-maker, Joost Bürgi, Ursus said, who devised a geometrical proof for that identity. (Emphasis in the original, p. 33)

In approximately 1586 or 1587, Bürgi designed and constructed a three-dimensional planetarium (i.e., a planetary model) for Reimers, of his "Tychonian" world model (Staudacher 2014, p. 119). The Tychonian model of the universe was a hybrid model of Ptolemy's geocentric model, where the sun and planets orbit around the Earth, and of Copernicus' heliocentric world model, which places the sun at the center. The hybrid model had the support of the Jesuits and also had two inventors, Reimers and Tycho Brahe, each of whom fought hard for his own priority until the death of Reimers in 1600. The hybrid world model shows the Earth in the center, surrounded by the moon and the sun. The other planets revolve about the sun, and all together they revolve around the Earth. Bürgi then constructed a second version of the planetary model at the request of Wilhelm IV and which incorporated feedback from Rothmann. In 1587, Reimers translated Copernicus' *De revolutionibus orbium coelestium* into German for Bürgi. Despite Bürgi's lack of Latin ability, his friend Reimers—imperial mathematician to Emperor Rudolf II—also likened Bürgi's abilities to those of Euclid an Archimedes (Gaulke 2015).

Afterwards (from 1587 until 1591), Bürgi began new work on the measurement of celestial bodies in order to define better orbital paths of the sun, Earth, and moon. And, in December 1590 until 1597, "Bürgi…regularly determined the angular distances of the planets and the Moon from those of the fixed stars recorded in the [*Grand Hessiae Register of Stars*] catalogue of 1587" (Gaulke 2015, p. 45). He needed these data for computations and to design a mechanically working device of Copernicus' moon theory to be integrated in the equation clock (or solar and lunar anomalies clock) of 1591.[9] This small table clock showed the mean moon and sun positions, as well as the highly accurate relative positions of the sun, the moon, (including eclipses), and the fixed stars (astrolabium dial) through the creation of elliptic movements of epicyclical and differential-epicyclical gearings. To integrate

[9] For a detailed discussion of this clock, see Gaulke (2015).

various paths, Bürgi selected the form of an elliptical movement, which is the same progression of the planets that Kepler discovered 15 years later. Thus, Bürgi's measurements and calculations would have required precision, and consequently, Bürgi needed methods for which he could carry out the computations. As an already skilled instrument maker, he needed mathematical tools to complete the work.

In 1588, Reimers published part of Bürgi's new mathematical methods in *Fundamentum Astronomicum*; however, Reimers published perhaps more than Bürgi would have actually agreed to—leading to a slightly strained relationship between the two men—and an unwritten or unspoken publication agreement of sorts was part of the problem. To prevent this undesirable outcome from happening again, Bürgi asked his friend and colleague Reimers to swear to keep quiet all of Bürgi's developments and innovations in future.[10] This misunderstanding (about what could and could not be published by Reimers) between Bürgi and Reimers in 1588 may have led to Bürgi being overly cautious about writing down his mathematical innovations and sharing them with others. For example, Bürgi's "Kunstweg" was a method that dealt with interpolation, and it was included in *Arithmetica Bürgii*, which was edited by Kepler in 1603.[11] Staudacher (2014), in following Ludwig Oechslin, is of the opinion that Bürgi had already prepared his *Aritmetische und Geometrische Progreß Tabulen* by this time, as he would have been able to create the tables and methods using his "Kunstweg," which included methods of interpolation.

German mathematics historian Menso Folkerts further supported this claim. Folkerts located a handwritten (allegedly by Bürgi himself) document titled *Fundamentum Astronomiae*—a document very similar to Reimers' *Fundamentum Astronomicum*—in the Biblioteka Uniwersytecka we Wrocławiu (Wroclaw University Library, Poland). The manuscript was personally given to Emperor Rudolf II as a gift 10 days after Bürgi's first audience with the emperor in June 1592.[12] The analysis and publication on the results of this Bürgi text on trigonometry, which includes algorithms for building sine tables and his "Kunstweg" method of interpolation, was published in 2015 (Folkerts, Launert, and Thom). The sine tables included in this document could be the same as shown to Brahe, which also took place in 1592.

Prior to Bürgi's first trip to Prague, he remained busy in Kassel, continuing to work on a system to measure planets, and he collects measurement data until 1597

[10] Reimers must have kept his promise; he refused to divulge information about Bürgi's "Kunstweg" (meaning artful (or skillful) method), because he had promised Bürgi to keep all of his (Bürgi's) information confidential (Staudacher 2014, p. 181).

[11] This work came to be known as Bürgi's *Coss*. The *Coss* manuscript was never delivered to a printer for publishing; it was finally edited and published in 1973 by List and Bialas. In 1604, Kepler wrote a letter to Fabricius, stating that he now had an understanding of the "Kunstweg" after having edited the *Coss* manuscript (Staudacher 2014, p. 181). However, Kepler did not mention his *Coss* editing work for Bürgi and therefore did not compromise the secrecy agreement he held with Bürgi.

[12] In the forward for *Fundamentum Astronomiae*, Bürgi gives the date "Prag, am Tage Mariae Magdalenae, Anno Christi 1592" (Folkerts 2015, p. 109), which corresponds to 22 July 1592.

on more than 1000 planet positions.[13] Bürgi built a silver and gold planetary globe in 1591–1592, which is considered one of the most highly developed automated models ever built. It is this planetary globe that Rudolf II asked Bürgi (through Wilhelm IV) to bring to Prague and which Bürgi personally delivered to Rudolf II in 1592. The construction of the globe required precise astronomical values for planetary positions, which Bürgi was able to compute in his own work as an astronomer and also as a mathematics expert (Staudacher 2014, p. 147). Bürgi returned to Prague in 1596, most likely for the purpose of checking and servicing the planetary globe and observatory clocks. Bürgi also met and spoke with Rudolf II during this visit regarding distances to planets and other astronomical interests. They also spoke about Bürgis' work in trigonometry, including the trigonometry document (*Fundamentum Astronomiae*) that he left with the Emperor during his last audience with him in 1592.

In addition to Bürgi's extensive work on celestial measurements and the design and construction of intricate instruments, he also worked to finalize a table of sines, *Canon Sinuum*, during this time. The table was probably completed at the end of the sixteenth century (Roegel 2010a), with List and Bialas (1973) and Staudacher (2014) giving the year 1598. However, as with every other mathematical endeavor of Bürgi's, coupled with his fear of others publishing without his permission, Bürgi most likely carried a copy of the *Canon Sinuum* on his person and used the tables for his own and Kepler's purposes and calculations.[14] Bürgi's *Canon Sinuum* contained sines calculated to eight (8) places, at intervals of 2″ (2 s).

Also at this time (1597–1599), Bürgi was completing the manuscript for the previously mentioned mathematical work, *Arithmetica Bürgii* (Staudacher 2014, pp. 185–186). Bürgi certainly felt at a disadvantage due to his poor knowledge of languages and his need to work more intently to read and understand the solutions of mathematical authorities. Thus, he searched for someone to improve and edit his draft of his *Arithmetica*. Bürgi's relationship with Reimers made him a candidate as editor of the manuscript; however, Reimers was himself writing a new book on mathematics and algebra. Also at this time, Reimers, Brahe, and Kepler's paths were converging, and strained relations in Prague were due to the priority fight between Reimers and Brahe (regarding their model of the universe), in which Brahe already asked Kepler to write a study of the subject. Brahe would eventually hire Kepler as an assistant at the observatory in Prague to help with analyzing data on Mars, although Kepler held ill feelings toward Brahe's dealings with Reimers (particularly since Kepler had only favorable dealings with Reimers). Eventually, Reimers handed Bürgi's draft of the *Arithmetica* over to Kepler for editing.

Soon after, in August 1600, Reimers died of tuberculosis while awaiting trial in a case that Brahe brought against him for allegedly stealing Brahe's idea for a hybrid model of the universe. Brahe had the support of Rudolf II, and Brahe expected

[13] The data was accessible to Kepler from 1603 until 1612, when both Kepler and Bürgi were in Prague.

[14] The *Canon Sinuum* was never published and most likely remains lost. However, it makes sense that if Bürgi kept it on his person, others would have seen it and stated that it did exist.

Reimers to be found guilty, the punishment for which would have entailed being "publicly beheaded, drawn, and quartered" (Staudacher 2014, p. 210).

## Prague: 1603–1631

Upon arriving in Prague, Bürgi continued to produce specialized mathematical instruments and Kepler finalized his edited draft of Bürgi's *Coss*. Additionally, Bürgi's astronomical data, which had been recorded over a period of 12 years in Kassel, became available to his friend (and now Imperial Court Astronomer) Kepler in Prague from 1603 until 1612. Bürgi's strong need for secrecy (as agreed upon between Kepler, Bürgi, and Bürgi's brother-in-law, Benjamin Bramer (1588–1652)) was a major factor for his work and name as an astronomer to be all but forgotten and eliminated from any mention by Brahe's successors. However, as Staudacher (2014) claimed, without Bürgi it would have been difficult or nearly impossible for Kepler to define and to verify the small elliptical deviation of an only eight (8) arc minutes from a circular path in his calculation of planetary motion. Bürgi provided to Kepler not only the most precise instruments for time-second and angle-minute part measurements but also the mathematical methods necessary to accommodate this mass of spherical data.

In December 1604, Bürgi was officially named imperial clockmaker. There he maintained a clock- and watch-making workshop, with two employees, in the same building as Rudolf II's alchemy laboratory and artist Adriaen de Vries' atelier with metal casting equipment. Beginning in 1608, Bürgi owned a private house in the downtown area close to the Powder Tower, and with a monthly salary of 60 guilders, he was the third-highest paid employee of Rudolf II. For the next dozen years or so, Bürgi continued to develop instruments, clocks, and watches in his workshop and to support Kepler as an astronomical observer. Furthermore, others applied Bürgi's mathematical methods in their own work. For example, in the 1608 edition of *Trigonometria*, Bartholomaeus Pitiscus (1561–1613) published brief excerpts of Bürgi's new algebraic methods, including how to determine the direction and magnitude of eccentricity of the Earth's orbit and finding the sine of half-angle from the sine of an angle. In this edition of his *Trigonometria* (a book with examples from Bürgi), Pitiscus called Bürgi an "ingeniosissimus Mathematicus," or "ingenious mathematician" (Staudacher 2014, p. 187). One of the main reasons for the publication of Bürgi's mathematical examples in Pitiscus' books is the secrecy agreement between Bürgi and Kepler. That is, Kepler could publish Bürgi inventions in his own publications only after Bürgi had previously presented it himself in another publication. Therefore, it was necessary for Bürgi to hand over an example or excerpt for publication before a Kepler example was shown in *Astronomia Nova*.

A great deal has been written about when Bürgi began his work to construct the tables of the *Aritmetische und Geometrische Progreß Tabulen*, and a brief step back is in order. Nový (1970) speculated that Bürgi began computing his tables of logarithms as early as 1584. Grattan-Guinness placed Bürgi's computation of tables of logarithms as early as 1590 (1997, pp. 180–181). Many sources, however, quote Bürgi's brother-in-law, Benjamin Bramer, for a firsthand account of when Bürgi must have computed his tables of logarithms (actually, tables of antilogarithms). In his testimony, Bramer stated in a book published in 1630 that:

> [It] is on these principles that my dear brother and master Jost Bürgi, calculated, twenty years ago and more, a beautiful table of progressions, …, calculated to nine digits, [and] he did not print the [tables] until 1620 in Prague, so the invention of logarithms is not by Napier, but was made by Jost Bürgi long before." (translated from Montucla 1758, p. 10)

This passage has influenced some to place Bürgi's construction of tables as a result of his invention around the year 1610 (Roegel 2010a).

Refining the time frame for which Bürgi completed the construction of his tables of logarithms may be possible with Folkerts' forthcoming analysis of Bürgi's *Fundamentum Astronomiae* (which is dated to 1592). In particular, the first of the two books of the *Fundamentum Astronomiae* includes an explanation of the four basic arithmetic operations and root extraction using sexagesimal (base 60) numbers, a 12-page multiplication table (again, with sexagesimal numbers), a chapter dealing with prosthaphaeresis, and the calculation of the sine value for each angle, in increments of 1 min and to six places. The sheer amount of calculation work in the *Fundamentum Astronomiae*, coupled with the underlying similarity among the various calculation techniques required to construct tables of sines and to make the accurate calculations required to construct the astronomical models, could place Bürgi's construction of his tables of logarithms prior to 1592. That is, his method for simplifying all manners of calculations using logarithms (like those eventually needed in the *Fundamentum Astronomiae*) may have been the precursor to Bürgi's more complex mathematical texts.

Kepler, as his friend and colleague, urged Bürgi to print and disseminate his tables and instructions for their use as "an efficient method to carry out multiplications and divisions" (Waldvogel 2012, p. 13). Some time between 1600 and 1603 and in an effort to avoid a similar situation that Bürgi experienced with Reimers publishing his work without first establishing a proper agreement with Kepler, Bürgi arranged a secrecy agreement with him. Consequently, along with handing over of Bürgi's *Coss* draft to Kepler, Kepler and Bürgi swore to not betray each other and to keep the methods and innovations in mathematics of the other secret until he published them himself (Staudacher 2014).

Yet Kepler knew and worked with Bürgi's *Aritmetische und Geometrische Progreß Tabulen* while editing Bürgi's *Coss*, and from 1603 onward, Kepler worked in silence with both of Bürgi's innovative tables, the *Canon Sinuum* and the *Aritmetische und Geometrische Progreß Tabulen*, in order to calculate with a vast amount of observation data collected by Tycho Brahe. Then, in 1609 both Kepler and Bramer were convinced that Bürgi would bring both manuscripts to the printer. Unfortunately, Bürgi's first wife (Bramer's sister) died in 1609, and this, along with the growing trouble in Prague between Catholic League soldiers and of the people of Old Town Prague, made the eventual printing of Bürgi's manuscripts difficult. Bürgi would not start publication until 1620, and even then only the actual tables were printed as proofs and in small quantity and without the instructions necessary for their use. Whatever copies of the tables existed in 1620 were most likely lost during the Thirty Years' War. One battle—the Battle of the White Mountain—was fought just outside of Prague in November 1620 and 7000 men lost their lives there (González-Velasco 2011, p. 101).

The subject of assigning a timeframe or year to Bürgi's construction of his tables of logarithms is often due to the question of priority with regard to the invention of logarithms. In 1614, John Napier (1550–1617) published his *Mirifici Logarithmorum Canonis Descriptio* (or the *Descriptio*), officially earning publication priority with regard to the invention of logarithms. However, for some, the priority issue is about more than the moment of publication. González-Velasco (2011) stated that "for the sake of fairness that the earliest discoverer of logarithms was Joost, or Jobst, Bürgi (1552–1632), a Swiss clockmaker, about 1588" (p. 100).

As was the case with Bürgi, Napier began working on his conception of logarithms some years before his first publication in 1614. Napier stated in his *Descriptio* that he worked some 20 years on the tables he presented within it, which would place the beginning of his work on logarithms in 1594. Interestingly and perhaps out of respect for his colleague and friend, Kepler did not show an official interest in Napier's logarithms since he had been urging Bürgi to publish the *Aritmetische und Geometrische Progreß Tabulen* for many years. In 1619, Kepler would have known that Bürgi's tables were being typeset for publication, and since they would soon be printed and distributed, Kepler no longer felt he was bound to secrecy. And his reaction was to not maintain allegiance to Bürgi but to align with Napier's (and, consequently, Briggs') tables of logarithms and, eventually, his own. In 1627 Kepler famously wrote in the foreword to *Tabulae Rudolphinae*: "Der zaudernde Geheimniskrämer liess sein Kind im Stich, anstatt es zum allgemeinen Nutzen grosszuziehen"[15] (Staudacher 2014, p. 206).

The discussion about assigning the title of inventor of logarithms to Bürgi or Napier is now over 400 years old. If we only consider publication date as the defining metric for priority, then Napier is the clear winner. Another dimension to the discussion, however, is to recognize that the parallel insights of both Napier and Bürgi occurred at approximately the same time. In the late sixteenth century and early seventeenth century, both Bürgi and Napier, in two different locations and engaged in very similar life's work (the need to perform a vast amount of difficult calculations, particularly with respect to astronomical computation applications), came to develop a mathematical method that enabled them to improve their own work and the work of others. Whereas Napier's original conception of the logarithmic relationship was dependent upon a kinematic argument (Appendix B), and which required complex calculations to construct his table of logarithms, Bürgi's original conception was algebraic in nature and much simpler in construction. It is unfortunate that because of Bürgi's need for secrecy to protect his innovations and methods until he believed them to be ready for publication and the events of the time (e.g., the worsening political conditions in Prague and the start of the Thirty Years' War),

[15] "*The hesitant secretive [man] abandoned his child instead of raising it for the general benefit.*"

the *Aritmetische und Geometrische Progreß Tabulen* would not be published and enter into mainstream use as Napier's conception of logarithms did.

There are several resources that describe Napier's conception of the logarithmic relationship, as well as the method used to construct his tables, including Havil (2014), Katz (2009), and Roegel (2010b).

## Return to Kassel: 1631–1632

In 1631, just before his death, Bürgi left Prague for the last time to return to Kassel. He died just 4 weeks shy of his 80th birthday on 31 January 1632, and without children of his own, his legacy died there as well. Although the grave no longer exists, a plaque was placed to commemorate his contributions:

> Auf diesem Friedhof liegt begraben
> der landgräflich- hessische und
> kaiserliche Uhrmacher sowie Mathematiker
> Jost Bürgi
> geb. 28.2.1552 in Lichtensteig, Schweiz
> gest. 31.1.1632 in Kassel.
> 1579–1604 und in späteren Jahren tätig in Kassel
> als genialer Konstrukteur von Messinstrumenten
> und Himmelsgloben, Erbauer der
> genauesten Uhren des 16. Jahrhunderts,
> Erfinder der Logarithmen. (Volk 2009)[16]

[16] *On this cemetery lies buried/the Landgrave of Hessen and/the Emperor's watchmaker and mathematician/Jost Bürgi/born February 28th, 1552 in Lichtensteig, Switzerland/died January 31st, 1632 in Kassel/ingenious designer of measuring instruments/and celestial globes, builder of the/most precise clocks of the 16th century,/inventor of the logarithms.*

# Chapter 2
# Details of *Aritmetische und Geometrische Progreß Tabulen*: Printed Tables, Manuscripts, and Mathematical Details

## Introduction

Two extant *Aritmetische und Geometrische Progreß Tabulen* manuscripts were considered for the commentary that appears in Chapter 3. In this chapter, the two copies (the Gdańsk (Gk) manuscript and the Graz (Gz) manuscript), as well as an example of a copy that contains only the title page and Bürgi's tables (e.g., the printed copy in München (Mn)), are briefly described. Then, further details of the Graz copy are given so as to inform the transcription, English translation, and commentary presented in Chapter 3.

## Brief Descriptions of Extant Prints and Manuscripts

### *The München Print (Mn) of 1620*

One copy of the *Aritmetische und Geometrische Progreß Tabulen*, comprising only the printed title page and 58 pages of tables, can be found in the Universitätsbibliothek of the Ludwig-Maximilians-Universität in München (Table 2.1). This was the first copy found by Rudolf Wolf in 1846 and was previously owned by Doppelmayer. As it does not contain any additional handwritten information (e.g., there is no accompanying written instruction manuscript), nobody understood it or could work with the tables. Furthermore, copies of this print containing only the title page and tables are also available online (http://daten.digitale-sammlungen.de/~db/0008/bsb00082065/images/). Analyses of the accuracy of Bürgi's tables are available (Roegel 2010a; Waldvogel 2012); however, there are print-quality discrepancies among the various copies used in the analyses.

© Springer Science+Business Media New York 2015

K. Clark, *Jost Bürgi's Aritmetische und Geometrische Progreß Tabulen (1620)*,
Science Networks. Historical Studies 53, DOI 10.1007/978-1-4939-3161-3_2

**Table 2.1** Description of copies of the *Aritmetische und Geometrische Progreß Tabulen*

| *Title (abbreviated)* and description | Current location | Date | Short label | Comments |
|---|---|---|---|---|
| *Aritmetische und Geometrische Progreß Tabulen*<br>Printed copy only: title page and complete tables | Ludwig-Maximilians-Universität, Universitätsbibliothek München (Germany) | 1620 | Mn | Discovered in 1846 by R. Wolf<br>Also available online from Bayerische Staatsbibliothek (Bavarian State Library), München, Germany[a] |
| *Aritmetische und Geometrische Progreß Tabulen*<br>Printed copy and manuscript: title page and complete tables, and the handwritten foreword and "Kurzer Bericht" ("short report," or instruction for using the tables) | Gdańsk Library of the Polish Academy of Sciences (originally discovered in the Stadtbibliothek in Danzig, Prussia/Germany, in 1855) | 1620 | Gk | Name of copyist not known; Published by Gieswald in 1856 |
| *Aritmetische und Geometrische Progreß Tabulen*<br>Printed copy and manuscript: title page and complete tables, and the handwritten foreword and "Kurzer Bericht" ("short report," or instruction for using the tables) | Universitätsbibliothek, Sondersammlungen, Graz, Austria (Department of Special Collections of the Library of the Karl-Franzens-University Graz, in Graz, Austria) | 1620 | Gz | Name of copyist not known; Found by Ernst Seidel in Guldin Archive (Graz, Austria) in 1982 |

[a]This is just one example of the print that is available. Other libraries such as Landesbibliothek in Coburg, Germany, Wissenschaftliche Stadtbibliothek in Mainz, Germany, and the Universitäts- und Forschungsbibliothek Erfurt/Gotha, in Erfurt, Germany link to the "full text" of *Aritmetische und Geometrische Progreß Tabulen*. However, the "full text" only includes the title page and the 58 pages of tables (i.e., the printed copy only), and each of these libraries links to the same copy (i.e., the one found at http://daten.digitale-sammlungen.de/~db/0008/bsb00082065/images/)

## *The Gdańsk Manuscript (Gk)*

The manuscript used by H.R. Gieswald for his edition resides in the historical manuscripts collection of the Gdańsk Library of the Polish Academy of Sciences. The manuscript is stitched between two cardboard covers, and the inside front cover contains the following notation (translated into English[1]) from the library's cataloging system:

Germ.; ca. 1620; 18 × 15.5 cm; 88 pages; cardboard binding.
Byrg, Justus, Logarithm tables.

1. Print: *Aritmetische und Geometrische Progress Tabulen, sambt gründlichem Unterricht, wie solche nützlich in allerley Rechungen zugebrauchen, und verstanden werden sol*, Prag 1620 pp. 1–59
2. "Vorrede an den Treuhertzigen Leser" pp. 61–62
3. "Kurtzer Bericht der Progress Tabulen, wie dieselbigen nutzlich in allerley Rechnungen zugebrauchen" pp. 63–80

Thus, only 80 of the 88 pages contain content; that is, the final eight pages are blank.

There is an additional comment on the catalog information provided, stating that:

On the title page (written by A. Engelckego?) name of the author and note: "(Dieser—nicht gedruckte—Unterricht ist im Manuscripte beigefügt[2])."

This note corresponds to the handwritten content on the title page of the Gk manuscript (see Figure 2.1). The sentence "Dieser … im Manuscripte beigefügt," and the completion of "Justus Byrg" that appears inside the circular representation of Bürgi's logarithms, is written by someone's hand—which is hypothesized to be A. Engelckego (or, possibly, A. Engelcke).

H.R. Gieswald first published the handwritten foreword and instruction for how to use the tables in 1856 (Gronau 1996; Lutstorf 2005; Waldvogel 2012), and Gieswald's transcription was based on the manuscript he found in possession of the Stadtbibliothek in Danzig, Prussia, in 1855. Gieswald stated in the introduction of his 36-page essay ("Justus Byrg als Mathematiker und dessen Einleitung in seine Logarithmen"[3]) that the intent and purpose of it was to publish the not previously printed "instruction" (the "vnterricht" (or "unterricht") as announced in Bürgi's title) that Bürgi himself gave for his logarithms (Gieswald 1856, p. 1). However, since the first known copy of the printed *Aritmetische und Geometrische Progreß Tabulen*—the Mn copy—did not contain the handwritten foreword and instruction, this made the use of the tables difficult if not impossible (Waldvogel 2012). As Staudacher (2014) stated, Bürgi distributed only a few nonprinted "proof copies" of the title page and the tables; all other information such as the foreword or instruction had to be copied in handwriting by a professional copyist or by the receiver of the proof copy.

Gieswald's essay contains a very short biography of Bürgi (less than one page) and also details Bürgi's accomplishments in geometry (10 pages) and in algebra

[1] Translation assistance provided by Ewa Lichnerowicz, Gdańsk Library of the Polish Academy of Sciences.

[2] *"This—not printed—instruction is attached to the manuscript."*

[3] *"Justus Bürgi as Mathematician and his Introduction to his Logarithms."*

Aritmetiſche vnd Geometriſche Progreſs
Tabulen/ſambt gründlichem vnterricht/wie ſolche nützlich
in allerley Rechnungen zugebrauchen/vnd verſtanden werden ſol.

Die gantze Rote Zahl
230270022.

Die gantze Schwartze Zahl
1000000000.

Gedruckt/ In der Alten Stadt Prag/ bey Paul
Seſſen/der Löblichen Univerſitet Buchdruckern/ Im Jahr/ 1620.

ZUR DANZIGER STADT-BIBLIOTHEK

**Figure 2.1** Title page of the Gk manuscript (Waldvogel 2012, p. 10)

(approximately 12 pages) or, rather, algebra as was known at the time. In this section of his essay, Gieswald also detailed methods of both Bürgi's and Napier's logarithms, using nineteenth-century mathematical notation. Finally, the transcription of the Gk copy of the *Aritmetische und Geometrische Progreß Tabulen* manuscript text begins on page 26 of Gieswald's essay. Gieswald included only a few footnotes and the reproduction of necessary excerpts of the tables to clarify examples, and he stated that he rectified any errors found in the "instruction" of the Gk copy within his transcription. In order to support reader comparison, a transcription of the Gk manuscript—and not the corrected Gieswald text—is provided along with the transcription of the Gz manuscript copy in Chapter 3.

Some scholars attempt to identify the copyist of the Gk manuscript and conjecture that Bürgi himself wrote it. Lutstorf and Walter (1992) and, later, Lutstorf (2005) offered the evidence that "J B" appears on the printed title page and that there are several uses of the first person. Lutstorf (2005, p. 102) also used some of the verbal cues found in the foreword of the "Kurzer Bericht" ("short report") to speculate on attributing the authorship to Bürgi, including the personalized salutation ("to the truehearted reader"), use of the German first person "ich," and the orientation of the instructional examples. However, a facsimile of the Gk manuscript was reproduced in Folta and Nový (1968), and the handwriting of the "gründlichem unterricht" of the Gk copy shown in their reproduction (Figure 2.2) is not the same as the sample of Bürgi's handwriting given in Staudacher (2014). The question about the order in which the copies were generated is discussed in the section "Detailed Description of the Gz Manuscript" of this chapter, as well as in Chapter 4.

## *The Graz Manuscript (Gz)*

The manuscript detailed herein and which will be referred to with the abbreviation "Gz" is the *Aritmetische und Geometrische Progreß Tabulen* housed in the Department of Special Collections of the Library of the Karl-Franzens-University Graz. The manuscript contains the shelf mark I 18600–18601 and is bound together with Johannes Krabbe's *Newes Astrolabium* (Frankfurt: Becker, 1609). The Gz manuscript contains the printed title page, 58 pages of printed tables, the 2-page handwritten foreword, and the 21-page handwritten "Kurzer Bericht" (or "short report" on instruction for how to use the tables).

The Gz manuscript was previously owned by Paul Guldin (1577–1643). Guldin was born in Mels, Switzerland, just 40 km from Lichtensteig. His early training as a goldsmith would have been similar to Bürgi's own early training. He later studied mathematics and became a Jesuit and amassed in his lifetime a large collection (some 300 titles) of sixteenth- and seventeenth-century volumes, manuscripts, and correspondence, which are now part of the Department of Special Collections of the Library of the Karl-Franzens-University Graz. The Gz manuscript, most likely collected by Guldin in response to the Counter-Reformation, in which scientific texts were acquired and kept from the mainstream, is in very good condition. No portion of the handwritten text appears to have been affected by damage. Furthermore, the manuscript contains no commentary or marginal notes, other than the text intended for the manuscript. "Justis Byrg[i]y" is handwritten in red ink above the first phrase of the title on the printed cover page. Also, and as previously mentioned, the initials "J B" are printed above the phrase "Die ganze Rote Zahl" ("The whole red number") on the same.

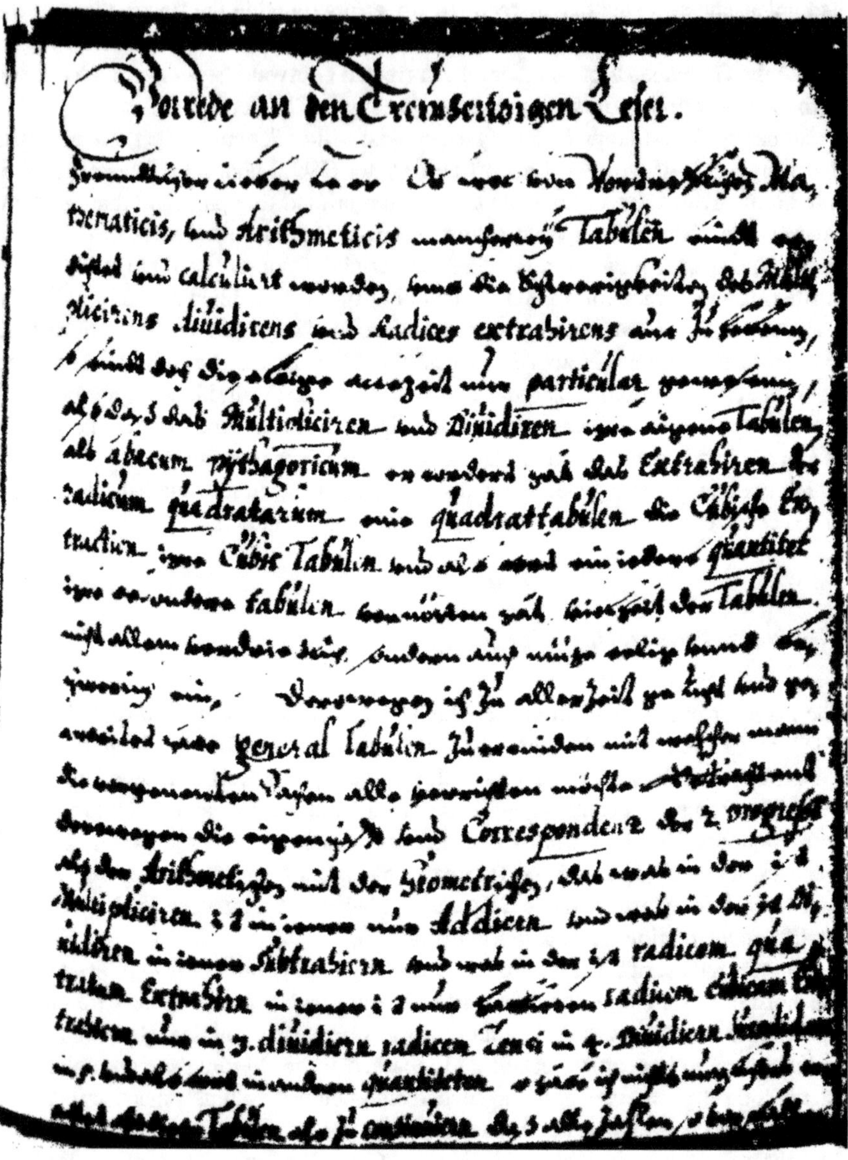

**Figure 2.2** Page 1 of the foreword from the Gk manuscript (Folta and Nový 1968)

## Detailed Description of the Gz Manuscript

The Gz manuscript is composed of three parts: the printed title page, printed tables of logarithms (printed by Paul Sess in Prague, 1620), and a handwritten "Kurzer Bericht" ("short report"), which also includes a two-page foreword. It is printed and written on paper with pages of size 19 cm by 16 cm and bound using flexible parchment. The tables fill 58 unnumbered pages, with labels centered at the bottom of every second and eighth page and which most likely formed a quire or four sheets of paper that formed eight pages when folded or stitched together. In the Gz manuscript, the second page is labeled "A 2" and the eighth page is labeled "B," and the tenth page is labeled "B 2." Next, the 16th page is labeled "C," the 18th page is labeled "C 2," the 24th page is labeled "D," and so on. The combination of Krabbe's and Bürgi's books together shows that the stitching was completed at a later time and that Bürgi's printed proofs were distributed only in the form of folded sheets. Finally, the odd-numbered pages of the 21-page handwritten "Kurzer Bericht" are so indicated, beginning with "1" and ending with "21." The foreword does not contain page numbers.

The use of red and black ink is a key element throughout Bürgi's *Aritmetische und Geometrische Progreß Tabulen*. Red ink was used in mathematical treatises of the time (e.g., during the sixteenth and seventeenth centuries), particularly for those containing tables and for the purpose of "better distinction" (Hutton 1811) among the different tabulated values such as Viète's (1540–1603) trigonometrical tables (1579), which contained "differences of the sines, tangents, and secants" (p. 4). Similarly, the use of color was fundamental for Bürgi to express the relationship between the two progressions of numbers. The notion of using the color red to better distinguish within a table continued in printed tables; there are several references to tables (especially tables of logarithms) that were produced in black and red ink some 200 years after Bürgi's tables including a reference from 1877: "The part left should contain the first and last numbers on the page in black ink, and their logarithms in red ink, or *vice versa*" (Buchheim 1877, p. 70).

The first and second-to-last line of the title page of the Gz manuscript are also printed in red ink, along with the appropriate values of his logarithms and "Die ganze Rote Zahl/230270022" ("the whole red number/230270022"), and two different instances of either Bürgi's name or his initials. The tables themselves are printed with the logarithms (the top row and left-most column) in red and the antilogarithms[4] in black. Finally, red ink is used throughout the "Kurzer Bericht" whenever elements of an arithmetic sequence or the red numbers (logarithms) from the tables are used, or operations on the red numbers are performed.[5]

There are notable uses of Latin, and in some cases, Latin-German hybrids, for mathematical terms in both the handwritten foreword and the handwritten "Kurzer Bericht." In most instances of the use of Latinized or hybrid terms there is a distinct change in the handwritten script. Examples of these terms in the "Kurzer Bericht" include *Fundamenta* (page 1), *Radicem Quadratum* (page 10), and *Medio proportional* (page 14),[6] to name just a few.

[4] In the modern sense, then, Bürgi provides a table of antilogarithms because these are the values of the body of the tables.

[5] However, since the "Kurzer Bericht" is handwritten, there are several inconsistencies, which are noted in the commentary that appears in Chapter 3.

[6] Latinized terms or the Latin part of hybrid terms will appear in italics or a combination of italics and normal font, respectively, in the transcription of the foreword and "Kurzer Bericht" in Chapter 3.

Lutstorf (2005) stated that the *Aritmetische und Geometrische Progreß Tabulen* manuscript is written in High German, with typical Swiss-German spelling conventions. Lutstorf also provided a sort of reading guide, in order for the reader to recognize the forms of words used by Bürgi (or his copyist) and the expected High German forms. Examples cited by Lutstorf (p. 98) include:

| Forms found in Bürgi | High German equivalent |
|---|---|
| allein | nur |
| allezeit | immer |
| nit | nicht |
| sein/seindt/sindt | sind |
| sein (as possessive) | sein, dessen, deren |
| worden (infinitive) | werden |

There are other characteristics of the manuscript that represent notable differences from what may be expected. For example, in the Gz manuscript, there is/are:

- Variations of the word "zwei" ("two"), such as "zwo Zahlen" and "zwüschen zwaÿen"
- Use of the dative and possessive for the description of the genitive case, as in "der rothen Zahl ihr schwarze Zahl"
- Archaic use of "so" for conjunctions such as "wenn" or as the relative pronoun
- Use of gap prepositions, as in "gegen derselbigen über" instead of "gegenüber derselben"
- Punctuation of cardinal numbers (as in "3." for three) and written forms of names of the ordinals (as in "dritte" for "third"), whereas the opposite is true today (in German)

There are also several variations in the spelling of often-used words in the manuscript. Examples of such words (in the correct, modern spelling) include gebührende (desired, due), rote or roten (red), and Zahl or Zahlen (number or numbers). These grammatical and, as will be shown in Chapter 3, content differences indicate that the Gk and Gz manuscripts were not copied from the same parent copy, nor were they recorded at the same lecture (e.g., an owner of a proof copy dictating to a copyist).

It is most likely not possible to determine the copyist of the Gz manuscript. Although several passages are written in the first person, it seems unlikely that Bürgi wrote this particular copy because of the numerous errors, particularly related to the examples and computations. As mentioned in regard to the identification of the copyist of the Gk manuscript, Staudacher (2014) and Waldvogel (2012, 2014) reproduced a handwritten letter by Jost Bürgi, and the handwriting of this letter does not resemble the handwriting found in the Gz copy.

In Chapter 4, I provide conjectures about the order in which the two manuscript copies may have been produced and provide evidence from the analysis of the Gk manuscript provided by Lutstorf (2005) and of the Gz manuscript conducted for the purpose of this book.

## The Content of Bürgi's "Kurzer Bericht" (As given in the Gk and Gz Copies)

### *The Foreword to the "Truehearted Reader"*

Bürgi announced clearly that the intention behind the construction of his special tables was to "remove the difficulties involved in calculating multiplications, divisions, and extractions of roots"[7] (Foreword, page 1). He continued, stating that: "I therefore searched for all time and worked to invent general tables with which you would like to do all the above[-mentioned] things"[8] (Foreword, page 1). Thus, Bürgi's key motivation was to construct special tables that could be used for a variety of calculations rather than needing collections of various tables, each of which aided the user to perform a particular operation. Indeed, Bürgi noted that having "the multitude of tables" for multiplication, division, square roots, and cube roots was "not only annoying but also cumbersome and difficult"[9] (Foreword, page 1).

It is here in the foreword that Bürgi stated that he was able to create one table for a multitude of calculations by considering two progressions: one arithmetic and the other geometric. He closed the foreword by noting that he would most likely work with the tables for years to come and promised another work for those readers who desired a deeper understanding of the tables. Sadly, this grand explanation, the "gründlichem vnterricht,[10]" or literally, the "detailed instruction," promised in both the title of the *Aritmetische und Geometrische Progreß Tabulen* and in the foreword, was never delivered in Bürgi's time, and its omission rendered the tables essentially useless. Instead, the "Kurzer Bericht" (which is the "gründlichem vnterricht" or "detailed instruction" in the title of the manuscript) did not become available until 1856, after Dr. Hermann R. Gieswald discovered, transcribed, and published it. The "Kurzer Bericht" contains a brief introduction to the relationship between an arithmetic progression and a geometric progression (with eight examples of calculations using the whole numbers and the nonnegative powers of 2) and some 26 examples of calculations using the tables Bürgi computed (see Table 2.2).

### *The Tables*

Unlike the handwritten "Kurzer Bericht," the tables in Bürgi's *Aritmetische und Geometrische Progreß Tabulen* are printed. The most significant and immediately apparent feature is that Bürgi's tables give antilogarithms, or powers, of the base 1.0001,

[7] "... um die Schwerigkeidten deß Multiplicierenß, Diuidierenß, und *Radices* Extrahierenß aufzuheben..."

[8] "Derowegen Ich zu aller zeit gesucht und gearbeitet habe, General Tabulen zu erfinden, mit welchen man die vorgenändten sachen alle verrichten müchte."

[9] "...villheidt aber der Tabulen nicht allein verdrießlich sondern auch muhseelig und beschwerlich seindt."

[10] The Gz and Gk manuscripts spell "unterricht" as "vnterricht" in the title page.

**Table 2.2** Distribution of examples in the *Aritmetische und Geometrische Progreß Tabulen* (Gz manuscript)

| Page number ("Kurzer Bericht") | Topic | Content |
|---|---|---|
| 1 | Introduction | Arithmetic $(n+1)$ and geometric progressions $(2^n)$, $n \geq 0$ |
| 2–4 | Definition of operations with examples using $2^n$ | 8 examples |
| 4–5 | Introduction to using the tables | 2 examples |
| 5–6 | Determining nontabulated values | 1 example |
| 7 | Multiplication | 2 examples |
| 8 | Division | 2 examples |
| 8–10 | Rule of three ("*Regula Detri*") | 3 examples |
| 10–11 | Extracting the square root | 2 examples |
| 11–12 | Extracting the cube root | 3 examples |
| 13 | Extracting the fourth root | 1 example |
| 13–14 | Extracting the fifth root | 1 example |
| 14–18 | Finding a single mean proportional[a] (geometric mean) between two boundary numbers of either equal or unequal magnitude | 6 examples |
| 19–21 | Finding multiple mean proportionals (MPs) between two boundary numbers of equal magnitude | 1 example finding two MPs<br>1 example finding three MPs<br>1 example finding four MPs |

[a]Bürgi's description, "mean proportional," will be used throughout this book

multiplied by $10^8$. Thus, as tables of antilogarithms,[11] the arguments are the logarithms themselves (the red numbers in the *Aritmetische und Geometrische Progreß Tabulen*), and the antilogarithms (the black numbers) are retrieved in the body of the table.

Bürgi employed several techniques to make his tables more usable and comprehensible. As previously mentioned, Bürgi consistently[12] used red and black ink throughout the "Kurzer Bericht," which served to emphasize the difference between the antilogarithms and logarithms and which served to emphasize the relationship between arithmetic and geometric progressions. So as not to overcrowd the tables, for each new page of the tables, only the first row of the body of the table always includes all nine digits for each entry. The red numbers increase by 10 for each row; however, there is also an implied scale factor of 10. By referencing both the left-hand column and the top row, the exact logarithm (red number) and its corresponding

[11]The fact that Bürgi produced a table of antilogarithms has been described as being an important "marketing" device (Folta and Nový 1968, p. 98). Also, it is worth noting that Bürgi would not have used the term "base," as such a mathematical description was not in use at the time.

[12]There are, of course, exceptions to this consistency.

base number can be retrieved. The values within the tables are divided into 17 clusters of three rows each, and this serves as a visual aid when reading a page of 408 nine-digit numbers.

Despite the fact that Bürgi presented the theoretical motivation for his tables via the comparison of arithmetic and geometric progressions (or sequences) based on the powers of 2, it was obviously necessary to use a different numerical parameter to construct his tables. Bürgi knew of the work of the German reckoning masters Simon Jacob (d. 1564) and Moritius Zonz (or Moritz Zons, dates unknown) and was therefore familiar with the fact that a geometric progression with a common ratio of 2 (or any value much larger than 1) would produce terms that became too large too quickly to be useful. For example, Zonz (Figure 2.3) displays the juxtaposition of a geometric progression (powers of 3) and a corresponding arithmetic progression.

To produce a table of values that did not progress as quickly (in the geometric progression), Bürgi required the common ratio of the geometric progression to be close to 1. Thus, he selected a common ratio of 1.0001 for constructing his tables, and this common ratio choice created a smaller gap between any two successive black numbers, enabling Bürgi to employ linear interpolation to determine close approximations for a black number (or red, if using the tables in that direction) corresponding to any red number (or black, depending upon the use) resulting from calculations.

The first value in the body of Bürgi's table (in black) is 100000000 (Figure 2.4). Its corresponding logarithm (red number) is 0. Modern reconstructions (e.g., Katz 1998) show that subsequent values are generated via

$$B = 10^8 (1.0001)^{R/10} \tag{2.1}$$

where $B$ is the antilogarithm, or the black number from the table, and $R$ is the logarithm, or the red number from the table. Again, the black numbers form a geometric progression with ratio $r = 1.0001$. It is important to note that Bürgi did not reveal any such details that underlie the construction of his tables.[13]

When using a modern lens, we see that straightforward indexing techniques were used to tabulate the logarithmic values found in Bürgi's tables. To use the left and top edge of the table, simple addition provides the logarithm value (or 10 times the logarithm value). For example, to find the logarithm of 100400781 (near bottom, left corner of Figure 2.4), the column value (0) is added to the row value (400), and the result is divided by 10. Thus, the logarithm of 100400781 is $\frac{400}{10}$ or 40 in Bürgi's system. The modern calculation (rounded to the nearest whole number) corresponding to this use of the tables is confirmed using (2.1):

$$\begin{aligned} 10^8 (1.0001)^{40} &= 100400780.989 \\ &= 100400781. \end{aligned}$$

[13] For descriptions of how Bürgi might have constructed his tables, as well as error analysis of the printed tables, the reader is encouraged to consult Roegel (2010a), Waldvogel (2012, 2014).

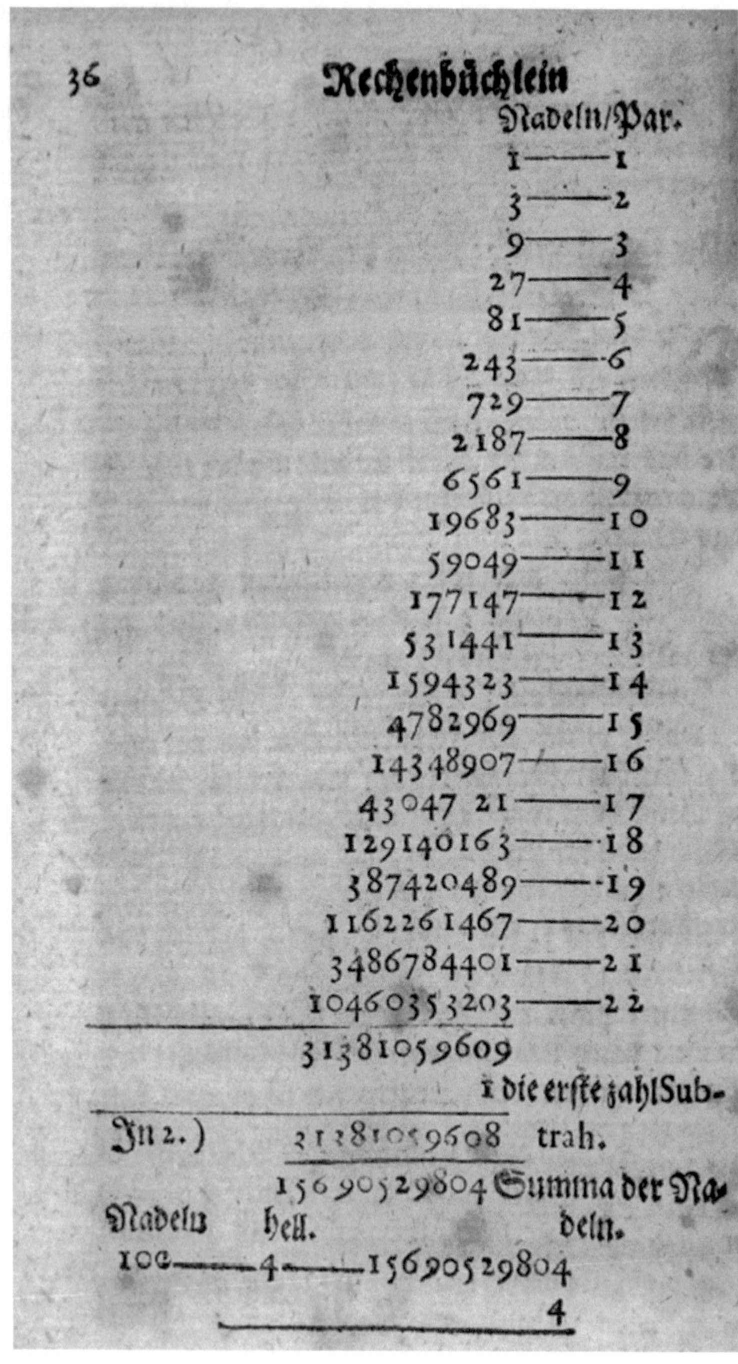

36 Rechenbüchlein

| Nadeln | Par. |
|---|---|
| 1 | 1 |
| 3 | 2 |
| 9 | 3 |
| 27 | 4 |
| 81 | 5 |
| 243 | 6 |
| 729 | 7 |
| 2187 | 8 |
| 6561 | 9 |
| 19683 | 10 |
| 59049 | 11 |
| 177147 | 12 |
| 531441 | 13 |
| 1594323 | 14 |
| 4782969 | 15 |
| 14348907 | 16 |
| 43047 21 | 17 |
| 129140163 | 18 |
| 387420489 | 19 |
| 1162261467 | 20 |
| 3486784401 | 21 |
| 10460353203 | 22 |

31381059609
1 die erste zahl Sub-
In 2.) 31381059608 trah.
15690529804 Summa der Nadeln.

Nadeln Hell.
100——4——15690529804
4

**Figure 2.3** Example of two juxtaposed progressions (Zons 1616, p. 36)

| | 0 | 500 | 1000 | 1500 | 2000 | 2500 | 3000 | 3500 |
|---|---|---|---|---|---|---|---|---|
| 0 | 100000000 | 100501227 | 101004966 | 101511230 | 102020032 | 102531384 | 103045299 | 103561790 |
| 10 | ····10000 | ····11277 | ····15067 | ····21381 | ····30234 | ····41637 | ····55603 | ····72146 |
| 20 | ····20001 | ····21328 | ····25168 | ····31524 | ····40437 | ····51891 | ····65909 | ····8250[illegible] |
| 30 | ····30003 | ····31380 | ····35271 | ····41087 | ····50641 | ····62146 | ····76216 | ····92861 |
| 40 | ····40006 | ····41433 | ····45374 | ····51841 | ····60846 | ····72402 | ····86523 | 103603221 |
| 50 | ····50010 | ····51487 | ····55479 | ····61996 | ····71052 | ····82660 | ····96832 | ····13581 |
| 60 | ····60015 | ····61543 | ····65584 | ····72153 | ····81259 | ····92918 | 103107142 | ····23942 |
| 70 | ····70021 | ····71599 | ····75691 | ····82309 | ····91467 | 102603177 | ····17452 | ····34305 |
| 80 | ····80028 | ····81656 | ····85799 | ····92468 | 102101676 | ····13438 | ····27764 | ····44068 |
| 90 | ····90036 | ····91714 | ····95907 | 101602627 | ····11887 | ····23699 | ····38077 | ····55033 |
| 100 | 100100045 | 100601773 | 101106017 | ····12787 | ····22098 | ····33961 | ····48391 | ····65395 |
| 110 | ····10055 | ····11834 | ····16127 | ····22949 | ····32310 | ····44225 | ····58705 | ····75765 |
| 120 | ····20066 | ····21895 | ····26239 | ····33111 | ····42523 | ····54489 | ····69021 | ····86132 |
| 130 | ····30078 | ····31957 | ····36352 | ····43274 | ····52738 | ····64755 | ····79338 | ····96501 |
| 140 | ····40091 | ····42020 | ····46465 | ····53438 | ····62953 | ····75021 | ····89656 | 103706871 |
| 150 | ····50105 | ····52084 | ····56580 | ····63604 | ····73169 | ····85289 | ····99975 | ····17241 |
| 160 | ····60120 | ····62150 | ····66696 | ····73770 | ····83386 | ····95557 | 103210295 | ····27613 |
| 170 | ····70136 | ····72216 | ····76812 | ····83938 | ····93605 | 102705827 | ····20616 | ····37986 |
| 180 | ····80153 | ····82283 | ····86930 | ····94106 | 102203824 | ····16097 | ····30932 | ····48360 |
| 190 | ····90171 | ····92351 | ····97049 | 101704275 | ····14045 | ····26369 | ····41261 | ····58734 |
| 200 | 100200190 | 100702420 | 101207168 | ····14446 | ····24266 | ····36642 | ····51585 | ····69110 |
| 210 | ····10210 | ····12491 | ····17289 | ····24617 | ····34488 | ····46915 | ····61910 | ····7[illegible]4[illegible]7 |
| 220 | ····20231 | ····22562 | ····27411 | ····34790 | ····44712 | ····57190 | ····72237 | ····8[illegible]865 |
| 230 | ····30253 | ····32634 | ····37533 | ····44963 | ····54936 | ····67466 | ····82564 | 103800244 |
| 240 | ····40276 | ····42707 | ····47657 | ····55138 | ····65162 | ····77742 | ····92892 | ····10624 |
| 250 | ····50300 | ····52782 | ····57782 | ····65313 | ····75388 | ····88020 | 103303221 | ····21005 |
| 260 | ····60325 | ····62857 | ····67908 | ····75490 | ····85616 | ····98299 | ····13552 | ····31387 |
| 270 | ····70351 | ····72933 | ····78035 | ····85667 | ····95845 | 102808579 | ····23883 | ····41770 |
| 280 | ····80378 | ····83011 | ····88162 | ····95846 | 102306074 | ····18860 | ····34216 | ····52155 |
| 290 | ····90406 | ····93189 | ····98291 | 101806075 | ····16305 | ····29142 | ····44549 | ····62540 |
| 300 | 100300435 | 100803168 | 101308421 | ····16206 | ····26536 | ····39425 | ····54883 | ····72926 |
| 310 | ····10465 | ····13248 | ····18552 | ····26387 | ····36769 | ····49708 | ····65219 | ····83313 |
| 320 | ····20496 | ····23330 | ····28684 | ····36570 | ····47003 | ····59993 | ····75555 | ····93702 |
| 330 | ····30528 | ····33412 | ····38817 | ····46754 | ····57237 | ····70279 | ····85893 | 103[illegible]04091 |
| 340 | ····40562 | ····43496 | ····48950 | ····56939 | ····67473 | ····80566 | ····96232 | ····14481 |
| 350 | ····50596 | ····53580 | ····59085 | ····67124 | ····77710 | ····90855 | 103406571 | ····24873 |
| 360 | ····60631 | ····63665 | ····69221 | ····77311 | ····87947 | 102901144 | ····16912 | ····35265 |
| 370 | ····70667 | ····73752 | ····79358 | ····87499 | ····98186 | ····11434 | ····27254 | ····45659 |
| 380 | ····80704 | ····83839 | ····89496 | ····97687 | 102408426 | ····21725 | ····37596 | ····56053 |
| 390 | ····90742 | ····93927 | ····99635 | 101907877 | ····18667 | ····32017 | ····47940 | ····66449 |
| 400 | 100400781 | 100904017 | 101409775 | ····18068 | ····28909 | ····42310 | ····58285 | ····76846 |
| 410 | ····10821 | ····14107 | ····19916 | ····28263 | ····39152 | ····52604 | ····68631 | ····87243 |

**Figure 2.4** Excerpt from the first page of Bürgi's tables (Gz manuscript)

The first 57 of the 58 pages of tables each contain 8 columns and 51 rows that produce 408 entries per page, for a total of 23256 entries. Except for the first entry of the first page, each final column entry is also the first entry of the next column. This means that of the 23256 entries, 22801 are distinct. Finally, page 58 of the tables includes 233 additional entries, of which 229 are distinct. Thus, Bürgi's tables are composed of 23489 entries, 23030 of which represent distinct calculations.

Although the construction of the tables appears to be driven by scale factors (either 10 and $10^8$) for the improvement in precision without the need for decimals, it is also clear from Bürgi's note at the conclusion of the tables that he saw the tables being easily used for any number desired. On the 58th sheet of the tables, Bürgi stated (see Figure 2.5):

So ends the sum of two numbers in 9 digits/and the Red [numbers up to][14]

**230270022 –**
**230270023 +**

[14]Texts within [...] are editorial insertions.

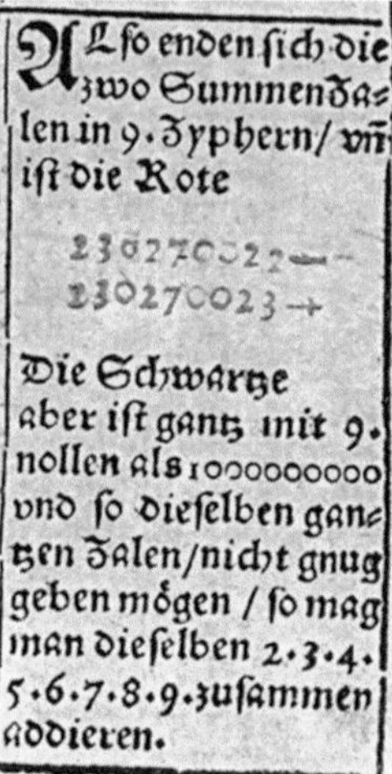

Also enden sich die
zwo Summen Za=
len in 9. Zyphern/ vn̄
ist die Rote

230270022—
230270023+

Die Schwartze
aber ist gantz mit 9.
nollen als 1000000000
vnd so dieselben gan=
tzen Zalen/nicht gnug
geben mögen / so mag
man dieselben 2.3.4.
5.6.7.8.9. zusammen
addieren.

**Figure 2.5** Excerpt from the last page of Bürgi's tables (bottom-right corner, page 58 of the tables)

The black [numbers], however, with only 9 zeros as in 1000000000 may not be enough and/you can add the same as 2, 3, 4, 5, 6, 7, 8, 9 [-digit numbers], together.

Thus, Bürgi, with this note at the end of the tables, declared that the "whole red number" (or, rather, the greatest logarithm he has calculated) is between 230270.022 and 230270.023 and that the black numbers can either be taken as multiples of or as parts thereof.

In summary, Bürgi based his system on a geometric sequence, with ratio 1.0001 and first term $10^8$. He then tabulated values from 100000000 to 999999999, and in modern notation (with the Bürgi logarithm taken as "$\log_{BG}$"), the calculation is

$$log_{\mathrm{BG}} 999999999.7 = 230270.022.$$

The black number reported in the body of the tables is just given as 999999999. Finally, Bürgi must have also calculated to obtain the relation (again, in modern notation)

$$log_{\mathrm{BG}} 1000000010 = 230270.023,$$

which yielded a value that was too large (i.e., greater than 1000000000). Thus, the logarithm of a number that was tenfold different from the starting value of the tables (100000000) was equivalent to adding or subtracting 230270.022 to or from its logarithm.

## Graphical Depiction of the Tables and Relation

In addition to his traditionally presented tables, Bürgi included a graphical table that summarized his system and that displayed a syncopated presentation of his tables. This circular representation includes increments of 5000 for the red numbers (the logarithms) and the corresponding black numbers (the antilogarithms), as well as a truncated version of the whole red number, 230270.022 (see Figure 2.6; this value is given (in red) as 230270022, with a small red circle appearing about the second "0" in the value). This representation appears on the title page, and the

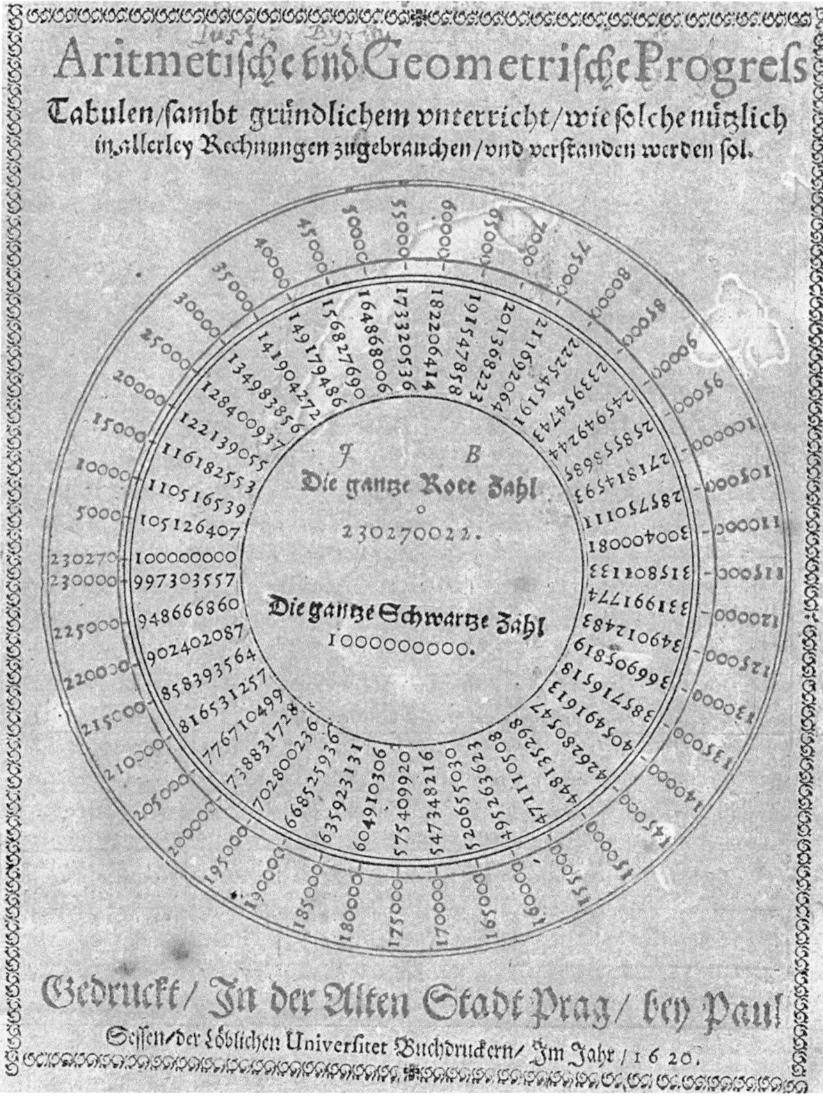


**Figure 2.6** Printed title page of Bürgi's *Aritmetische und Geometrische Progreß Tabulen* (Gz manuscript)

graphical display consists of two concentric circles.[15] Therefore, with this graphical rendering on a single sheet and the appropriate interpolation pattern, users could compute using Bürgi's system. Whether or not it was in fact used in place of the tables, this graphical image is emblematic of Bürgi's system in several ways: it highlights the cyclical nature of his system, it conveys the actual numerical relationships, and it captures the red–black numerical relation. Waldvogel (2012) observed that the "arrangement of the entries in a circular dial clearly shows Bürgi's genius since it documents his insight that the next decade, e.g., [10,100) is a mere repetition in tenfold size of the current one, e.g., [1, 10)" (p. 9).

The title page of the Gz manuscript contains two errors in the circle of values depicted. First, the black number corresponding to the red number 5000 should be 105126847, not 105126407 as given on the title page. Then, proceeding to the final value in the syncopated representation, the rounded value for the red number 230270 should be given as 1000000000, not the first value that appears in Bürgi's table (100000000). The remaining values appear correctly in the circular arrangement.

## The *"Kurzer Bericht"*

Bürgi produced tables that could be used for a variety of calculations, as well as brief instruction on how to use them. Table 2.2 provides each type of calculation found in the "Kurzer Bericht," and for each calculation (multiplication, division, square root, cube root, fourth root, fifth root, and mean proportionals) Bürgi's examples increase in complexity. The range of examples includes those that are (1) a straightforward use of the tables (e.g., black numbers and an operation are given; the corresponding red numbers are found and associated "simple" operation is performed; resulting black numbers are determined from the table); (2) interpolation (e.g., resulting red values that do not appear in the table that require linear interpolation between two that do appear); (3) the need for adding or subtracting the value of the "whole red number," or 230270.022 (e.g., a resultant red number larger than 230270.022 required that the whole red number be subtracted before determining the associated black number); and (4) a combination of a subset of the first three types.

The "Kurzer Bericht" contains 21 pages of worked examples and corresponding explanation or instruction on how to use the tables. Bürgi began the written instruction on how to use his tables by introducing the reader to two types of numbers: red numbers that are elements of an arithmetic progression[16] and black numbers that are elements of a corresponding geometric progression. Except in very few instances, the commentary in Chapter 3 will retain Bürgi's terms of "red number(s)" and "black number(s)" since he did not use the modern terms of "logarithm(s)" and "antilogarithm(s)," respectively.

[15] In the true geometric sense of the word, the circles are not exactly concentric, as this would have been a very difficult page to typeset in the sixteenth century.

[16] Bürgi used for the term "progression" for what is often referred to as a sequence. "Progression" and "sequence" will be used interchangeably throughout the commentary (Chapter 3).

# Chapter 3
# *Aritmetische und Geometrische Progreß Tabulen*: Edition, Translation, and Commentary

## Introduction

In the tercentenary memorial volume commemorating the invention of logarithms, or more precisely, John Napier's invention of logarithms, Florian Cajori (1915) observed: "In the history of science it is the rule, rather than the exception, for two or more men independently to develop the same idea" (p. 93). In the same publication, Cajori also stated that "[f]ew inventors have a clearer title to priority than has Napier to the invention of logarithms" (p. 93), and he continued to highlight men who contributed preliminary ideas for and simultaneous (though independent) conceptions of the logarithmic relation. Subsequent to the tercentenary memorial volume, modern scholarship has highlighted Bürgi's contributions to the development of logarithms. Boyer (1991), for example, noted that:

> Napier was indeed the first one to publish a work on logarithms, but very similar ideas were developed independently in Switzerland by Jobst Bürgi at about the same time. In fact, it is possible that the idea of logarithms had occurred to Bürgi as early as 1588, which would be half a dozen years before Napier began work in the same direction. (p. 314)

The remainder of this book focuses on a different conception of the logarithmic relation that appeared at roughly the same time as Napier's and was proposed by Jost Bürgi in 1620.

This chapter first presents a complete facsimile of the non-table pages of the Gz manuscript of the *Aritmetische und Geometrische Progreß Tabulen*. Following the reproduction, a complete transcription of the handwritten text is provided. To facilitate comparison with the other known extant text, I provide a copy of the Gk manuscript alongside the transcription of the Gz manuscript. Finally, the translation and accompanying commentary are divided into seven subsections according to related examples. The commentary includes a description of the calculations carried out by Bürgi, identification of notable errors, and discrepancies between the Gz and Gk manuscripts.

© Springer Science+Business Media New York 2015

K. Clark, *Jost Bürgi's Aritmetische und Geometrische Progreß Tabulen (1620)*,
Science Networks. Historical Studies 53, DOI 10.1007/978-1-4939-3161-3_3

## A Guide to Reading the Manuscript Transcription and Translation

There are several conventions used in the pages that follow that will assist the reader with the edition and translation of this nearly 400-year-old handwritten text. So that the reader may follow the text of the Gz copy of the *Aritmetische und Geometrische Progreß Tabulen*, First, the complete 24 pages of the manuscript are given, which includes the printed title page, the 2-page foreword, and the 21-page "Kurzer Bericht" (these pages may also be downloaded from www.springer.com/us/book/9781493931606). Next, the corresponding transcription is provided, as it was written, in order to preserve the original text (along with errors and inconsistent spelling conventions) as well as Bürgi's tone and style. A transcription of the Gk copy of the *Aritmetische und Geometrische Progreß Tabulen* appears in tandem, which was also recorded as it appears in the original text.[1] Finally, the translation and commentary are divided into seven subsections (Table 3.1), according to the purpose of the text and the type of examples discussed. In preparation for the segmented translation and commentary, the transcription is divided in the same manner.

The translation seeks to preserve the original narrative style of the manuscript; however, there are exceptions to this preservation. For example, very little punctuation exists in the Gz manuscript, and so that the reader does not have to suffer through very long passages, punctuation is included in the translation. Whenever possible, the pages of transcription and translation end as they are given in the Gz text, and beginning with page 2 of the "Kurzer Bericht," the lines of text in the translation follow as closely as possible to the lines of text in the transcription.

**Table 3.1** Transcription, translation, and commentary subsection descriptions

| Subsection | Corresponding *Aritmetische und Geometrische Progreß Tabulen* pages | Description and/or calculations |
|---|---|---|
| 1 | Title page and two-page foreword | |
| 2 | 1–4 | Introduction to the tables; eight examples of calculations using powers of 2 |
| 3 | 4–6 | Locating black and red numbers in the tables; linear interpolation |
| 4 | 7–10 | Multiplication, division, and regula detri |
| 5 | 10–14 | Extraction square, cube, and fourth roots |
| 6 | 14–18 | Determining two mean proportionals between two boundary numbers of either equal or unequal magnitude |
| 7 | 19–21 | Determining two, three, or four mean proportionals between two boundary numbers of equal magnitude |

[1] There are some minor exceptions to this. The transcription of the Gk manuscript here differs slightly from the Gieswald (1856) edition.

The important conventions used within the transcription and translation are:

- The Gz manuscript of the *Aritmetische und Geometrische Progreß Tabulen* is transcribed as it was written, including spelling errors (i.e., spelling idiosyncrasies are retained), omitted punctuation, and strikethroughs. The one exception is a leading ‘v.’ Words such as “vnd” are transcribed as “und.”
- Ink color is retained in the Electronic Supplementary Material (ESM); however, all text that appeared in red in the Gz copy appears in boldface in the edition (transcription), translation, and commentary.
- Latin terms are italicized; and, for hybrid terms that contain both Latin and German, the Latin part of the term is italicized.
- Words or parts of words that could not be determined in the Gz manuscript are given as <…>, and if the undetermined word(s) impacted the translation, a similar <…> appears within it, as well.
- Editorial insertions in the transcription and translation appear as [...].
- Since the Gz copy is transcribed “as is,” footnotes are used to explain errors to aid in reading the translation. For example, the term “souil” is used in the Gz copy and is understood as “so viel” or “soviel.” Indeed, this word illuminates two conventions: (1) the copyist printed the letter “u” (instead of “v” as we would expect in modern times) when not appearing as the first letter of a word, and (2) the copyist often dropped “e” in the “ie” letter combination. This convention of transcribing “as is” is also applied to incorrect values recorded by the copyist.
- As an aid to a smoother reading of the translation, all values are corrected and appear within square brackets [...].

The edition, transcription, translation, and commentary of the Gz copy of Bürgi’s *Aritmetische und Geometrische Progreß Tabulen* (1620) comprise the remainder of Chapter 3.

## Gz Manuscript of *Aritmetische und Geometrische Progreß Tabulen* (1620)

The Gz copy of *Aritmetische und Geometrische Progreß Tabulen* is reproduced in this chapter in the following order:

**Printed title page**

**Handwritten foreword (two pages), with transcription of the Gk manuscript on the facing page**

**Handwritten "Kurzer Bericht" ("Short Report"; 21 pages), with transcription of the Gk manuscript on the facing page**

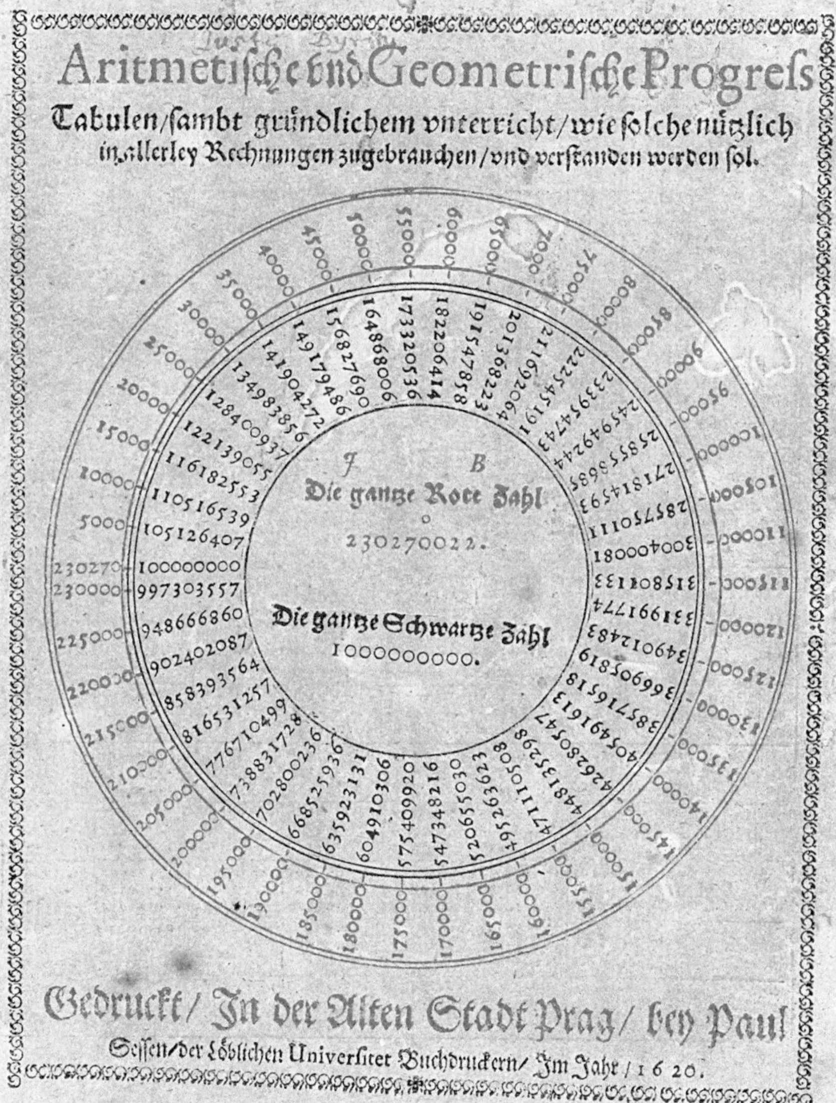
Aritmetiſche vnd Geometriſche Progreſs
Tabulen/ſambt gründlichem vnterricht/wie ſolche nützlich
in allerley Rechnungen zugebrauchen/vnd verſtanden werden ſol.
Die gantze Rote Zahl
230270022.
Die gantze Schwartze Zahl
1000000000.
Gedruckt/ In der Alten Stadt Prag/ bey Paul
Seſſen/der Löblichen Univerſitet Buchdruckern/ Im Jahr / 1620.

# Vorrede An den Trewhertzigen Leser.

Freundtlicher lieber Leser, obwohl von Vortrefflichen Mathematicis vnnd Arithmeticis Mancherley Tabulen seint verdichtet vnnd Calculirt worden, vm die Schwerrigkeiten deß Multiplicierens, diuidierens, vnd Radices Extrahierens Aufzuheben, so sein doch dieselbigen Allzeit nur particular gewesen, Also daß das Multiplicieren vnd Diuidieren Ihre eigene Tabulen, Als Abacum Pythagoricum erfordert hat, das Extrahiren der Radicum quadratarum seine quadrat Tabulen, die Cubische Extraction ihre Cubic Tabulen, vnnd also fort eine Jede quantitet ihre besondere Tabulen vonnötten hatt, Vil seindt aber der Tabulen nicht allein verdrießlich sondern auch mühselig vnnd beschwerlich seindt, derowegen Ich zu allerzeit gesucht vnd gearbeitet habe, General Tabulen zu erfinden, mit welcher man die vorgenandten sachen alle verrichten möchte, Betrachtendt derowegen die Aigenschafft vnnd Correspondentz der 2 Progressen, Als der Arithmetischen, mit der Geometrischen, das was in der ist Multiplicieren ist in Jener nur Addiren, vnd was ist in der diuidiren, in Jener Subtrahieren, vnd was in der ist Radicem quadratam Extrahieren, in Jener ist nur halbiren, Radicem Cubicam Extrahiren, nur in 3 diuidiren, Radicem Zensi in 4 diuidiren, Sursolidam in 5. vnd Also fort in Andern quantitaten, So habe Ich nicht Nutzlichers Erachtet, dan diese Tabulen Also zu Continuiren, das Alle Zahlen, so vorfallen, in derselben mögen gefunden werden,

Auß welcher Continuation dise Tabulen erwachsen, durch welche man nicht allein die schwerigkeiten, des Multiplicierens,
dividierens, und allerlej Radices Extrahierens, welche bej der
Algebra oder Coß einen trefflichen Vortheil und nutzen hat, ver-
hüetet werden; sondern auch das einschießen zwischen 2. gegebenen
Zahlen soviel media proportionales als man begehrt, mügen
gefunden werden, welches wie schwer es ohne dise Tabulen zugehet, denen bewußt ist, so sich einichmalig mit disen publice Exer-
cirt haben, und ob wohl ich mit disen Tabulen vor etlichen Jahren umbgangen bin, so hatt doch mein beruff von der Edition
derselben mich aufgehalten, wolle derowegen der guthertzige
Leser dises Ihme also gefallen lassen, und die Tabulen mit vol-
gender Underweisung des Verstandts, durch und mit etlichen Ex-
emplen erklert, günstig annehmen.

Wie hernach folgt.

Kurtzer Bericht der Progress Tabulen wie dieselbige nützlich in allerlay Rechnung zu gebrauchen.

In disen tabulen findet man zwayerlay zahlen, Eine mit roten Characteren, welche wir einem jeden künfftlich zu stehen, nichts anders dann ein Arithmetischer Progress, die andere aber mit schwartzen, anders dann ein Geometrischer progress ist, und auf daß wir in disen desto kürtzer durchgehen, wollen wir forthin den Arithmetischen progress die rothe, und den Geometrischen progress die schwartze zahl nennen, damit auch ein jeder die fundamenta diser tabulen gründtlicher fassen, und dieselbe desto bässer gebrauchen mag, so wollen wir in volgendem begriff die aigenschafft diser beeden progressen fürtragen stellen und dieselbigen mit etlichen Exempeln erklären.

| | | | | | | | | | | | | | |
|---|---|---|---|---|---|---|---|---|---|---|---|---|---|
| Arithmetisch | 0 | 1 | 2 | 3 | 4 | 5 | 6 | 7 | 8 | 9 | 10 | 11 | 12 |
| Geometrisch | 1 | 2 | 4 | 8 | 16 | 32 | 64 | 128 | 256 | 512 | 1024 | 2048 | 4096 |

Wir haben in der Vorrede angezaigt, wie auch vor etlichen

Arithmeticis Simon Jacob, Moritius Zons, Vnnd Andern ist
erweisst worden, daß was in den Geometrischen Progressen
oder in der Schwartzen zahl Multipliciert, das selbig ist
in den Arithmetischen Progressen oder in der Rothen zahl
Addiren.

Alß zum Exempel man soll Multipliciren 8 mit 64 die
Rothe zahl von 64 ist 6 Und von 8 ist 3 der Summa ist 9
von 6 Vnnd 3 ist dessen schwartze zahl ist 512 Und souil
kombt auch 6 man 8 mit 64 Multipliciert,

Item man soll Multipliciren 32 mit 256 Ihre Rothe
zahlen seindt 5 und 8 thut zusamen 13 dessen schwartze
zahl ist 8192 Und souil kombt so man 32 mit 256
Multipliciert,

Item man soll diuidiren 16384 durch 512 Ihr Rothe zall
seindt 14 und 9 Subtrahir den weniger 9 von 14 bleibt 5 sein
schwartze zahl ist 32 Und souil kompt 16384 durch 512
diuidiert.

Wiler dan die Regula detri nichts anders als Multipliciren
und diuidiern bedarff, so volgt das die Regula detri auch
zuuorderlich durch diese Tabulen verrichtet muge werden Als
zum Exempel,

8 · geben · 128 · mal geben · 32 gibt der Zahl ihr gebürende

Rote Zahl 3 · 7 · 5 addir und zusammen,

7
5
12 darnach Subtrahir die erst Rote
3 Zahl ist 3 bleibt der oder ihr schwarze
9 Zahl ist 512 welches ist der begehrten
Zahl facit genant,

Item man will Radicem quadratum aus 256 Extrahirn, sein Rote Zahl ist 8 · die halbier kombt 4 dessen schwarze Zahl ist 16 welches ist Radix quadratus aus 256 ·

Item man will Radicem Cubicam aus 512 Extrahirn sein Rote Zahl ist · das in 3 · dividirt, kombt · sein schwarze Zahl ist 8 und ist Radix Cubica aus 512 ·

Item man will Radicem Zensi Zensicum Extrahirn aus 4096 · sein Rothe Zahl ist · diß dividirt in 4 · kombt dessen schwarze Zahl ist 8 · welches Radix Zensi Zensica ist aus 4096,

Item man will zwischen 4 und 64 die mitler proportional finden ihre rothe Zahlen seindt 6 · und 9 6 man eine von der andern Subtrahirt bleibt 3 · diese in 3 · dividirt kombt 1 · diese 1 addire ich die erste der zweyen Mittlern proportionaln und man die 1 widerumb zu · 7 addiert kombt 8 dessen schwarze Zahl ist 256 die andern mitler proportional, und also fort, wie hernach vil angezaigt werden, und ist dise aigen,

hafft, aber nicht allein die zuvor abgesezte Proportion miteinander, sondern alle andere, so wir sie wollen
man der Aritmetische bey 0 und der Geometrische bey 1
anfangt, wie dan auch die folgende Tabulen, nichts
anders als solche Progressen seindt, und dieselbe seyn
gerechnet nur allein aus der obgesezten Progression, so
wollen wir zu dem gebrauch der Progress
Tabulen schreitten, und derselben lehrnen.

I Wie einer jeden schwarzen Zahl, so in den Tabulen under
schwarzen gefunden wirdt, ihr correspondirende
Rotte zufinden sey, als zum Exempel, man soll diese Zahl
133373810 Rote Zahl suechen, diese Zahl findet man in der
Tabulen am 8 blatt, in der Columna 28500 und an der
linckhen seiten under 300, diss addir dazu 300
macht 28800 welches ist also die Rote Zahl bey 133373810
und aus diesem weiss, das einer jeder Zahl, so in der
Tabulen zufinden, seine Rote Zahl erfunden werden.

II Wie einer jeden Roten Zahl, so in der Tabulen zufinden
ist, ihr gebürende schwarze Zahl soll gefunden werden
so wolle begehrt werden zum Exempel zu wissen,
welche schwarze Zahl dieser Rothen bey 28800

gebürendt, dieselben zu erforschen, so such ich under den Rotten Zahl die oben beim Anfang stehet, eine dergleichen, oder so nechst kleiner, als die fürgelegte ist, diese findt ich am 8 blatt in der Columnæ 28500 an welchem noch 300 manglen, such derowegen die 300 auf demselbigen blatt in der ersten Columnæ begegnen derselbigen über in der Columnæ under der 28500 werden gefunden 133373810 welche ist die begerte schwarze von 28500 und also handelt man auch mit den anderen, dann man findet der Rotten Zahl alle von 0 bis auf 230270 ihre gebürende schwarze Zahl auf obgemelte weise.

Wie aber, man eine Zahl fürstelte, so in der Tabula nit just zu finden were, khan man villen Rechnungen da vornehmen die rothe Zahl, welche der fürgebenen Zahl am nechsten ist, wer ihm aber damit nicht vergnügen ließ, khan auf folgende weise, seine begehrte Rotte Zahl erforschen.

Man soll zum Exempel die wahre Rott Zahl von 36 suchen, so setzt man noch sieben 0 für damit ich neun Ziffern bekomme, dan alle schwarze Zahlen haben in Vnnserer

Tabula, nicht weniger dan 9. differentz haben, derohalben
diese schwartze Zahl 36000000 darnach suche man in
der Tabula under der schwartzen Zahl die 2 nechst klein-
neren, und nechst grösseren ist das 36000000 diese finde
ich am 33 blat in der Columnæ 28000 auf der linken
seitten under: mir die schwartzt als 36000000 zwischen,

90 — diese hat schwartz — 359964763 — diese ist zu klein
der 10 die differentz · 35996 die differentz
100 — diese hat schwartz — 360000759 diese ist zu gross

diese kleinere zall von 90 ist ihr schwartze 359964763 subtra-
hen meinen gegebenen zall — — — — — 36000000

Restat 00035237 die dritte differentz

Wie sich helt die differenz zu der roten, also helt sich die 3 zur 4.

35996 | 10000 | 35237 also 9789

diese vierte addier zu der kleinern roten zall

die kleine rote zahl ist ——— 90

die zall der Columnæ ——— 128000

dis ist der schwartzen zall von 36000000 ihr rote 128099789

so soll gleich wol verstanden werden 36 haben ihr rote 128099 $\frac{789}{1000}$

Und werden alle weil bis under die 0 ganz verstanden und die folgen der bruch

Wie zwo Zahlen mit ein ander zu Multiplicieren seindt
Alß man soll Multiplicieren die Zahl. 15403018 mit 20551812
Und ihre Correspondierende Rothe Zahl ist 43200 und 72040
Die zwo Rote Zahl addire zusamen 43200
72040
Kombt diese Rote Zahl —— 115240
Von der schwarzen in 9 Ziffern. 36559928 und diese
seindt die Neün erste Ziffern deß products, an welches wir
bei diser Tabulen nur Neün Ziffer haben, Und die Letzte oder
Neündte nur vor ain brauch geben wöllen, dieweil vil Irrati-
onal Zahl vorfallen
Item man soll Multiplicieren. 551192902 mit 709153668
Ihre Rote Zahlen seindt. 170700 195900
Die zwo Rote Zahl addir zusammen 170700
195900
Tabula nicht so groß zu finden so Subtra: 366600 diese Rote Zahl
230270022 ist in der, das ist
bleibt die Rote diser rothen Zahl 136329978 die groß rote, such ihre
Schwarze Zahl ist — — — — 3908804680 welches
seindt die 9 ersten Ziffern deß begehrten products,
Allhier ist zu merckhen, daß ich dises Exempel zu endt ein
Ziffer mehr deß im vorigen manglet, das die Tabulen haben
nit mehr dan 9 Ziffern, und solten wohl 10 sein, daß ist die
Ursach, daß wir die ganze Rote Zahl Subtrahiern
müessen, welches nachfolgendts weiter erklert
soll werden,

Wie man eine Zahl durch die andere dividiren soll,
alß man soll dividiren 316559928 durch 20551812 und ist
die rote Zahl. 115240 und 72040 Subtrahirt man
des divisoris rote Zahl von des roten des dividendi alß 72040 von
115240 bleibt dise rote Zahl 23200 dessen schwarze
Zahl ist 154030185 oder 1 $\frac{54030185}{100000000}$

Item man soll dividiren 154030185 durch 20551812
ihr rote Zahlen sein - - 43200 und 72040
Subtrahirt man des divisoris rote Zahl von der roten des
dividente alß,
72040 von 43200 dieweil aber weniger ist so addirt man
die ganze rote Zahl 230270022
kombt - - - - 273470022 davon Subtrahir des
divisoris rote Zal - - 72040000
201430022 such dise rote Zahl ihr gebürende
Zall ist - - - - 749472554 schwarz, und biß kombt so
man 154030185 durch
20551812 dividiret, welches doch keine ganze, sondern
lautter bruch von ganzen alß 0 749472554 oder
$\frac{749472554}{1000000000}$

V

Wie man auch drey bekhandter Zahlen die vierte proporzional finden soll welches man gemeinlich die Regulæ detri zu nennen pflegt,

Alß zum Exempel,

Die erste — Die ander — Die dritte — Die vierte

wie sich 154030185 helt zur 205518112 also 399854564 zur vierten

Ihr Correspondierende rothe Zall: 43200 — 72040 — 138600

Addier die ander und dritte Rote Zahl zusammen als 138600.
72040

dis ist die Rote Zahl der vierten die — 210640 aus der Multiplicatio ...

Subtrahir darnach die erste rote Zahl — 43200

dis ist die Rote Zahl der vierten Zahl — 167440 die nun begert ...

Zeit — 533514619 dis ist die vierte Zahl

das Ander Exempel

Die erste — Die andere — Die dritte — Die vierte

Wie sich 100160120 helt zu 889122800 als 945919848 zur vierten.

Ihr Correspondieren de Rothe Zahl 160 — 21750 — 224710

227500

Addier die ander und dritte Rote Zal zusamen als

weil aber diese Zal grösser ist, das die ganze Rothe Zahl — 452210

So subtrahirt man die erste ganze rothe Zahl darnach — 160

Weil aber diese Zahl grösser ist das die ganze Rote Zahl — 452050
230270022

Subtrahirt man die ganze rote Zahl darnach als so bleibt — 211779978

Dieses vierte Zahl ist — 831194715

Dis ist, so man die endt noch ein 0 darnach setzt, die vierte begerte proportional und das darumb das ich die ganze Rote Zahl einmahl davon hab müssen subtrahiren, welches aber zuviel ist als mit 10 dividieren desshalben das hernach wider mit 10 muess Multipliciert werden,

Das dreitte Exempel

I II III IIII

Wie sich ·945919848 helt zur 100160120 also 880122800 zu der Vierten
diß sindt 224710 ihre Rote 160 Zahl 217500
Addier die Rote zweite und dritte Zall zusamen, 160
und welle durch die erste darvon subtrahiren, 217660
dieweill aber weniger ist, so addier darzu die gantze Rote 230270022 Zall
darnach subtrahir die erste Rothe Zall darvon 447930022
224710
so bleibt diese Rote Zahl und ist derselben - - - - - - - - 223220022
Schwartze Zahl ist ·931931024 welche, so man die letzte Ziffer abschneidt,
so darumb geschicht, daß die gantze rothe Zahl einmahl zum aggregat addirt ist,
die Vierte gesuchte proportional,

Auß einer gegebnen Zahlen Radicem quadratam zu extrahiren, wan soll zum
Exempel Radicem quadratam auß 4015374 extrahiren, wirdt also erstlich
punctirt, wie bey der Extraction breuchlich ist und stehet also ·4015374·
und weiln alhie vier Puncten seind, so wirdt seine Radix auß vier Ziffern
haben, die Rothe Zahl diser abgesetzten ist 139020 diß halbiert kombt
69510 dessen schwartze Zall ist 200383982, oder baß verstanden werden,
$20038\frac{3982}{10000}$.

Wan soll zum andern Exempel Radicem quadratam auß ·22033094· Extrahiren, wirdt also erstlich Punctirt, wie bey der Extaction breuchlich ist, und
stehet also ·22033094 und weiln alhie fünf Puncten kommen, so werden mi
Radix auß 5 Ziffern kommen, die noch den 5 punct bringt, seine Rote Zahl ist
79000 dieweil aber der letzte Puncten nit auf die letzte Ziffer felt
in der schwartzen Zahl, als im vorgenanten Exempel, sondern er felt
auf die zweite Ziffer, darumb muß die gantze Rothe Zall darzu
addiret werden und halbiert, also folgt, es sey seine Rote Zall also
folgt, 79000.
darzue addier die gantze Rote Zahl - - 230270022
diser Rothe Zahl halbier ———— 309270022
und derselben schwartze Zahl von diser Roten - - - - - - - - 154635011

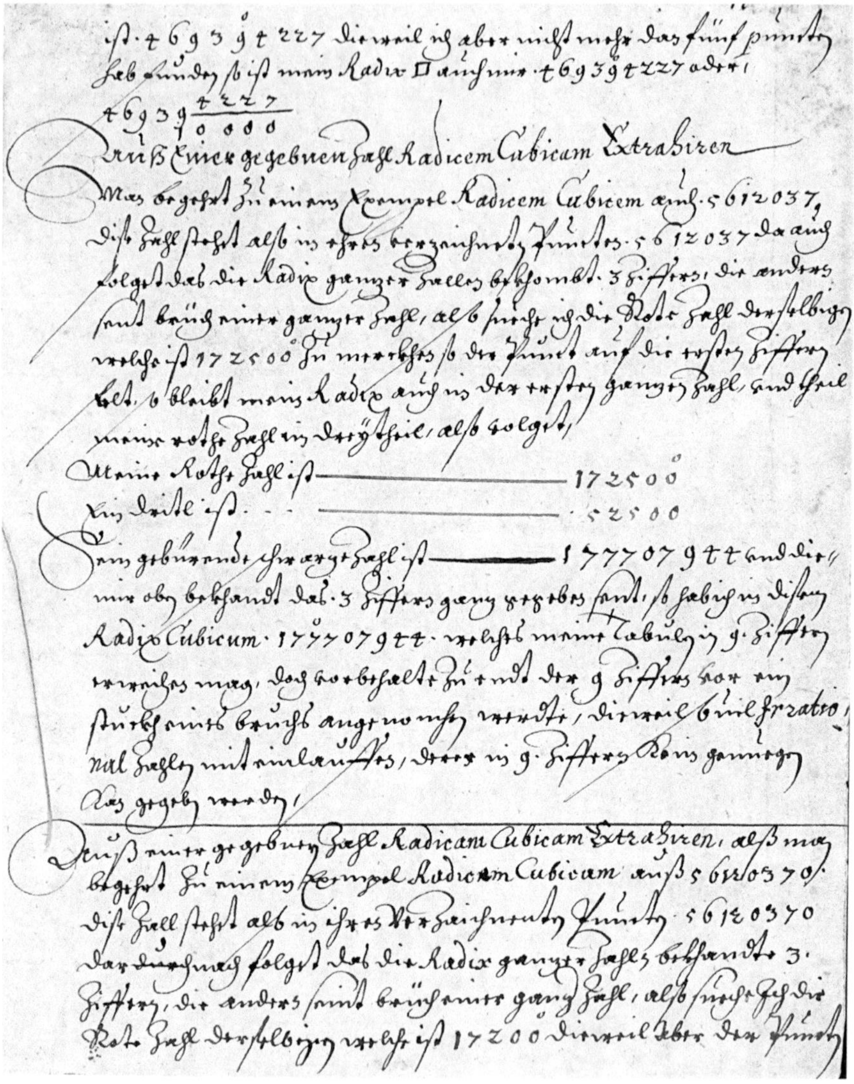

nit auf die erste Ziffer felt, sondern auf die ander, so würdt
zu der Rothen Zahl welche ist vorgegeben noch eine gantze Zahl addiert
Thut also zusamen 172500 und die gantze Zahl 172500
230270022

Cubus der 3 quantitet ist das

Dis theil in drey theil, dieweil der — 402770022
ein drittheil ist in Rothen — — — — — — — 134256674
Durch derselben schwartzes Zall ist — — — — — — 38286015 9

Radix Cubica

Auß einer gegebnen Zahl Radicem Cubicam Extra hiren mag
begehrt zu einem Exempel Radicem Cubicam auß 561203700
dise Zahl stehet als nichts bezeichnet Puncten. . . 561203700 aller
Zall, auch 3 puncten aber der letzte puncten felt auf die dritte
Ziffern obwohl dieselbe Zall der vorigen Exempel Rote Zal gebürt
als so werden doch noch zwo gantze Zahlen darzu addiret 172500
und ist das die ursach, die letzte 5 sambt der andern 230270022
Ziffern gebürt die Rote Zahl, dieweil aber der puncten 230270022
nit auf der ersten als 5 auch nit auf die andere als 633040044
0 sondern felt auf die dritte, so hat die erste 5 mit der andern Ziffern — 172500
und die 0 darnach eine gantze Rote Zahl — 230270022
Mehr die dritte ~~Zahl die erste~~ stehet 1 darauf der punct felt auch. 230270022
also hab ich der 3 ersten Ziffern ihr rote Zahl zusamen.
dieweil der Cubus die dritte quantitet ist so nimb ich von der
selben roten Zahl dein drittheil ist — — — . 211013346

Dis drittheil ist die Rote Zahl, die schwartze Zall ist Radix. - 824847192

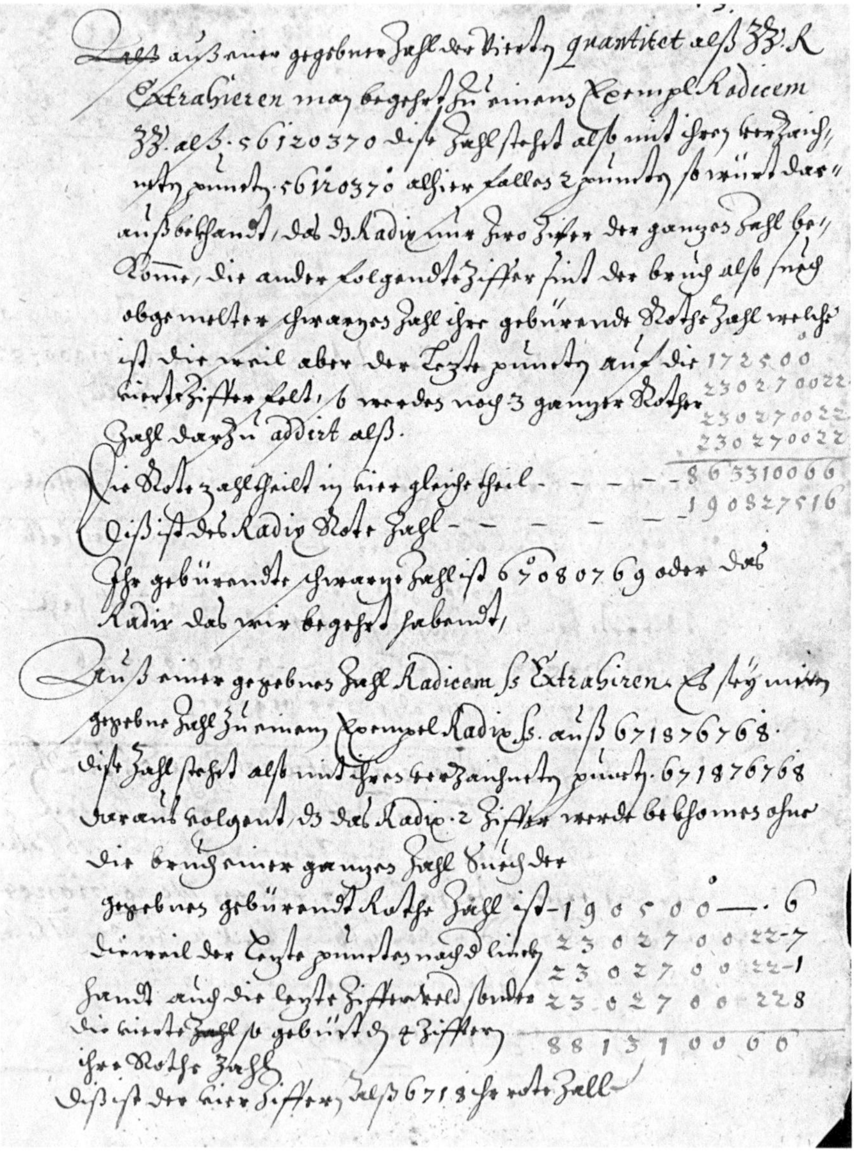

Dieselbe Theil in 5 gleiche theil kombt $\frac{1}{5}$ — 176262015 $\frac{1}{2}$ das
ist die rothe Zahl derselben gebürende schwarze Zahl des Radius
so bey 671876768 als 582717328 oder 58 $\frac{2717328}{10000000}$

Erstlich zwischen zwayen bekandten Zahlen ein Media pro-
I. portional Zahl zu finden, ob seien die 2 Zahlen 119004521
und 89342383. Ihre gebürende Rote Zall,
ist - - - - - - - 17400 und 219000
die differenz der roten Zahl ist ——— 201600 die theil
in zway gleiche theil od halbiert ist — 100800 das halb
Addir zu der kleinen roten Zahl ist — 17400
Diß ist die rote Zal der Medio proportio 118200 nal Zahl
Und Ihre schwarze ist die - - - 326069676
Medio proportional Zahl die wir begehren,

Zum andern, 2 Medio proportional Zahl zu finden
Theil die obgemelte rote differenz in 4 gleiche theil, und
II Addir der theil einen zu der kleinen roten Zahl, so haben
wir die erste rote Zahl, derselbigen Medio proportional Zahl, oder addir derselbigen theil 2 zu der kleinen
roten Zahl, so haben wir die andere rothe Zahl, derselbigen schwarzen Medio proportional Zahl,

III Zum dritten 3 Medio proportional zu finden, theil die ob gemelte differenz in 4 gleiche theil, und addier der theil eins zu der kleinern Roten Zahl, so haben wir die erste Rote Zahl derselbigen schwarzen Medio proportional Zahl, oder addier derselben theil 2 zu derselben kleinern Roten Zahl, so haben wir die andere Rote Zahl derselbigen schwarzen Medio proportional Zahl, oder addier derselben 3 zu der kleinen Roten Zahl, so haben wir die dritte Rote Zahl derselbigen Medio proportional Zahl, 15.

Weil diser weg kommen alle Medio proportional Zahl gefunden werden, so die 2 gegebenen Zahlen gleiche Numera Ziffern haben, also dass weiter in folgenden Exempel zu sehen,

Zwischen diesen 2 Zahlen ein Medio proportional Zahl zu finden, ist aber die 2 gegebenen Zahlen nit mit gleichen Numeren Ziffern, das die erste hat 7 Ziffern die andere 8 und stehendt also 2447471 und die andere 33053604, welche ihr gebührende Rote Zahl ist 89510 und 119500.

die addier zusammen ——— 89510

gibt diese rothe Zahl ---- 209010 dieweil aber die eine rote eine ziffer mehr hat dann die andere 230270022 wird die ganze Rote Zahl darzu addiert ist ——— 439280022 dise rote Zahl ist halb ——— 219640011 der gebürende schwarze ist diese Medio proportional 899159541 Zahl.

Zwischen 2 Zahlen ein Medio proportional Zahl zu finden
so sein aber die 2 Zahl nit mit gleichen numero Ziffern, da
die erste Zahl 7 Ziffern, die andere Zahl 8 hat und steht also.
// - - - - // 2447471 und die andere 33033604 ihre ge//
bürende Zahl rote ist 89510 die andere 119500
die addier zusammen — 89510
thut zusammen — — 209010 darzue
addier 2 ganze rote Zahl die weil die grössere 230270022
die kleinere mit 2 Ziffern übertrifft so kombt 230270022
669550044
diese rote Zahl halbir ist die Rote Zahl d— 334775022
gebürende zweyen Zahl, die weil aber
grösser ist das die ganze Rothe Zahl, so nimb
die ganz rote Zahl davon: so bleibt die Rothe 230270022
Zahl, der Medio proportional Zahl —— 104505000
welche ist —— 284339213

Dieweil ich hab die ganz Rothe Zahl, von der halbiren Rothen Zahl
subtrahiren können, so das sie auch ein Ziffern mehr haben das
die erste, als 8.
Zwischen zweien Zahlen ein Medio proportional Zahl zu finden, so
sein aber die zwo Zahlen, die mir vorfallen, als volgt.

Die erste mit 6 Ziffern, die aber mit 9 Ziffern

I · 303419 - - - II - 304939818
ihr gebührende rote 111000 - - Zahl 1115000
1118000

Addir zusamen dise bede --- 222500

darzu addir 3 ganze rote Zahl, die 230270022.
weil eine Zahl die ander mit 230270022.
2 Ziffern übertrifft ---- 230270022.

So kombt die Rote Zahl die halbier -- 913310066.
von diser halben Zahl subtrahir die ganz Rot Zahl 456655033.
230270022.

So bleibt dise rothe Zahl die gebürende Medio 226385011.
proportional Zahl welche ist ---- 961415942

Und ist nur umb ein Ziffer mehr dan die erste, und das ist der beweiß
das ich die ganze Rote Zahl nit mehr dan einmahl von der halben
halbierten Rote Zahl hab nehmen mügen,

Zwischen 2 Zahlen ein Medio proportional zu finden.

Zwischen 2 Zahlen ein Medio proportional Zahl zu finden,
Setz aber die 2 Zahlen die mir fürfallen, als welche,
die erste mit 5 Ziffern, die andere mit 9. und ist die erste,
-- 32891 -- die andere ist -- 454907654
ihr 119067351 gebürende rote Zahl . 151500000
addir zusamen ---- 119067351 die
ihre diese rote Zahl ------- 270567351.
darzu addir 4 ganze rote Zahlen die 230270022
weil eine die andere mit vier Ziffern 230270022
übertrifft. 230270022.
230270022.

So kombt dise rote Zahl, die halbier --- 1191647439.
von der halben rote subt: die ganze rothe Zahl 595823719 $\frac{1}{2}$.

Und so offt ich dieselbig mag, so viel ziffer wirdt die Medio proprotional
Zall 2 mahl mehr haben das die erste, das ich mag die gantze rothe Zahl
2 mahl, und bleibt mir über die rote Zal der Medio proprotional
welche ist - - - - - - - - - - - - - - - 1 3 5 2 8 3 6 7 5
Also ist die Medio protional Zahl - - - - - - - 3 8 6 8 1 2 9 9 8
so die mir begehrt haben,

Zwüschen 2 Zahlen ain Medio proprotional Zahl zu finden,
es sein aber die zwo Zahlen die mir vorkommen als.
die erste mit 4 ziffern, die ander mit 9 ziffern, und stehen als
als 5 7 6 4 . die andere — 2 8 7 6 4 9 8 3 3 .
ihr — 17517064 0 gebürende Rote 1 3 5 5 0 0 0 0 0 Zall die
addier zusammen - - - - - - - - - - 1 7 5 1 7 0 6 4 0
macht diese Rote Zahl - - - - - - - - 3 1 0 6 7 0 6 4 0 .
darzu 5 gantz Roter Zallen die weil
eine die andere mit 5 ziffern über
trifft - - - - - - - - - -

2 3 0 2 7 0 0 2 2 .
2 3 0 2 7 0 0 2 2 .
2 3 0 2 7 0 0 2 2 .
2 3 0 2 7 0 0 2 2 .
2 3 0 2 7 0 0 2 2 .

diese addierte Rote Zahl halbier - 1 4 6 2 0 2 0 7 5 0 .
ist diese rothe Zahl - - - - - - - - 7 3 1 0 1 0 3 7 5
darnach subtrahier die gantze Rote Zall so offt als ich mag in diesem
Exempel 3 mahl, darumb wirdt der Medio proprotional Zahl 5.
ziffern mehr haben dan die erste und bleibt ihr rothe Zahl
- - - - - - - - - - - - - - - 4 0 2 0 0 3 0 9
dieser gebürende schwartze Zahl ist die Medio — 1 4 9 4 7 8 5 9 1 .
proprotional Zahl.

19.

Zwischen 2 Zahlen die Medio proportional Zahl zu finden,

Ist auf unsere meinung eine geringe verenderung, 2 3 4
oder mehr Medio proportional Zallen zwischen 2 bekhandten Zallen zu finden,
darumb wollen wir die verenderung bekhandt machen, durch ein
Exempel, welches Irer [?] durch bekhandte Zallen geben ist,
und seien die 2 Zallen 119004521 und 893423483
ihr gebürende rote Zahl ist — 17400 und 219000
die differenz der Roten Zahl ist ——— 201600 die

theilen 3 theil ist ——,
ein drite addier zu der kleinern Zahl roth

So ist die Rothe Zahl der ersten proportional Zahl
ihr gebürende schwarze Zall ist die —

Zwey drite der differenz der roten Zal ist
und die kleiner rote Zahl addier darzu,

dis ist die rote Zahl der andern proportional Zahl
ihr gebürende schwarze Zahl ist die

| A: | B: | C: | D: |
|---|---|---|---|
| 119004521. | 23020839. | 45932698. | 893423483. |
| 17400 | 84600 | 151800 | 219000. |

wie sich helt A zu B: also helt sich B zu C und C zu D:

Zwischen 2 Zahl drey Medio proportional Zal zu finden/
nemblich die zwo bekhandten Zahlen 119004521 und 893423483
ihre gebürende rote Zall ist — 17400 die ander 219000
Ihre differenz ist ———————— 201600.
Die theil in vier gleiche theil, ist ein theil 50400.
17400.

Der theil auch addir zu der kleinen Roten 67800 Zahl die ist
die gebürende rote Zahl der schwarzen 196986715 diß ist
die erste Medio proportional Zall
Zum andern addir $\frac{2}{4}$ der roten differenz zu der kleinen roten Zahl
als. ———————— 50400 — die $\frac{2}{4}$
50400.
Und die kleiner rote Zahl ———— 17400.

Gibt die rote Zall die andere proportional 118200 Zahl
welches ist ihre gebürende schwarze Zahl 32606976.
die ander begehrte.
Zum dritten addir $\frac{3}{4}$ der roten differenz 50400.
50400.
und der kleiner roten Zahl — — — 50400.
17400.

diß ist die rote Zahl die dritte proportional 168600 Zahl,
welches ist ~~die rote~~ ihre gebürende schwarze Zahl 539735109
die dritt begehrte/

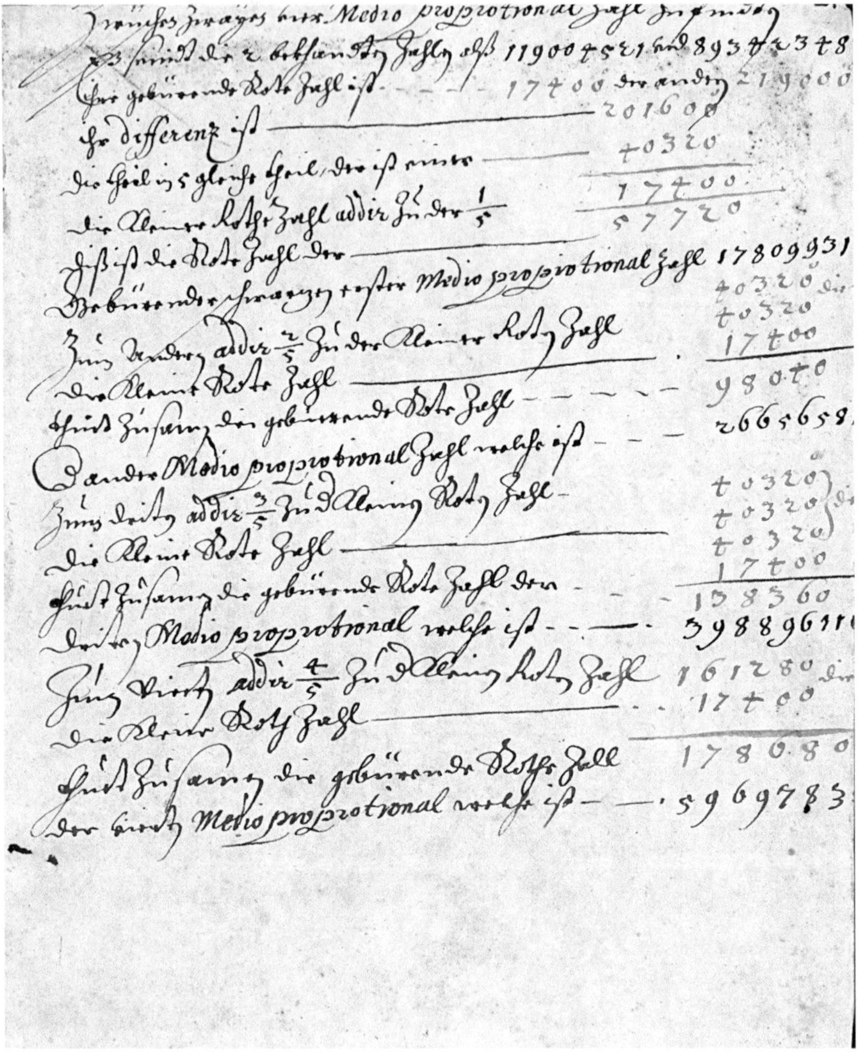

# Transcription

## Title Page and Two-Page Foreword

***Aritmeti*sche *und Geometri*sche *Progreß***

**Tabulen/sambt gründlichem unterricht/wie solche nützlich**
**in allerley Rechnungen zugebrauchen/und verstanden werden sol.**

**Gedruckt/In der Alten Stadt Prag/bey Paul**

**Sessen/der Löblichen Universitet Buchdruckern/Jm Jahr/1620.**

## Vorrede an den Treuhertzigen Leser.

Freundlicher lieber Leser, obwohl von Vortrefflichen *Mathematicis* und *Arithmeticis* Mancherleÿ Tabulen Sindt erdichtet und *Calculirt* wordten, um die Schwerigkeidten deß Multiplicierenß, Diuidierenß, und *Radices* Extrahierenß aufzuheben, so sein doch dieselbigen Allezeit nur *particu-*[2] *lar* gewesen. Also das das Multiplicieren und *Diuidieren* Ihre eigene Tabulen, Als *Abacum Pythagoricum* erfordert hat, das Extrahiern der *Radicum quadratarum* seine *quadrat* Tabulen, die Cubische *Extraction* ihre Cubic Tabulen, unnd also fort eine jedere *quantitet* ihre besondere Tabulen von nötten hatt, villheidt aber der Tabulen nicht allein verdrießlich sondern auch muhseelig unnd beschwerlich seindt. Derowegen Ich zu aller zeit gesucht und gearbeitet habe, General Tabulen zu erfinden, mit welchen man die vorgenändten sachen alle verrichten müchte. Betrachtendt derowegen die Aigenschafft unnd *Correspondenz* der 2 progressen, Als der Arithmetischen, mit der Geometrischen, das was in der ist Multiplicieren ist in jener nur *Add*iren, und was ist in der diuidieren, in Jehner Subtrahieren, und was in der ist *Radicem quadratam* Extrahieren, in Jener ist nur halbieren, *Radicem Cubicam* Extrahieren, nur

[2] For clarity throughout the edition and translation, the modern hyphen (-) will be used in this edition of the Gz manuscript instead of the notation (//) found in the Gz manuscript to designate the division of a word.

**Vorrede an den Treuhertzigen Leser.**

Freundlicher lieber Leser, Ob wol von Vortrefflichen *Ma*

*thematicis*, und *Arithmeticis*, mancherleÿ *Tabulen* seindt er-

dichtet, und *calculiert* worden, umb die Schwierigkeiten des *Multi*

*plicirens diuidirens* und *Radices extrahirens* auf zu heben,

so sindt doch dieselbige allezeit nur *particular* gewesen,

also daß das *Multipliciren* und *Diuidiren* ihre eigene *Tabulen*

als *abacum pÿthagoricum* erfordert hat das *Extrahiren* der

*radicum quadratarum* seine *quadrat tabulen* die cubische *Ex-*

*traction* ihre *Cubic Tabulen* und also fort ein iedere *quantitet*

ihre besondere *tabulen* vonnöten hat, vielheit der *Tabulen*

nicht allein verdrießlich, sondern auch müheselig und be-

schwerlich sein. Derowegen ich zu aller Zeit gesucht und ge-

arbeitet habe *general Tabulen* zu erfinden, mit welchen mann

die vorgenannten Sachen alle verrichten möchte. Betrachtent

derowegen die eigenschafft und *Correspondenz* der 2 *progreßē*

alß der *Arithmet*ischen mit der *Geometri*schen, das was in der ist

*Multipliciren* ist in iener nur *Addiern*, und was in der ist *Di-*

*uidiren* in iener *Subtrahiern* und was in der ist *radicem qua-*

*dratum Extrahirn* in iener ist nur halbiren *radicem cubicam Ex-*

in 3 diuidieren, *Radicem Zonsi* in 4 diuidieren, *Sursolidam*
in 5. Und Also fort in Andern *quantiteten*, so habe Ich nichts
Nutzlichers erachtet, dan dise Tabulen Also zu *continuiren*,
das Alle Zahlen so vorfallen, in der selben mugen gefunden werden,[3]
Auß welcher *Continuation* dise Tabulen erwachsen, durch welche man nicht allein die Schwerligkeiden des Multiplicierenß,
diuidierenß, und Allerleÿ *Radices* Extrahierenß, welches in der
*Algolia*[4] oder Coß einen trefflichen Vorthel und nutzen hatt, ver-
huetet werden, sondern auch das <suchrieß>[5] zwischen 2 gegebnen
Zahlen sovil *media porportionales* Alß man begehrt, mugen
gefunden werden, welches wie schwehr es ohne dise Tabulen zu
gehet, deenen bewust ist, so sich ein wenig mit disen *publice Exor-
cirt* haben. Und ob wohle Ich mit disen Tabulen vor ettlichen Jahren umbgangen bin, so hatt doch mein Beruff von der *Edition*
derselben mich enthalten, wolle derowegen der gutherzige
Leser dises Ihm Also gefallen lassen, und die Tabulen mit vol-
gender Underweisung des Verstandts, durch und mit ettlichen Ex-
emplen erklehrt, günstig Annehmen.[6]

Wie hernach folgt.

[3] Page 1 of the Foreword ends here.

[4] This is an odd spelling of "Algebra," but it is clear in the copy that it is given as "Algolia."

[5] This word is difficult to determine from the manuscript. If correct, it cannot be found in any German language resource. Gieswald (1856) transcribed "mehr ist," which does not make sense here.

[6] The final two words of this sentence are not found in Gieswald's transcription (1856).

*trahiern* nu in 3. diudiern, *radicem Zensi* in 4. *Diuidiern Sursolidam* in 5. und also fort in andern *quantiteten*, so habe ich nichts nutzlichres erachtet alß diese *Tabulen* also zu *continuiern* dass alle Zahlen so vorfallen in derselben mögen[7] gefunden werden, auch welcher *continuation* dieße *Tabulen* erwachßen, durch welche man nicht allein die schwerlichkeiten des *Multiplicierens Diuidierens* und allerleÿ *Ra dices Extrahierens*, welches in der *Algolia* oder *Cos* ein trefflichen Vortheil und nuzen hat, verhütet werden, sonder auch das mehr ist Zwischen 2. gegeben Zahlen so viel *media proportionalis* alß mann begert, mögen gefunden werden, welches wie schwer es ohne dieße *Tabulen* zugehet, ist denen bewust, so sich ein wenig in dießem *puluere exerciert* haben. Und ob wol ich mit dießen *Tabulen* vor ettlichen Jahren bin umbgang so hat doch mein Beruff von der *Edition* derselben enthalten, wolle derowegen der Guttherzige Leser dieße ihm also gefallen laßen, und die *Tabulen* mit volgenden Unterweisung, des Verstandes mit ettlichen Exempel erklärt[8]; wie folgt[9];

---

[7] Only one dot of umlaut is clear in the manuscript.

[8] There is an unusual character here, which looks like a subscripted script "P." For the transcription, I have used a semicolon (;) as it appears in Gieswald (1856).

[9] The same subscripted script "P" character appears here.

**Kurzer Bericht der *Progress Tabulen***

wie dieselbige nutzlich in Allerlaÿ

Rechnung Zu gebrauchen.

Zu dißen Tabulen findtet man Zwaÿerlaÿ Zahlen Eine mit rothen *Charact*ern, welche wie einem jeden Leüchtlich zu sehen, nichts anders dann ein *Arithm*etischer *progress*, die ander aber mit schwarzen, [10] anders dan ein Geometrischer *progress* ist, und auf das wir in disen desto kürzer durch gehen, wollen wir forthin den *Arithmeti*schen *progress* die rothe, und das *Geomet*rischen *progress* die schwarze Zahl Nennen, dar mit auch ein Jede die *Fundamenta* diser Tabulen grundtlicher Fassen und dieselbe desto beß-ser gebrauchen Mag so wollen wir in folgenden Be-griff die Aigenschafft diser Zweien *progress*en für Augen stellen und dieselbigen mit ettlichen Exempeln erkhleren.

| ***Arithmet*isch** | **0·** | **1·** | **2·** | **3·** | **4·** | **5·** | **6·** | **7·** | **8·** | **9·** | **10·** | **11·** | **12·** |
|---|---|---|---|---|---|---|---|---|---|---|---|---|---|
| *Geometr*isch | 1. | 2. | 4. | 8. | 16. | 32. | 64. | 128. | 256. | 512. | 1024. | 2048. | 4096. |

[10] There is a white space (i.e., a large gap in the text) following "schwarzen," which has the same dimension as the word "nichts" in the line before, and it should appear in this white space as well.

**Kurzer Bericht der *Progress Tabulen*,** Wie dieselbigen nutzlich in allerleÿ Rechnungen zu gebrauchen.

Zu diesen *Tabulen* findet mann Zweÿrleÿ Zahlen, Eine mitt rothen *Caractren*, welche wie einem ieden leichtlich zu sehen nichts anders dann ein *Arith*metischer *progress*, die ander aber mit schwarzen nichts anders dann ein *Geome*trischer *progress* ist, und auf daß wir in dießem desto kurzer durchgehen, Woll wir dorthin den *Arith*metischen *progress* die rothe, und den *Geomet*rischen *progress* die schwarze Zahl nennen, damit auch ein ieder die *fundamenta* dießer *Tabulen* grundlicher faßen, und dieselbigen desto beßer gebrauchen mag, so wollen wir in folgenden Begriff die Eigenschafft dießer 2. *progressen* fur Augen stellen und dieselben mit etlichen Exempeln erklären.

| Aritmeti*sch* | **0** | **1** | **2** | **3** | **4** | **5** | **6** | **7** | **8** | **9** | **10** | **11** | **12** |
|---|---|---|---|---|---|---|---|---|---|---|---|---|---|
| | 1. | 2. | 4. | 8. | 16. | 32. | 64. | 128. | 256. | 512. | 1024. | 2048. | 4096. |

Wir haben in der Vorrede Angeregt, wie Auch von etlichen[11]

*Arithmeticis Simon Jacob Moritius Zons*, und andern ist

berührt worden, das was in dem *Geomet*rischen *progress*en

oder in der schwarzen Zahl *Multipliciert*, das selbig ist

in den *Arithmeti*schen *Progress*en oder in der Rothen Zahl

*Addiren.*

Alß zum Exempel man soll *Multipli*ciern .8.[12] mit 64. Die

Rothe Zahl von .64 ist **6**. Und von 8. ist **3**. Der Summa ist **9**.

denn .**6** und **3**. ist[13] Deßen schwarze Zahl ist 512. und souil[14]

kombt eüch so man 8 mit 64 *Multipli*ciert.

Item man soll Multiplicieren .32. mit 256. Ihre Rothe

Zahlen[15] Seindt .5 und 8. Thuet Zusammen .13. Dessen schwarze

Zahl ist .8192 und souil khompt so man .32. mit 256.

Multipliciert.

Item man soll Diuidieren .16384. durch .512 Ihr Rote Zall

seindt .**14** und **9**. *Subtrahir* derowegen **9** von **14** Bleibt **5** sein

schwarze Zahl ist 32 und souil khompt .16384 durch .512

diuidiert.

---

[11] Page 1 of the "Kurzer Bericht" ends here.

[12] The punctuation that appears before and after numbers in the Gz edition does not serve as sentence punctuation. Instead, the notation is used to highlight numerical values in the text. There are inconsistencies in how the numerical punctuation is applied in the Gz edition.

[13] The copyist has forgotten the red number **9**, which should appear here. Additionally, this is a prime example of where end-of-sentence punctuation should appear; however, since the text of the manuscript may have been dictated to the writer or copyist, punctuation may not have been a primary concern.

[14] This is a version of "soviel," where as in many other instances "v" is written as "u" when it is not the first letter of the word.

[15] The copyist has failed to use red ink for the red numbers in this example or he or someone else entered these later.

Wir haben in der Voredt angeregt, wie auch von etlichen *Arithmeticis Simon Jacob Moritius Zons* und andere ist berürt worden, daß was in der *Geometri*schen *Progress* oder in der Schwarzen Zahl *Multipliciert* daßelbige ist in der *Arithmeti*schen *Progress,* oder in der rothen Zahl *addier*en, Alß zum Exempel mann sol *multipliciren* 8. mit 64. die rothe Zahl von 64. ist **6** und von 8. ist **3.** Der Summa ist **9**, dann **6** und **3** ist **9** dieße schwarze Zahl ist .512. und soviel kombt auch, so mann 8. mit 64. *multipliciert.*

Item mann soll *multipliciern* 32 mit 256. ihre rothe Zahl sindt [16] und [17] thuet zursamben, deße schwarze Zahl ist .8193. und soviel kombt, so mann 32. mit 256. *multipliciert.*

Item mann sol *Diuidirn* .16384. durch 512. ihre rothe Zahlen sind **14.** und **9.** Subtrahire derowegen **9** von **14** bleibt **5** sein schwarze Zahl ist 32. und soviel kombt .16384. durch 512. *Diuidiert.*

---

[16] No red number value is given.

[17] No red number value is given.

Weilen dan die Regula detrÿ nicht anders als Multipliciern

und diuidiers bedarff, so folgt das die Regula detrÿ auch

fürderlich durch dise Tabulen verricht müge werden. Als

zum Exempel,[18]

8 . geben. 128. was geben. 32 gib <…> der Zahl ihr gebürende

Rote Zahl **3**.[19] **7** **5** addir und Zusammen.

**7**

<u>**5**</u>

**12** Darvon Subtrahir die erste Rote

<u>**3**</u> Zal ist **3** bleibt der oder ihr schwarze

**9** Zahl ist .512 welches ist der begehrten

Zahl *facit* genant.

Item man will *Radicem quadratum* auß .256 *Extrahirn.* Sein

Rotte Zahl ist **8**. dis halbier khombt **4** Deßen schwarze Zahl ist

16 welches ist *Radix quadrata,* auß .256.

Item man will *Radicem Cubicam* auß .512 *Extrahirn.* Sein Rote

Zall ist .[20] Das in 3. diuidiert, kompt. [21] Sein schwarze Zahl ist

8 und ist *Radix Cubica* auß .512.

---

[18] Page 2 of the "Kurzer Bericht" ends here.

[19] In this example, the black numbers and red numbers are aligned vertically. For example, **3** appears directly below 8.

[20] The copyist has forgotten the red number **9**, which should appear here.

[21] The copyist has forgotten the red number **3**, which should appear here.

Weil dann die *Regula Detri* nichts anders als *Multi-*

*pliciern* und *Diuidierns* bedarff, so folget daß die *Regul Detri*

auch fürderlich durch dieße *Tabula* erreicht mag werden, alß zum

Exempel .8. geben .128. was geben 32. gib der Zahl ihre gebürende

**ll**

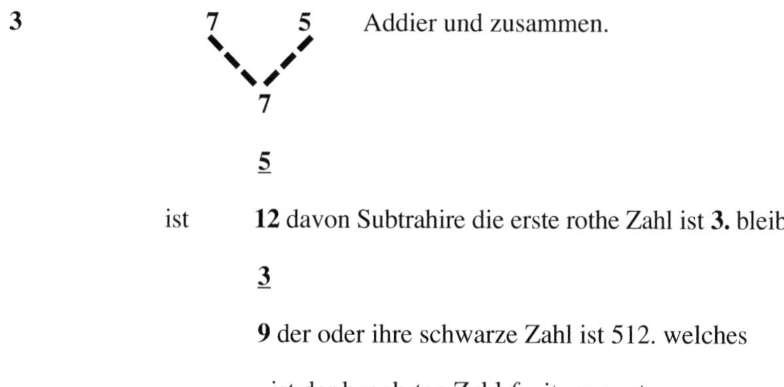
**3**
**7** **5**
Addier und zusammen.
**7**
**5**
ist
**12** davon Subtrahire die erste rothe Zahl ist **3.** bleibt
**3**
**9** der oder ihre schwarze Zahl ist 512. welches

ist der begehrten Zahl *facit* genannt.

Item mann wil *Radicem quadratam* auß .256. *Extrahieren* sein

rothe Zahl ist **8.** dis halbire kombt **4.** deße Schwarze Zahl ist

16. welches ist *Radix quadrata* auß 256.

Item mann wil *Radicem Cubicam* auß 512. *Extrahiern* sein rothe

Zahl ist **9.** das in 3 diuidiert kombt **3.** sein Schwarze Zahl ist 8.

und ist *Radix Cubica* auß 512.

Item man will *Radicem Zonsi Zonsicum*[22] *Extrahirn* auß .4096.

Sein Rothe Zahl ist ____.[23] Diß diuidiert in 4. kombt ___[24] Deßen schwarze

Zahl ist 8. welches *Radix Zonsi Zonsica* ist auch 4096.

Item man will Zwischen .4 und 64 die mitler *proportional*

finden ihre rothe Zahlen seindt 6. und 9[25] So man eine von der

andern *Subtrahirt* bleibt .3 Diße in 3. diuidiert kombt 1.

Dise 1 *addire* ich die erste der Zweÿen Mitlern *proportionalen*

und so man die 1 wiederumb zur .7 *addiret* kombt 8 Deßen

schwarze Zahl ist .256 Die andern mitler *proportional*, und

Also fort, wie hernach soll angezeigt werden, und ist Dise Eigen- [26]

schafft haben nicht allein die zwoe obgesezte *propro*[27]

mit einander sondern alle andere so sein wie sie wollen

wan der *Arithm*etische von .0 und der *Geomet*rische von .1.

anfangt, wie dan auch die folgende Tabulen nichts

anders als .2 solche *Progress*en seindt, und dises seÿ

geredt nur allein von den obgesezten *Progress*en. Jezo

wollen wir zu dem gebrauch unser *Progress*

Tabulen schreitten, und Erstlich Lehrnen.

[22] For the abbreviation of this root, *ZZR* (which appears on page 13 of the Gk edition), Gieswald (1856, p. 32) identifies this as "die Zahl 4 in der *Coss*," meaning it is the fourth root of a given number.

[23] The copyist has forgotten the red number **12**, which should appear here.

[24] The copyist has forgotten the red number **3**, which should appear here.

[25] These numbers are not written in red in the text. Furthermore, the two given numbers (i.e., black numbers) should be 64 and 512 for this corresponding pair of red numbers.

[26] Page 3 of the "Kurzer Bericht" ends here.

[27] This appears to be left over from the "*proportional* Zahlen" of the previous examples—and is not completely written. The copyist most likely meant "Progressen" instead, in order for the following three lines to make sense.

Item mann wil *Radicem Zensi Zensicum Extrahiern* auß

4096. sein rothe Zahl ist **12.** diß *Diuidiert* in 4. kombt **.3.**

deßen Schwarze Zahl ist 8. welches *Radix Zensi Zensico* ist auch 4096.

Item mann wil zwischen 4. und 64. die mittler *Propor-*

*tional* finden, ihre rothen Zahlen seindt **2** und **6** dieße *addiert*

geben **8.** deßen helfft ist **4.** sein schwarze Zahl ist 16. und

dießes ist die *Media proportionalis* zwischen 4 und 64.

Item mann wil 2. *media proportionalia* zwischen 64 und 512.

finden, ihre rothen seindt **6** und **9** so mann die eine von

der andern *subtrahiert* bleibt **.3.** dieße in 3. *diuidiert* kombt

**1.** diß **1** *addiere* ich zu der **6.** kombt **7.** sein schwarze Zahl ist

128. welches ist die erste der Zweÿen mittlern *proportionalen*

und so mann die **1.** wiederumb zu **7.** *addiert*, kombt **8.** deßen

schwarze Zahl ist .256. die ander mittler *proportional*, und

also fort, wie hernach sol angezeiget werden, und ist dieße

Eigenschafft haben nicht allein die 2. abgesezten *Progressen*

mit einander, sonder alle, sie sein, wie sie wollen, wenn der

*Arithmeti*sche von **0**. und der *Geometri*sche von 1. anfanget, wie

dann auch die folgendt *Tabulen* nichts anderß alß 2. solche

*Progressen* sindt. Und dießes seÿ geredt wir allein von der

obgesezten *Progressen*. Jetzo wollen wir zu dem gebrauch

unsre *Progress Tabulen* schreiten und Erstlich Lehren.

**I.**[28] Wie einer Jeden schwarzen Zahl so in den Tabulen under-

schwarzen gefunden wurdt, ihr *Correspondi*rendte

Rott zu finden seÿ. Als zum Exempel, man soll diser Zahl

133373810 Rote Zahl suechen. Dise Zahl findet man in der

Tabulen am 8 blat, in der *Columnæ* **.285$\dot{0}$0** [29] und an der

linekhen seiten under .**300** Diß *addir* darzue .**300**

macht **28800** welches ist also die Rote Zahl von .1333738$\dot{1}$0

und auf dise weise, kan ein jedwetern Zahl, so in der

Tabulen zu finden, seine Rote Zahl erfundten werden.

**II.** Wie einer jeden Rott Zall so in der Tabulen zu findten

ist, ihre gebürende schwarze Zall soll gefundten werden.

Eß wolle begehrt werden zum Exempel zu wüssen,

welcher schwarzen Zahl diser Rothen von **28800**[30]

gebürren. Dißes zu erforschen, so suech under der Rothen

Zahl die oben verzeichnet seint eine der gleichen, oder so

nechst kleiner, als die fürgelegte ist. Dise finde ich am

8 blat in der *Columnæ* .**28500** am welchen noch **300** manglen

suech derowegen die **300** auf demselbigen blatt in der

ersten *Columnæ* von gegen derselbigen über in der *Columnæ*

under der **28500** werden gefunden .1333738$\dot{1}$0 welche ist

---

[28] These section indicators appear in the margin to the left of the text.

[29] In the manuscript transcriptions, the "decimal zero" (i.e., decimal point) is superscripted above the last digit of the whole number, in the form of a small dot (as it is here) or small circle (in later examples).

[30] Page 4 of the "Kurzer Bericht" ends here.

**I**. Wie einer ieden schwarzen Zahl, so in den *Tabulen* Unter Schwarzen gefunden wirdt, ihre *correspondirende* rothe zu finden seÿ; alß zum Exempel.

Mann sol dießer Zahl 133373810. rothe Zahl suchen, dieße Zahl findt n in der *Tabulen* am 8. blat in der *columna* **28500** und an der linken seiten, under **300**. die *addier* darzu **300.** macht **28800**. welches ist also die rothe Zahl von 133373810. und auf dieße weis kann eines iedern Zahl, so in der *Tabul* zu finden, sein rothe Zahl erfunden werden.

Wie einer iedern rothen Zahl, so in der *Tabulen* zu finden ist, ihr gepürende schwarze Zahl soll gefunden werden.

Es wolle begehret werden zum Exempel zu wissen, welcher schwarzen Zahl dießer rothen von **28800** gebüeren, dießes zu erforschen, so such unter den rothen Zahlen, die oben vorzeichnet sein eine dergleich, oder so nahe kleiner alß die fürgegebene ist

Dieße finden ich am 8. blat in der *columna* **28500** an welchem noch **300**. mangelt such derowegen die **.300.** auf denselbigen blat in der ersten *columna* und gegen derselbigen über in der *columna* unter der **28500.** werden gefunden .133373180.

die begerte schwarze von **28500** und also handelt man auch

mit den anderen, dann man findet der Rothen Zahl alle

von **0** bis auf **230270** ihre gebürendte schwarze Zahl auf

obgemelte weise.

Wie aber man eine Zahl für fiel, so in der Tabula nit

just zu finden wëer, khan man viller Rechnungen da-

uor[31] nehmen die rothe Zahl, welche der für gebenen Zahl

am nechsten ist, wer ihm aber darmit nicht vergnuiegen

ließ, khan auf folgendte weise, seine begehrte Rote

Zahl erforschen.

**III.** Man soll zum Exempel die wahre Rote Zahl von 36

suechen, so sezt man noch siben 0 für damit ich neun Ziffern

bekhomme, den alle schwarze Zahlen haben in unßerer[32]

Tabula, nicht weniger dan .9. Zifferen haben, derohalben

diße schwarze Zahl .360000000 Darnach sueche man in

der Tabula under der schwarzen Zahl. Die .2. nechst klei-

nern, und nechst größere ist dan 360000000 diße finde

ich am 33 blat in der *Columnæ* .28000[33] auf der linekhen

:falt

seitten under :mir die schwarze als 360000000 zwischen,

[31] As in previous cases, this serves as a "v," as in "davor."

[32] Page 5 of the "Kurzer Bericht" ends here.

[33] Not printed in red in the Gz manuscript and this value should be **128000**.

welche ist die begehrte schwarze von **28800** und also handelet

mann auch mit den anderen, dann mann findt der rothen Zahl

alle von **0** biß auf **230270** ihm gebüerendt schwarze Zahl

auf obgemelten weg.

Wiedann eine Zahl für fiele, so in der *Tabul* nicht just zu finden weer

kann mann in vielen Rechnungen davor nemen die rothe Zahl welche

der fürgebene Zahl am nechsten ist, vor ihm aber darmit nicht vor-

gnügen ließ, kann auf folgende weise seine wahre rothe Zahl erforschen.

**II.** Mann soll zum Exempel die wahre rothe Zahl von 36. suchen, so setzet

mann noch Sieben 0. für damit ich 9. Ziffern bekomme, denn alle

schwarze Zahlen haben in unser *Tabula* nicht weniger

[alßo .9. Zifferen haben, derohalben diße schwarze Zahl][34]

360000000. darnach sucht mann in der *Tabul* unter der

schwarzen Zahl die 2. nechst kleiner, und nechst größer ist

dann 360000000 . diß finde ich am 33. Blat. in der

*columna* **12800** und auf der linkhen seite, nun felt

mir die schwarze alß 360000000. Zwischen.

[34] This final line of page 4 of the Gk manuscript is cut off at the bottom of the bottom of the page. The line that appears here is taken from Gieswald (1856).

| | |
|---|---|
| | ° zu |
| **90̇** – dißе hat schwarz – | 359964763. – Dißе ist: klein |
| den **.10 die differenz** | .35996 die differenz |
| **100** – dißе hat schwarz | ° .360000759 dißе ist zur groß |
| Dise kleinere Zall von **90** ist ihr schwarze | ° .359964763 Suptra: |
| von meiner gegeben Zahl _ _ _ _ _ _ _ | ° .360000000 |
| *Restat* | 000035237 die- drite differenz |

| Wie sich helt die | *Differenz* | zu der Roten, also helt sich die 3 zur 4. |
|---|---|---|
| | 35996 | **10̇000** 35237 alß **9789** |

Dißе vierte *addier* zu der kleinern Roten Zall

Die kleine Rote Zahl ist: ____________________________ **9°0**

**Die Zall der *Columnæ*** ___________________________ **128000**

---

Diß ist der schwarzen Zahl von 36°0000000 ihr Rote **128099°789**

Eß soll gleich wol so verstanden werden, 36 haben ihre Rote **128099° $\frac{789}{1000}$**

Und werden Alle Zeit biß under die o ganz verstanden und die folgen der bruch[35]

[35] Page 6 of the “Kurzer Bericht” ends here.

**90̊** . dieße hat schwarz – 359964763. diese ist zu klein <…>[36]

Den **10 die Differenz** 35996. die Differenz

**100** dieße hat schwarz – 3̊60000759 . diß ist zu groß <…>[37]

Dieße kleiner Zahl von ist ihr schwarz 3̊59964763. Subtrahire

von meiner gegebenen Zahl. 3̊60000000

000035237. Die <deit…>[38]

| Wie sich helt die | Differenz | zu der | rothen | also helt sich die 3 zur 4. |
|---|---|---|---|---|
| | 35996 | | **10̊000** | 35237 alß **9789** |

Diße Viert Vierte *addier* zu der kleinen rothen Zahl

Die kleiner rothe Zahl ist ________________________ **90̊**

**Die Zahl der *columnæ* ________________________ 128000.**

Dieß ist der Schwarzen Zahl von 3̊60000000 . ihr rote **128099̊789**

Es sol gleichwol so verstand worden .36. haben ihre rothe **128099̊ $\frac{78}{1000}$**

Und werden alle Zeit biß unter die ° ganze verstanden und

die folgen der bruch.

[36] There are several lines on this page where the words or numbers are cut off on the right edge. Where the rest of the page cannot be read, and if Gieswald's transcription provided editorial insertions, they are included in square brackets.

[37] Words and/or numbers are cut off from the right edge.

[38] Words and/or numbers are cut off from the right edge.

Wie zwo Zahlen mit ein ander zur Multiplicieren seindt

alß man soll Multiplicieren die Zahl .154030185 mit 205518112

Such ihre *Correspondir*endte Rothe Zahl ist **43200** und **72040**

Die zwo Rote Zahl *addire* zusamen **43200**

**72040**

Kombt diße rote Zall ______________ **115240**

Von der schwarzen in 9. Ziffern. 36559928[39] und diße

seindt die Neün erste Zifferen des *Products*, an welchen wir

unßer Tabulen nur Neün Ziffer haben, und die Lezte oder

Neündte nur vor ein bruch geben wöllen, dieweil vil *Irrati-*

*onal*[40] Zahl vorfallen.

Item man soll *Multiplicier*en .551192902 mit 709153668

Ihre Rote Zallen seindt **170700** **195900**

**170700**

Die Zwo Rote Zahl *addir* zusammen **195900**

**366600** Diße Rote Zahl ist in der

Tabula nicht so gross zufinden so *Subtra*: **230270022**

bleibt die Rote dißer rothen Zall **136329978** Daß ist Die große rote, suech ihre

Schwarze Zahl ist __ __ __ __ __ 3908804680[41] welches

---

[39] This is an error; the number should be 316559928.

[40] Gieswald (1856) has "*ihr rational*," but it is clear, as Lutstorf (2005) also indicated, that the Gz manuscript has "*Irrational.*"

[41] There is an extra digit here (the final "0") so that a 10-digit number is reported and not the typical 9-digit number.

Wie Zwo Zahlen mit einander zur *multipliciren* seindt alß mann

sol *multipliciern* die Zahl ____ 154030185 mit 205518112.

Such ihre *correspondierende* rothe Zahl ist **43200** und **72040**

Die zwo rothe Zahlen *addir* zusammen **43200**

<u>**72040**</u>

Kombt dieße rothe Zahl ____. **115240**

Von der schwarzen in 9. Ziffern ____ 316559928. und dieße

sindt die 9. ersten Ziffern des *product*s an welchen wir unser *Tabulen* nur 9. Ziffern haben und die letzte oder Neundte nun vor ein bruch geben wollen, dieweil viel ihr *rational* Zahlen vorfallen.

Item man sol *multiplicirn* 551192902. mit 709153668.

ihre rothe Zahl sein . . . . . **170700** und **195900**

**170700**

Die 2. rothe Zahlen *addier* zusammen <u>**195900**</u>

so kombt dieße rothe Zahl ____ **366600** dieße rothe Zahl ist in der

*Tabula* nicht so groß, so subtrair <u>**230270022**</u> es ist die größte rothe

bleibt die rothe dießer rothen Zahl. **136329978** such ihrer

schwarze Zahl ist ___________ 3908804680. welches seindt die seindt die .9 ersten Ziffern des begehrden *Products*

Alhier ist zumerekhen, das ich dißen Exempel zu endt ein

Ziffer mehr das im vorigen manglet, das die Tabulan haben

nit mehr das .9. Ziffren, und solten wohl 10 sein, das ist die

Ursache, das wir die ganze Rote Zahl *Subtrahirn*[42]

muessen 0 welches nach folgends weiter erkhleret

soll werden.[43]

Wie[44] man eine Zahl durch die Andere *diuidiren* soll,

alß man soll *diuidiren* **.**316559928 durch **.**205518112 und ist

die Rote Zahl.[45] **11524$\overset{\circ}{0}$ und 7204$\overset{\circ}{0}$** *Subrahirt* man

dn[46] *diuisoris* Rote Zahl von des Rothen[47] des *diuidendi* alß **72040** von

**11524$\overset{\circ}{0}$** so bleibt diße Rothen Zahl **23200**[48] deßen schwarze

Zahl ist $\overset{\circ}{1}$54030185 oder $1\frac{54030185}{100000000}$ [.]

Item man soll *diuidiren.* 154030185 durch 205518112

ihr Rote Zahlen sein - - **4320$\overset{\circ}{0}$ und 7204$\overset{\circ}{0}$**

---

[42] A larger than usual gap appears between the two words, "Zahl" and "*Subtrahirn.*"

[43] Page 7 of the "Kurzer Bericht" ends here.

[44] Section IV should be noted here, although the notation, as was given for I, II, III, and V, is not given in the margin for section IV.

[45] Again, this is not grammatical punctuation. Instead, "." is often written before a number, though, in this case, there is a large gap between "." and the red number.

[46] In numerous examples throughout the manuscript, a specific letter "d"' is used in place of the full article (e.g., der, die, das, den, dem). The symbol is derived from the fourth lower case letter in the Greek alphabet (d = δ). It is also similar to the letter "d" in the alphabet of the proper old German handwriting typeface.

[47] There is an error before the "R" which is difficult to represent here. It appears that something in red (possibly a number) was marked out and is now in the shape of an "O."

[48] This value should be **43200**, as was also reported in Lutstorf (2005).

9. ersten Ziffern des begehrten *products*.

Alhier ist zu merkhen, daß in dießem Exempel zu endt ein Ziffer mehr dann im vorigen manglet, dann die *Tablen* haben nit mehr dann 9. Ziffren, und solten wol 10 sein, das ist die Ursach, dß wir die ganze rothe Zahl haben *Subtrieren* müssen, welches nach'n obgendt weiter erklärt sol werden.

Wie mann ein Zahl durch die ander *Diuidiern* soll.

Alß mann sol *diuidiern* 316559928. durch 205518112 und ist

Ihr rothe Zahl . . . . **115240** und **72040** *subtrahiert*

Mann das *diuisoris* rothe Zahl von das rothen des *diuidendi* alß

**72040** von **115240** <…>[49]

Item mann sol *diuidiern* .154030185. durch 205518112.

ihre rothe Zahl sein **43200** und **72040.** *subtrahiert*

[49] Words and/or numbers are cut off from the bottom edge of the page. Also, it appears that Gieswald (1856) has omitted this example altogether.

*Subtrahirt* man des *diuisoris* Rote Zahl von der Roten des

*diuidenti* alß,

**7204$\mathring{0}$ von 4320$\mathring{0}$** dieweils aber weniger ist so *addirt* man

die ganze rote Zahl **230270022**

kombt _ _ _ _ **273470022** darvon *Subtrahir* das

*diuisoris* rote Zal - -**72040000**

**201430022**[50] suech diß rote Zahl wir gebürente

Zall ist - - - - 749472554 schwarz, und so vil kombt so

man .154030185 durch

205518112 *diuidiret*, welches doch keine ganze, sondn[51]

lautter bruch von ganzen alß $\mathring{0}$749472554 oder

$0\frac{749472554}{1000000000}.$

**V.** Wie man auch dreÿ bekhandten Zahlen die vierte *Proportional* finden soll welches man gemeinlich die *Regula Detri* zunenen pflegt.

Alß zum Exempel,[52]

Die erste Die Ander Die drite die vierte

Wie sich .154030185 helt zur 205518112 Also .399854564 zur 4 Za[hl][53]

Ihr *Correspondieren:* **4320$\mathring{0}$ 7204$\mathring{0}$ 93860$\mathring{0}$**[54]

---

[50] The first "**0**" appears as a blob of red ink.

[51] The word, "sondn," appears to be a shortened version of "sondern."

[52] Page 8 of the "Kurzer Bericht" ends here.

[53] The "-hl" of "Zahl" is cut off from copy on the edge of the page.

[54] The value, **938600**, should be **138600**.

mann des *diuisoris* rothe Zahl von der rothen des *diuidendi* alß

$\mathbf{7204\overset{\circ}{0}}$ von $\mathbf{4320\overset{\circ}{0}}$. Dieweils aber weniger ist, so *addiert* man die

ganze rothe Zahl **230270022** davon *subtr*ire des *diuisoris*

**273470022**

kombt

**72040000**

Rothe Zahl **201430022** such dießer rothen Zahl ihr gebürendt

schwarze Zahl ist .749472554. und soviel kombt so mann 154030185 durch

205518112 *diuidiert*, welches doch keine ganze, sonder lauter bruch

vom ganzen alß $\overset{\circ}{0}749472554$ oder $0\frac{749472554}{1000000000}$.

**V.** Wie mann auß 3 bekandten Zahlen die Vierdte *proportional* finden

sol, welches mann gemeinlig die *Regul detri* zu nennen pflegt.

alß zum Exempel,

| | die Erst | die ander | | die drit[t]e | die Vierte |
|---|---|---|---|---|---|
| Wie sich | 154030185 helt zwo | 205518112 | also | 399854564 | zur 4 Zahl ihre |
| ihre *correspondirende* rothe Zahl | $\mathbf{4320\overset{\circ}{0}}$ | $\mathbf{7204\overset{\circ}{0}}$ | | $\mathbf{13860\overset{\circ}{0}}$ | |

*de* rothe Zall

*Addier* die ander und drite Rote Zahl zusamen alß **138600.**

**72040**

Dis ist die Rote Zahl der schwarzen die _ _ _ _ _ **210640** auß der Mult-

Subtrahir darvon die erste rote Zahl _ _ _ _ **.43200** plicatio erwa[55]

ist

Diß ist die Rote Zahl der Vierdten schwarzen _ _ _ _ _ **167440.** die wir beger[56]

habe

Alß ________________________________ 533514619 diß ist die 4 Za[hl].[57]

---

Das Ander Exempel

| Die erste | Die andere | Die dreite | Die Vierte |
|---|---|---|---|

Wie sich 100160120 helt zu 889122800[58] also .945919848 zur der 4.

Ihre *Correspondieren*

*de* Rothe Zahlen

**160** **21750**[59] **224710**

**227500**[60]

*Addier* die ander und drite Rote Zal zusamen als ________.

[55] The rest of this word cannot be read from the edge of the page.

[56] The remainder of this word was cut off from the edge of the page; however, from the pattern within other examples, it makes sense that it is a version of "begehrten."

[57] "-hl" cut off from end of page.

[58] This is an error. The value should be 880122800.

[59] As a corresponding error to the given black number, this value should be **217500** (also reported in Lutstorf 2005).

[60] This is also an error. If using the value above, this should be **217500** (also reported in Lutstorf 2005).

Addier die ander und drit[t]e rothe Zahl zusamen alß**138600**

**72040**

Diß ist die rothe Zahl der Vierten Schwarzen die . . **210640** auß d Multi<…>[61]

<… …>[62]

*subtrier* darvon die Erst rothe Zahl . . . . . . **43200**

diß ist die rothe Zahl der Vierten Schwarzen **167440** die wir beg <…>[63]

alß . . . . . . . . . . . . . . . . 533514619 diß ist die <…>[64]

Das ander Exempel.

---

| | die Erst | die Ander | | die dritte | die Viert[e] |
|---|---|---|---|---|---|
| Wie sich | 100160120 helt zwo | 889122800.[65] | also | | 945919848 zur der 4 |
| ihre correspondirende | | | | | |
| rothe Zahl | **160.** | **21750** | | | **224710** |
| | | | | | **217500** |

[61] Words and/or numbers are cut off from the right edge.

[62] Words and/or numbers are cut off from the right edge.

[63] Words and/or numbers are cut off from the right edge.

[64] Words and/or numbers are cut off from the right edge.

[65] Should be 880122800 and red number is 217500.

**weil aber dißе Zal groser ist, dan die ganze** **452210**

**Rothe Zahl** _ _ _ _ _ _ _ _ _ _ _ _ _ _ _ _ _ _[66]

So subtrahirt man die erste ganze rothe Zahl darvon _ _ _ _ **160**

Weiln aber dise Zahl gröser ist den die ganze Rote Zahl **452050**

So subtrahirt man die ganze rote Zahl darvon als **230270022**

so bleibt ________________________ **211779978**

Dessen schwarzen Zahl ist _ _ _ _ _ _ _ 831194715

Dises ist, so man zu endt noch ein 0 darvon sezt, die vierte begerte *propor: tional* und das darumb das ich die ganze Rote Zahl einmahl davon hab mueßen Subtrahirn, welches aber souil ist als mit 10 diudiren deß halben das Facit weider mit 10 numrs Multiplicirt werden.

Das Dreitte Exempl[67]

| | I | | II | III | IIII[68] |
|---|---|---|---|---|---|
| Wie sich | .945919848 | helt zur | 100160120 Also | 880122800 | Zu der Vierten |
| Diß sindt | **224710** | ihre Rote | **160** | Zahl **217500** | |

*Addier* die Rote Zweite und drite Zall zusamen, **160**

**217660**

und solts durch die erste darvon Subtrahiren,

Dieweils aber weniger ist, so *addir* darzu die ganze Rote **230270022** Zall

**447930022**

Darnach Subtrahir die erste Rothe Zall darvon **224710**

so bleibt dise Rote Zahl und ist derselben _ _ _ _ _ _ _ **223220022**

Schwarze Zahl ist .931931024 welches ist so man die Lezte Ziffer abschneidt, so darumb geschieht, das die ganze rothe Zahl enmahl zum *aggregat addirt* ist, die Vierte gesuechte *propartional.*[69]

---

[66] Short red line drawn from here to point to the value 452210 in the line above.

[67] Page 9 of the "Kurzer Bericht" ends here.

[68] An arched-type symbol lies above each of the Roman numerals.

[69] Here, "proportional" is incorrectly spelled.

Addier die ander und dritt Zahl zusammen alß **442210**

So subtriert mann die erste Zahl darvon ....... **160**

Weil aber dieße Zahl größer ist dann die grosse rothe Zahl **442050**

Dießes ist so mann zu Endt noch ein 0 davon sezt, die Vierte begehrte *proportional* und das darumb, das ich die ganze rothe Zahl einmahl davon hab müßen *Subtrahiern* welches aber souiel ist als mit 10. *Diudiern*, deß halben das *Facit*, weider mit 10. nums *multipliciert* werden.

Das Dritt Exempel.

I II III IIII

Wie sich 945919848. helt Zwo 100160120. also 880122800. zu der Vierten.

diß seindt **224710.** ihr rothe **160** Zahl **217500**

*Addier* die rothe Zweÿte und dritte Zahl Zusamen **160**

**217660**

und solst die Erste darvon Subtriren dieweils

aber weniger ist, so *addier* darzu die ganze rothe **230270022** Zahl

**447930022**

Darnach *subtrier* die erste rothe Zahl darvon **224710**

so bleibt dieße rothe Zahl und ist derselben **223220022**

schwarze Zahl ist .931931024. welches ist so mann die letzt Ziffer abschneidt, so darumb geschieht, daß die ganze rothe Zahl einmal zum *aggregat addiert* ist, die Vierte gesuchte *proportional*.

Auß[70] einer gegebnen Zahlen *Radicem quadratam* Zu Extrahiren man soll Zum exempel *Radicem quadratam* auch 4015374. Extrahiren, wirdt also erstlich *puctirt*, wie beÿ der Extraction breüchlich ist und stehet also $.\dot{4}0\dot{1}5\dot{3}7\dot{4}$. und weilen, alhir vier Puncten sein, so wirt sein *Radix* auch vier Ziffern haben. Die Rothe Zahl diser obgesetzten ist $\mathbf{\dot{1}3902\mathring{0}}$ dise halbiert kombt $\mathbf{6951\mathring{0}}$. Desen schwarze Zall ist $2003\mathring{8}3982$, oder soll so verstanden werden $20038\frac{3982}{10000}$.

---

Man[71] soll Zum andern Exempel *Radicem quadratam* auch .22033094. Extrahiren, wirdt also erstlich *punctirt*, wie beÿ der *Extraction* breüchlich ist, und stehet, also $2\dot{2}0\dot{3}3\dot{0}9\dot{4}$ und weilen alhir fünf Puncten komen, so werden *im Radice* auch .5 Ziffern komen, die nach den 5 sindt bruch, sein Rote Zahl ist $\mathbf{7900\mathring{0}}$. Dieweil aber der Lezte Puncten nit auf die Lezte Ziffer felt in der schwarzen Zahl, als im vorgenanden Exempel, sondern er felt auf die Zweite Ziffer, darum muß die ganze Rothe Zall darzur *addiret* werden und halbiert, alß folgt, es seÿ sein Rote Zall alß -

| | |
|---|---|
| - folget | $\mathbf{7900\mathring{0}}$. |
| darzur *addier* die ganze Rote Zahl _ _ | $\mathbf{23027\dot{0}022}$ |
| diße Rothe Zahl halbier ______ | **309270022** |
| Suech derselben schwarze Zahl von diser Roten _ _ _ _ _ _ _ | **154635011**[72] |

---

[70] There is a symbol in the left margin which is difficult to determine; however, it could be an "8," as in section 8.

[71] Here again, there is a symbol in the left margin, which could be a "9."

[72] Page 10 of the "Kurzer Bericht" ends here.

Aus einer gegebenen Zahlen *Radicem quadratam extrahirn.*

Mann sol zum Exempel *Radicem quadratam* auß 4015374. *extrahiern*, wirdt also erstlich punctiert, wie beÿ der *extraction* breuchlich ist und steht also $\dot{4}0\dot{1}5\dot{3}7\dot{4}$ und weil alhier fünff puncten seindt, so wirdt sein *Radix* auch 5 Ziffern haben, die rothe Zahl dießer obgesetzten ist **139020.** dieße halbiert kombt **6951**$\mathring{\mathbf{0}}$**.** deßen Schwarze Zahl ist

$2003\mathring{8}3982$. oder soll so verstanden werden . $20038\frac{3982}{10000}$.

---

Mann sol zum andern Exempel *Radicem quadratam* auß 22033094. *extrahiern* wirt also erstlich punctiert, wie beÿ der *Extraction* bräuchlich ist, und steht also $2\dot{2}0\dot{3}3\dot{0}9\dot{4}$ und weil alhier 5. puncten kommen, so werden im *Radix* auch 5. Ziffern kommen, die nach den 5 sindt brüch, sein rothe Zahl ist **7900**$\mathring{\mathbf{0}}$**.** Dieweil aber der letzte puncten nit auf die erste Ziffer felt in der schwarzen Zahl alß im vorgenannten Exempel, sondern er felt auf die zweÿte Ziffer, darumb muß die ganze rothe Zahl darzu *addiert* werden und halbiret alß solche **7900**$\mathring{\mathbf{0}}$

darzu *addier* die ganze rothe Zahl . . . . . . . **230270022**

Dieße rothe Zahl halbier . . . . . . . . **309270022**

such derselben schwarze Zahl von dießer rothe . . . . . . **154635011.**[73]

---

[73] The final text at the bottom of this page is cut off; Gieswald (1856) has nothing further as well.

ist .4693$\overset{\circ}{9}$4227 dieweil ich aber nicht mehr dan fünf Puncten hab funden so ist mein *Radix* □ auch wir .46939$\overset{\circ}{4}$227 oder,

$$46939\frac{4227}{10000}.$$

Auß Einer Gegebnen Zahl *Radicem Cubicam* Extrahiren

Man begehrt zu einem Exempel *Radicem Cubicem* auch .5612037, Dise Zahl stehet also in ehren verzeichneten Puncten .5612037[74] da auch folget das die *Radice* ganzer Zallen bekhombt .3 Ziffern, die andern sent brüch einer ganzer Zahl, also sueche ich die Rote Zahl derselbigen welche ist **17250$\overset{\circ}{0}$** zu merekhen so der Punct auf die ersten Ziffern felt, so bleibt mein *Radix* auch in der ersten ganzen Zahl, und theil meine rothe Zahl in dreÿ theil, also volget.

Meine Rothe Zahl ist ________________ **17250$\overset{\circ}{0}$**

Ein drite ist _ _ ________________ **52500**[75]

Sein gebürende schwarze Zahl ist ________ 177707944 und diewie oben bekhandt das .3 Ziffern ganz gegeben sent, so hab ich in disen *Radix Cubicum* .17$\overset{\circ}{7}$707944 . welches meine Tabulen in 9. Ziffern erreichen mag, doch vorbehalte zu endt der 9 Zifferen vor ein stuekh eines bruchs angenomhen werdte, dieweil souil *Irrational* Zahlen mit einlauffen, deren in 9. Ziffern kein genuegen kan gegeben werden.

---

[74] The "dotting" procedure, that is, placing points above particular digits, is not indicated in this instance of the number, as was seen in the examples on page 10 of the "Kurzer Bericht."

[75] This is an error; the corresponding red number should be **57500**.

Auß einer geben Zahlen *Radicem Cubicam extrahieren.*

Mann begehrt zu einem Exempel *Radicem Cubicam* auß 5632037. diese Zahl steht also in ihren verzeichneten puncten $\dot{5}63\dot{2}03\dot{7}$ darauß folgert, daß die *Radix* ganzer Zahlen bekombt 3 Ziffern, die andern sindt bruch einer ganzen Zahl, also suche ich die rothe Zahl derselbigen, welche ist $\mathbf{17250\mathring{0}}$ zu merkhen so der puncten auf die erste Ziffer felt, so bleibt mein *Radix* auch in der ersten ganzen Zahl, und theil mein rothe Zahl in 3. theil, also volglich mein

rothe Zahl ist . . . . . . . . . . . . $\mathbf{17250\mathring{0}}$

Ein drittheil ist . . . . . . . . . . . . . . . . **57500**

die gebürendt schwarze Zahl ist . . . . . . . . . 177707944.

dieweil wir oben bekant, daß 3. Ziffern ganz gegeben seint, so habe ich in diesem *Radix cubicam* $17\mathring{7}707944$. Welches mein Tablen in 9. Ziffer erreichen mag, doch vorbehalten zu Endt der 9. Ziffern vor ein stukh eines bruches angenommen werde, dieweil souiel *ihrrational* Zahlen mit ein[lauffen], der in 9 Ziffern kein genügen kann gegeben werden.

---

Auß einer gegebnen Zahl *Radicam Cubicam Extrahiren*, alß man

begehrt zu einem Exempel *Radicam Cubicam* auß 56120370.

Dise Zall stehet also in ihren verzeichnenten Puncten 56120370[76]

dar ~~durch~~ nach folget das die *Radix* ganzer Zahlen bekhandte 3.

Ziffern, die andern seint brüch einer ganz Zahl, also sueche Ich die

Rote Zahl derselbigen welche ist **17200** [77] dieweil Aber der Puncten[78]

nit auf die erste Ziffer felt, sonder auf die ander, so wurdt

Zu der Rothen Zahl welche ist vorgegeben noch eine ganze Zahl *addiret*

Thuet also Zusamen **172500 und die** ganze Zahl **172500**

**230270022**

______ *Cubus* die

Diß theil in dreÿ theil, dieweil der _ _ _ _ _ **402770022** 3 *quant-*

ein drittheil ist in Rothen _ _ _ _ _ _ **134256674** *itet* ist

Suech derselben schwarzen Zall ist _ _ _ _ _ _ 382860159 das

*Radix Cubica.*

Auß einer gegebnen Zahl *Radicem Cubicam Extrahiren* man

begehrt Zu einem Exempel *Radicem Cubicam* auß 561203700

Dise Zahl stehet also in ihren verzeichneten Puncten _ _ _ 561203700[79] alhir

fallen auch .3 puncten aber der lezte punctn[80] felt auf die drite

---

[76] As in the previous example, the "dotting" procedure is not indicated here.

[77] This value is an error; it should be **172500**.

[78] Page 11 of the "Kurzer Bericht" ends here.

[79] The "dotting" procedure is not indicated on the associated digits of this number.

[80] In some cases, the ending "-en" appears just as "-n."

**10.** Auß einer geben Zahl *Radicem Cubicam extrahiern* Alß man begehrt zu einem Exempel *Radicem cubicam* auß 56120370. Darauß folget, daß die *Radix* ganzer Zahlen bekommen 3. Ziffern, die andern seindt bruch einer ganzen Zahl, also suche ich die rothe Zahl derselbigen, welche ist **172500 .** Dieweil aber der puncten nit auf die erste Ziffer felt, sonder auf die ander, so wirdt zu der rothen Zahl, welche ist vorgegeben, noch eine ganze Zahl *addirt*, thut also zursammen **172500 .**

und die ganze Zahl . . . . . . . . . . . **172500**

**230270022**

diß theil in 3. theil, dieweil der *Cubus* die 3 *quantitet* ist **402770022**

Ein drittheil ist im rothen . . . . . . . . **13425667[4]**

such derselben schwarze Zahl ist 382860159 . das *Radix Cubicam.*

Auß einer gegeben Zahl *Radicem Cubicam extrahiern.*

Man begehrt zu einem Exempel *Radicem Cubicam* auß 561203700.

dieße Zahl stehet also in ihr verzeichneten puncten 561203700 alhier fallen auch 3. puncte, aber der letzte puncten felt auf die dritte Ziffer, obwol

Zifern[81] obwohl dieselbe Zall der vorigen Exempel Rote Zal gebürt

| | |
|---|---|
| als so werden doch noch Zwo ganze Zahlen darzur *addiret* | **172500** |
| und ist dn[82] die Ursach, die erste .5 sambt den anderen | **230270022** |
| | **230270022** |
| Ziffern gebürt die Rote Zahl, die weil aber der puncten | **633040044** |
| nit auf den ersten alß .5 auch nit auf die andere als | |
| 6 sonder felt auf die dreite, so hat die erste 5 mit den andn[83] Zifern _ | **172500** |
| und die 6 darnach eine ganze Rothe Zahl ______________________ | **230270022** |
| Mehr die dritte ~~so hat die erste 5~~ stehet 1 daruf d[84] puncten felt auch | **230270022** |
| Also hab ich der 3. erstern Zifern ihre rothe Zahl zusammen. | |
| Dieweil der *Cubus* die dreite *quantitet* ist so nimmb von der- | |
| selben rothen Zahl dein dritheil ist__________________________ . | **211013346**[85] |
| Diß dritheil ist die Rothe Zahl, d[86] schwarzen Zal ist *Radix*_ _ | 824847192[87] |

---

[81] "Ziffern" is not often misspelled in the manuscript.

[82] A shortened version of the article "den".

[83] As with other truncated words in the text, "andern" is shortened to "andn."

[84] This is a shortened version, using the specific "d" letter, of the article.

[85] This value should be **211013348** (also reported in Lutstorf 2005).

[86] This is a shortened version of the article, possibly "den" or "der."

[87] This corresponding value is also inaccurate; it should be 824847208 (also reported in Lutstorf 2005). Also, page 12 of the "Kurzer Bericht" ends here.

| | |
|---|---|
| dieselbe Zahl des voreigen Exempels rothe Zahl gebürt, alß | **172500 .** |
| so werden doch noch zwo ganze Zahl darzue *addiert* | **230270022** |
| Und ist das die Ursach die ersten 5 sambt den anderen | **230270022** |
| Ziffern gebürt die rothe Zahl, dieweil aber der puncten nit | **633040044** |
| auf den ersten alß 5. auch nit auf die andere alß 6 | |
| sondern felt auf die dritte, so hat die Erste 5. mit den | |
| andern Ziffern . . . . . . . . . . | **172500** |
| und die 6. darnach ein ganze rothe Zahl . . . . . . | **230270022** |
| nachher die dritte steht .1. darauf der puncten felt auch | **230270022.** |
| Also hab ich der 3. erstern Ziffern ihr rothe Zahl zusammen | **633040044.** |
| Dieweil der Cubus die dritte *quantitet* ist, so nimb von | |
| derselben rothen Zahl, die drittheil ist . . . . | **211013346** |
| diß drittheil ist die rothe Zahl der schwarzen Zahl ist *Radix* . . | 824847192 . |

~~Alß~~ auß einer gegebner Zahl der Vierten *quantitet alß ZZ.R Extrahiren* man begehrt zu einem Exempl *Radicem ZZ.* alß .56120370 Dise Zahl stehet also mit ihren verzeichneten puncten . 56120370 alhier fallen 2 puncten so würt darauß bekhandt, das dn *Radix* nur zwo Zifer der ganzen Zahl bekomme, die ander folgendte Ziffer sint der bruch also suech obgemelter schwarzen Zahl ihre gebürende Rothe Zahl welche

ist die weil aber der Lezte puncten auf die **1725000**

vierte Ziffer felt, so werden noch .3 ganzer Rother **230270022**

Zahl darzu *addirt* alß. **230270022**

**230270022**

Die Rote Zahl theilt in vier gleiche theil _ _ _ _ _ **863310066**

Diß ist des *Radix* Rote Zahl _ _ _ _ _ _ **190827516**[88]

Ihr gebürendte schwarze Zahl ist 67080769[89] oder das *Radix* das wir begehrt habendt.

Auß einer gegebnen Zahl *Radicem Ss*[90] *Extrahiren.* Es seÿ meinen gegebne Zahl zu einem Exempel *Radix Ss* auß 671876768 .

Dise Zahl stehet also mit ihren verzeichneten puncten .671876768[91]

[88] This value should be **215827516** (the value also reported in Lutstorf 2005). However, it is correct if the dividend is **763310066** as given in the Gk manuscript.

[89] This number is also incorrect; it should be 8655260259 (the value also reported in Lutstorf 2005), given the corrected red number above.

[90] For "radix Sursolida," which is equivalent to the fifth radix (or root).

[91] The "dotting" procedure is not indicated on the associated digits of this number.

Auß einer geben Zahl der Vierten *quantitet alß ZZ R Extrahiern.*

Mann begehrt zu einem Exempel *Radicem ZZ* auß 56120370. Dieße Zahl steht also mit ihr verzeichneten puncten $561\dot{2}037\dot{0}$ alhier fallen 2 puncten, so wirt darauß bekant, daß das *Radix* nur 2. Ziffer der ganzen Zahl bekhome, die ander folgende Ziffer seindt der bruch, also such obgemelter schwarze Zahl ihr gepüerendt rothe Zahl welche

| | |
|---|---|
| ist . . . . . . . . . . | **$17250\mathring{0}$** |
| dieweil aber der letzte puncten auf die 4te Ziffer felt | **230270022** |
| so werden noch 3 ganzer rothen Zahlen darzu *addiert*, alß | **230270022** |
| dieße rothe Zahl theil in 4. gleiche theil . . . . . . . | **230270022** |
| | **763310066** |
| diß ist der *Radix* rothe Zahl . . . . . . . | **$190827\!\!\mathring{\phantom{0}}\!\!7516$** |

Ihr gebüerendt schwarze Zahl ist $6\mathring{7}408769$ od $67\frac{4080769}{10000000}$ das *Radix* das wir begehrt haben.

Auß einer gegeben Zahl *Radicem Ss extrahiern.*

Es zeige meine gegebene Zahl zu einem Exempel *Radix Ss* auß 671876768 dieße Zahl steht also mit ihr verzeichneten Puncten $67\dot{1}87\dot{6}768$ darauß

Daraus volgent, dn das *Radix* .2 Ziffer werde bekhomen ohne

Die bruch einer ganzen Zahl suech der

gegebnen gebürendt Rothe Zahl ist _ **$1905\overset{\circ}{0}0$** __ .6

Dieweil der lezte puncten nach d[92] linken **230270022** – 7

**230270022** – 1

handt auch die Lezte Ziffer veld sonder **230270022** 8

Die vierte ~~Zahl~~ so gebürt den 4 Ziffern ___________

ihre Rothe Zahl **881310066**

Diß ist der vier Ziffern Alß 6718 ihr rote Zall[93]

Dieselbe Theil in 5 gleiche theil seint $\frac{1}{5}$**$17626\overset{\circ}{2}015\frac{1}{2}$**[94] dn

ist die rothe Zahl derselben gebürende schwarze Zahl des *Radicis*

*Ss* von 671876768 alß 582717328[95] oder $58\frac{2717328}{10000000}$.[96]

[92] This is a shortened version of the article, using the specific "d" letter.

[93] Page 13 of the "Kurzer Bericht" ends here.

[94] This number is incorrect; it should be 176262013 $\frac{1}{5}$.

[95] This instance of the number is missing a "decimal zero," "o," which should appear above the first "8." Also, the value should be $5\overset{\circ}{8}2717318$.

[96] This value is incorrect. The fifth root of 671876768 is $5\overset{\circ}{8}2717318$.

folgen das, d *Radix* 2 Ziffer werde bekommen ohn die bruch einer

| | |
|---|---|
| ganzen Zahl, such der gegebenen gebürendt rothe Zahl ist | **19050$\overset{\circ}{0}$** |
| dieweil der letzte puncten nach der linkhen handt nicht | **2302700** |
| auf die letzte Ziffer felt, sonder auf die Vierdte so ge- | **2302700** |
| bürrt der 4. Ziffern alß 6718. ihr rothe Zahl . . . . | **2302700** |
| | **8813100** |
| dieselbe theil in 5 gleiche theil sindt $\frac{1}{5}$ | **170$\overset{\circ}{2}$620** |

das ist der rothen Zahl derselben gebüerende schwarzen Zahl

da *Radix Ss* von 671876768 alß $5\overset{\circ}{8}2717328$ od $58\frac{2717328}{10000000}$

**I**[97]

Erstlich zwischen Zwaÿen bekhandten Zahlen ein *Media proportional* Zahl zu finden, es seien die .2 Zahlen .119004521 und 893423483. Ihre gebürendte Rote Zall-

ist _ _ _ _ _ _ _ _ _ **17400** und **219000**.

Die *Differenz* d roten Zahl ist ______ **201600** die theil

---

In Zwaÿ gleiche theil od[98] halbiert ist __ **100800** das halb

*Addir* Zu der keinen[99] roten Zahl ist ___ **17400**

Dn ist die rote Zal der *Medio Proportio* **11820**.0 *nal* Zahl

Und ihre schwarze ist die _ _ _ _ .326069676

*Medio proportional* Zahl die wir begehren.

---

**II**[100]

Zum Andern .2 *Medio porportional* Zahl zu finden

Theil die obgemelte rote *Differenz* in 4[101] gleiche theil, und *Addir* der theil einer Zur der keinen[102] rothen Zahl, so haben wir die erste rote Zahl, derselbigen *Medio proportional* Zahl, oder *addir* derselbigen theil 2. zu der kleinen

---

[97] Section number (with a hat-shaped figure above) appears in left margin.

[98] This is a shortened version of "oder."

[99] Should be "kleinen," as in "add to the *smaller* red number" (translation).

[100] Section number (with a hat-shaped figure above) appears in left margin.

[101] This should be "3."

[102] The "l" is not apparent in the spelling of "kleinen" here, as in the previous instance.

Zwischen zweÿen bekannten Zahlen ein *Media Proportiona*l Zahl zu finden.

Es zeigen die 2. Zahlen 119004521. und 893423483.

ihre gebüerende rothe Zahl ist **1740**$\overset{\circ}{\mathbf{0}}$ und **21900**$\overset{\circ}{\mathbf{0}}$

die Differenz der rothen ist . . . <u>**201600**</u> die theil in

2 gleiche theil oder halbier ist **100800** Das halb

*addier* zu der kleinen rothen Zahl ist **17400**

diß ist die rothe Zahl d *medio proportional* **118200** Zahl

und ihre schwarze ist die . . . .326069676.

*medio proportional* Zahl, die wir begehren.

Zum Andern 2. *medio Proportional* Zahl zu finden.

Theil die obgemelte rothe Differenz in 3. gleiche Theil und *addier* der Theil

eines zu der kleinen rothen Zahl so haben wir die erste rothe Zahl derselbigen

*medio proportional* Zahl, oder *addier* derselbigen theil 2. zu der kleinen roten Zahl, so haben wir die andere rothe Zahl derselbigen schwarzen *Medio porportional* Zahl.[103]

---

**III**[104]

Zum dreiten .3 *Medio proportional* zu finden, theil die ob gemelte *Differenz* .in 4 gleiche theil, und *addir* der Theil einß zu der kleinen Roten Zall, so haben wir die erste Rote Zahl derselbigen schwarzen *Medio Proportional* Zahl, od[105] *addir* derselben theil 2 zu derselben kleinen Roten Zal so haben wir die andere Rote Zall derselbigen schwarzen *Medio proprotional*[106] Zall, oder *Addir* derselben 3 zu der kleinen Rothen Zahl, so haben wir die dreite Rote Zahl derselbigen *Medio Proportional* Zahl.

Auf dise weg können alle *Medio proprotional* Zall gefunden worden. So die 2 gegebnen Zahlen gleiche Summa Ziffern haben alß den weiter in folgenten Exempel zu ersehen.

Zwischen 2 Zahlen ein *Medio proprotional* Zahl zufinden. es[107] sein aber die 2 gegebene Zahlen nit mit gleichen Summan Zifferen, dan die erste hat .7 Ziffern die ander 8

---

[103] Page 14 of the “Kurzer Bericht” ends here.

[104] Section number (with a hat-shaped figure above) appears in left margin.

[105] This is a shortened version of “oder.”

[106] In many instances from this page forward, “proportional” is spelled as “proprotional.”

[107] “Es” is not capitalized here.

rothen Zahl, so haben wir die ander rothe Zahl derselbigen schwarzen *medio Proportional* Zahl.

Zum dritten 3 *Medio Proportional* zu finden, theil die obgemelte Differenz in 4. gleiche theil, und *addier* der theil eins zu der kleinern rothen Zahl so haben wir die erste rothe Zahl derselben schwarzen *medio Proportiona*l Zahl, oder *addier* derselben theil 2, zu derselben kleinern rothen Zahl, so haben wir die ander rothe Zahl derselbigen schwarzen *medio Proportiona*l Zahl oder *addier* derselben theil 3. Zu der kleinen rothen Zahl, so haben wir die dritte rothe Zahl derselben Schwarzen *medio proportiona*l Zahl.

Auf dieße weg können alle *medio proprotiona*l Zahlen gefunden werden, so die 2. gegeben Zahlen gleiche Summa Ziffern haben alß weiter in folgendem Exempel zu ersehen.

Zwischen 2. Zahlen ein *Medio Proportiona*l Zahl zu finden.

Es zeigen aber die 2. gegeben Zahlen nit mit gleichen Summen Ziffern, dann die Erste hat 7. Ziffern, die ander 8. und seindt also 2447471.

und stehendt also 2447471 und die ander 33033604.

Suech ihr gebührent Rote Zal ist **89510** und **119500**.

die *addir* zusamen. ___________________, **89510**

---

Gibt dise rothe Zahl _ _ _ _ _ _ _ **209010** die weil aber die

eine rote eine Zifer mehr hat dan die and[108] **230270022** so wirt die ganz

Rote Zahl darzu *Addirt* ist ___________ **439280022** dise rote Zal

ist halb ____________________________________ **219640011** d[109] gebürendte

Schwarze ist dise *Medio proprotional* 899159541 Zahl.[110]

---

Zwüschen .2 Zallen ein *Medio proportional* Zahl zu finden

es sein aber die 2 .Zahl nit mit gleichen Summa Ziffern, den

die erste hat .7 Ziffern, die andere hat 8[111] und stehet also.

„ - - - - „[112] 2447471 und die ander 330336040 ihre ge-

bürende Zahl rote ist **89510 die ander** **119500.**

Die *addier* Zusamen ______________________ **89510**

Thuet Zusamen _ _ _ _ _ _ _ _ _ _ _ **209010** darzue

*addir* .2 ganze rote Zahl die weil die grosere **230270022**

Die kleinere mit 2 Zifern übertrifft so kombt **230270022**

---

[108] This is a shortened version of the word “andere.”

[109] The article is shortened to the specific “d” character.

[110] The bottom-right corner of this page (page 15 of the “Kurzer Bericht” of the Gz manuscript) is torn. Also, page 15 of the “Kurzer Bericht” ends here.

[111] This is an error; it should be “9.”

[112] This collection of symbols designates that the first two words from above are repeated here.

und die ander 33033604. Such ihre gebüerende rothe Zahl

ist . . . **89510** Und **119500**

die Addier Zusammen **<u>89510</u>**

gibt dieße rothe Zahl . . . . . . . **209010** dieweil aber

eine Zahl ein Ziffer mehr hat dann die ander **230270022** so wird[t]

ganz rothe Zahl darzu *addiert* ist . . . . . **<u>439280022</u>** diese rothe

ist halb . . . . . . . . . . . . . **<u>219640011</u>** der <ge…>

schwarze ist dieße *Medio Proportional* Zahl . 899159541

Zwischen 2. Zahlen ein *Medio Proportional* Zahl zu finden.

Es zeigen aber die 2. Zahlen nicht mit gleichen Summen Ziffern, dann

die Erst hat 7. Ziffern, die ander hat 8. und stehendt

also. . . . . . 2447471. und die Ander 330336040. Ihre

gebüerende rothe Zahl ist **89510** die Ander **119500**

die Addier Zursammen . . . . . . . **<u>89510</u>**

Thuet zusammen . . . . . . . . . . **209010** darzu

*Addier* 2. ganze rothe Zahl dieweil die größer die **230270022** kleine <…>

**<u>230270022</u>**

| | |
|---|---|
| | **669550044** |
| Dise rote Zahl halbir ist die Rote Zal d[113] – . | **334775022** |
| gebürenden schwarzen Zahl, die weils aber | |
| größer ist dan die ganze Rothe Zahl so würt | |
| die ganz rote Zall Subt[114]: so bleibt die Rothe | **230270022** |
| Zahl, der *Medio proprotional* Zahl ______ | **104505000** |
| welche ist ________________________________ | 284339213 |

Die weil ich hab die ganz Rothe Zahl von der halbierten Rothen Zall *Subtrahirn* könen. so kan Ich auch ein Ziffern mehr haben dan die erste, also .8.[115]

Zwüschen zweien Zallen ein *Medio proprotional* Zall zu finden. es[116] sein aber die zwo Zahlen die mir vorfallen als volgt.

---

Die erste mit 6 Ziffern, die aber mit 9 Ziffern

| | | | | |
|---|---|---|---|---|
| **. I .**[117] | .303419 | _ _ _ _ | **II** _ _ | 304939818. |
| ihr gebürhede[118] rote | **111000** | _ _ | Zahl | **1115000** |
| | | | | **1110000** |

---[119]

[113] The article is shortened to the specific "d" character.

[114] Truncated version of "Subtrahirt" (or similar spelling of the word).

[115] Again, this should be "9."

[116] The "e" is not capitalized.

[117] This number is written with a hat-shaped figure above.

[118] This word is difficult to read, but it is most likely a version of "gebührende."

[119] Page 16 of the "Kurzer Bericht" ends here.

mit 2. Ziffer übertrifft, so kombt . . . . . **<u>669550044</u>**

dieße rothe Zahl halbier ist die rothe Zahl der **334775022.**

gebürenden schwarzen Zahl, dieweils aber

größer ist dann die ganze rothe Zahl, so wirdt

die ganze rothe Zahl *sub*: . . . . . . . **<u>230270022</u>**

so bleibt die rothe Zahl der *medio Proportiona*l Zahl **104505000**

welche ist . . . . . . . . . . 284339213.

Dieweil ich hab die ganze rothe Zahl von den

halbierten rothen Zahl … können, so kan

Ich auch ein Ziffer <…>[120]

Zwischen zweÿen Zahlen Ein *medio Proportional* Zahl zu finden.

Es zeigen aber die 2. Zahlen die mir vorfallen alß folget

die erste mit 6. Ziffern, die Ander mit 9. Ziffern

die Erste . 303419 . . . . die Ander . 304939818. ihr

gebüerendt roth **111000** . . . Zahl . . **11150**$\overset{\circ}{\mathbf{0}}$

**11150**$\overset{\circ}{\mathbf{0}}$

[120] The remainder of the text at the bottom of this page is cut off; Gieswald (1856) has nothing further after the line: “welche ist . . . . . . . . . .284339213.”

*Addir* zusamen Thuet sovil **22250 0̊**

Darzu *addir* .3 ganze rote Zahl, die **230270022 .**

weil eine Zahl die ander mit **230270022 .**

2 Ziffern ubertrifft _ _ _ _ _ **230270022 .**

---

So kombt die Rote Zahl die halbier _ _ **913310066 .**

von dise halben Zahl Subt die ganz Rot Zahl **456655033 .**

**230270022 .**

---

So bleibt dise rothe Zahl d[121] gebürende *Medio* **226335011 .**[122]

*proprotional* Zahl welche ist _ _ _ _ 9614159̊42 [123]

Und ist nur umb ein Ziffer mehr dan die erste, und das ist der beweiß das ich die ganze Rote Zahl nicht mehr dan einmahl von der halben halbierten Roten Zahl hab nehmen mügen.

Zwüschen .2 Zahlen ein *Medio proportional* Zahl Zuefinden.

---

Zwüschen .2 Zahlen ein *Medio proportional* Zahl Zu finden.

Eß sein aber die 2 Zahlen die mir vorfallen, alß volgt.

die erste mit 5 Ziffern, die andere mit 9. und ist die erste-

„ _ _ 32891. _ die andere ist _ _ 454907654,

---

[121] The article is shortened to the specific "d" character.

[122] This number should be **226385011** (also reported in Lutstorf 2005).

[123] This number should be 9618967̊44.

Addier zusammen thut soviel . . . **222500**

darzu Addier 3. ganze rothe Zahl die- **230270022**

weil ein Zahl die ander mit 3 Ziffer übertrifft, **230270022**

**230270022**

so kombt die rothe Zahl . . . . . . **913210066** die halbier.

Von dißer halben Zahl *sub*: die ganze rothe Zahl **456655033**

so bleibt dieße rothe Zahl . . . . . . **230270022**

**226335011.**

der gebüerende *medio proportional* Zahl welche ist .961415942 .

und ist nur umb ein Ziffer mehr dann die Erst, und das ist der beweiß

daß ich die ganze rothe Zahl nicht mehr dann einmahl von der halben

halbirten rothen Zahl hab nemmen mögen.

Zwischen 2. Zahlen ein *medio proportional* Zahl zu finden.

Es zeigen aber die 2. Zahlen, die mir vorfallen, alß folget.

Die Erste mit 5. Ziffern, die ander mit 9. Ziffern, und ist die Erste

32891. Die Ander ist . . . 454907654.

ihr **1190̊67351** gebürende rote Zal **151500000** [124]

*Addir* zusammen _____ _____ ______ **119067351** die

thuet dise rote Zall _ _ _ _ _ _ _ **270567351 .**

darzu *addir* .4 ganze Rote Zahlen die- **230270022 .**

weilen eine die andere mit vier Ziffern **230270022 .**

ubertrifft. **230270022 .**

**230270022 .**

---

So kombt dise rote Zall, die halbier _ _ _ _ **1191647439 .**

von der halben roten Subt: die ganze rothe Zahl **595823719$\frac{1}{2}$.**[125]

Und so offt ich desselbig mag sovil Ziffer wirdt die *Medio proprotional* Zall .2 mahl mehr haben dan die erste, dan ich mag die ganze roth Zahl .2. mahl und bleibt mir über die rote Zal der *Medio proprotional* welche ist _ _ _ _ _____ _ _ _ _ _____ _ _ _ **135283̊675 .**

Also ist die *Medio $\overline{\text{pro}}$tional*[126] Zahl _ _ _ _ _ _ 386812198[127] //[128]

// die wir begehrt haben.

---

[124] Most of this number is overwritten with red ink. Also, the first "0" appears to have been a "6" originally.

[125] Page 17 of the "Kurzer Bericht" ends here.

[126] The line above the three letters "pro" (e.g., " $\overline{\textit{pro}}$ ") means that the letters are repeated and should be read as "propro." As previously mentioned, the copyist continues to spell "proportional" as "proprotional."

[127] This number's decimal zero is difficult to discern in the Gz manuscript (if it appears at all); it should be placed over the second "1," corresponding to 3868121̊98 .

[128] Here the "//" does not appear to serve as hyphenating a word, so the symbols are retained here and in the next instance.

ihr gebüerende **119067351**. Rothe Zahl **151500000** die

Addier zusammen . . . . . . **119067351**

thuet dieße rothe Zahl . . . . . . **270567351**

darzu addir 4. ganze rothe Zahl **230270022**

dieweil eine die ander mit 4. Ziffern **230270022**

**230270022**

**230270022**

So kombt dieße rothe Zahl die halbier **1191647439**

von der halben rothen *Sub*: die ganze rothe Zahl **595823719** $\frac{1}{2}$

und so offt isch derselbigen mag so viel Zifffern wird die Zahl mehr haben dann die die erste.[129]

[129] The remainder of the text at the bottom of this page is cut off; Gieswald (1856) only states "die ganze rothe Zahl und such deren schwarze" after reporting the final value of the example.

Zwüschen 2 Zahlen ein *Medio proprotional* Zahl zu finden.

Eß sein aber die zwo Zahlen die mir vorkhomen alß.

die erste mit 4 Ziffern, die ander mit 9 Ziffern, und stehendt

also 5764. die Andere __ 287649833[130]

ihr ______ **175170640** gebürende Rote **135500000** Zall die

*addier* Zusamen _ _ _ _ _ _ _ _ **175170640**

---

macht dise Rote Zahl _ _ _ _ _ _ _ _ **310670640** .

Darzue 5 ganz Roter Zallen diewiel **230270022 .**

eine die andere mit 5 Ziffern über **230270022 .**

trifft _ _ _ _ _ _ _ _ **230270022 .**

**230270022 .**

**230270022 .**

---

Dise *addierte* Rote Zahl halbier **1462020750 .**

---

ist dise rothe Zahl _ _ _ _ _ _ _ _ _ _ **731010375**

darvon Subtrahier die ganze Rote Zall so offt als Ich mag in disem[131] Exempel .3 mahl, darumb wirdt die *Medio Proprotional* Zahl .3. Ziffern mehr haben dan die erste und bleibt ihre rothe Zahl

_ _ _ _ _ _ _ _ _ _ **40200309**

diser gebüren d[132] schwarzen Zahl ist die *Medio* – 149478591[133] *proprotional* Zahl.[134]

---

[130] This number should be 387649833.

[131] Or, “disen,” since the final letter is elongated.

[132] The specific “d” character is used for the article.

[133] This number should be 149479552 .

[134] Page 18 of the “Kurzer Bericht” ends here.

Zwischen 2. Zahlen Ein *Medio Proportiona*l Zahl zull finden.

Es zeigen aber die 2. Zahlen die mir vorkommen, alß die Erst mit 4. Ziffern, die ander mit 9. Ziffern, und stehende

| | | |
|---|---|---|
| also . . . . 5764. | die Ander . . . | 387649833. |
| ihre gebüe- **175170640** | rendt rothe Zahl. | **135500000** die |
| Addier zusammen . . . . . . | | **175170640** |
| macht dieße rothe Zahl . . . . . . | | **310670640** |
| darzu fünff ganzer rothen Zahl die- | | **230270022** |
| wiel eine die ander mit fünff | | **230270022** |
| Ziffern übertrifft | | **230270022** |
| | | **230270022** |
| | | **230270022** |
| Dieße addierte rothe Zahl halbier | | **1462020750** |
| ist dieße rothe Zahl . . . . . . | | **731010375** |
| Darvon Subtrire die ganze rothe Zahl so offt als ich mag, in dießem Exempel 3 mahl, darumb wirdt die *Medio proportiona*l Zahl 3 Ziffern mehr haben | | |
| dann die Erste, und bleibt ihre rothe Zahl | | **402000309** |

Dießer gebüerender Schwarze Zahl ist die *Medio proportional* Zahl . . . . . . . 149478591

Zwischen .2 Zahlen, die *Medio proprotional* Zahl zu finden.

Eß ist auf unsere meinung eine geringe veranderung ein 2 3 4

oder mehr *Medio proprotional* Zallen zwischen .2 bekhandten Zallen zufinden,

darumb wollen wir die veranderung bekhandt machen, durch ein

Exempel, welches zuvor durch bekhandte Zahlen gegebn ist,

und sein die .2 Zahlen .119004521 und 893423483

ihr gebürende rote Zahl ist __ **17400**. und **21900 0**

die *differenz* de Rothen Zahl ist ________ // **20160 0** die

---

theil in .3 theil ist _____________ .[135]

ein drite *addier* zu der kleinen Zahl rothen[136]

---

So ist die Rothe Zahl d[137] Ersten *proprotional* [large gap in text] Zahl

ihre gebürende schwarze Zall ist die __ .

---

Zwaÿ drite der *differenz* der Roten Zal ist.

und die kleiner Rote Zahl *Addir* darzur.

---

Diß ist die Rote Zahl der anderen *proprotional* [large gap in text] Zahl

ihr gebürende schwarze Zahl ist die

[135] There are no numbers provided from this point until the final lines of this example.

[136] A stray “2” and “1” appear above the words “Zahl” and “rothen,” respectively. This numbering of word order is most likely because the phrase should be “rothen Zahlen.”

[137] Again, the article is shortened to the specific “d” character.

Zwischen 2. Zahlen 2. *Medio Proportiona*l Zahlen zufinden.

Es ist auf unsere meinung Ein geringe verenderung Ein 234 oder mehr *Medio proportional* Zahlen, zwischen 2. Bekandten Zahlen zu finden, darumb wollen wir die Verenderung bekandt machen durch ein Exempel, welches zu vornen durch bekandte Zahlen gegeben ist, und zeigen die 2 Zahlen. 119004521. und 893423483.

ihre gebürendt rote Zal ist **17400** und **219000**

die Differenz der rothen Zahl ist . . 201600 die

theil in 3. theil ist . . . . . **67200**

Ein drittheil *addier* zu der kleinen rothe Zahl. **17400**

So ist die rothe Zahl der Ersten *Proportio*: **84600** Zahl

ihre gebüerende Schwarze Zahl ist die 23020839.

Zweÿ drittheil der differenz der roth Zahl ist **134400**

und die kleiner rothe Zahl addir darzur **17400**

diß ist die Rothe Zahl der Ander *Proportional* **151800** Zahl.

ihre gebüerende Schwarze Zahl ist die .459326198.

| *A*: | *B*: | *C*: | *D*: |
|---|---|---|---|
| 119004521. | 23020839. | 45932698.[138] | 893423483. |
| **17400** | **84600** | **151800** | **219000** |

Wie sich helt *A* zu *B*: also helt sich *B* zue *C*: und *C*: zue *D*:

[139]

Zwischen 2 Zahl dreÿ *Medio proprotional* Zal zu finden.

es sein die zwo bekhandten Zahlen 119004521 und 893423483

Ihre gebürende Rote Zahl ist ____ **17400 die ander .219000**

Ihre *differenz* ist ____ **201600.**

die theil in vier gleiche theil, ist ein theil **50400.**

**17400.**

Der theil eins *Addir* zu der kleinen Roten **67800** Zahl die ist

die gebürendte Rothe Zahl der schwarzen 196986715 diß ist

die Erste *Medio proprotional* Zall

Zum andern *addir* $\frac{2}{4}$ der Roten *differenz* zu der kleinen Roten Zahl

Alß . ____ // **50400 – die $\frac{2}{4}$**

**50400.**

Und die kleiner Rote Zahl ____ // **17400.**

[138]The values for *B* and *C* should be 233020839 and 456274358, respectively, since the given (determined) red numbers are correct (also reported in Lutstorf 2005).

[139]Page 19 of the "Kurzer Bericht" ends here.

| *A.* | *B.* | *C.* | *D.* |
|---|---|---|---|
| 119004521. | 23020839. | 45932698. | 893423483. |
| $\mathbf{1740\overset{\circ}{0}}$ | $\mathbf{8460\overset{\circ}{0}}$ | $\mathbf{15180\overset{\circ}{0}}$ | $\mathbf{21900\overset{\circ}{0}}$ |

Wie sich helt *A.* zu *B.* also helt sich *B* zu *C.* und *C.* zu *D.*

Zwischen 2. Zahlen 3 *Medio Proportional* Zahlen zu finden.

Es zeigen die zwo bekandte Zahlen .119004521. und 893423483.

ihre gebüerende rothe Zahl ist **. . .** $\mathbf{1740\overset{\circ}{0}}$ der ander $\mathbf{21900\overset{\circ}{0}}$

ihre Differenz ist . . . . . . . . . $\mathbf{20160\overset{\circ}{0}}$

die theil in 4 gleiche theil in ein theil. $\mathbf{5040\overset{\circ}{0}}$

<u>**17400**</u>

der theil eins addier zu der kleinen rothen Zahl $\mathbf{.6780\overset{\circ}{0}}$ die

ist die gebüerende rothe Zahl der Schwarz 196986715 diese ist

die Erste *Medio proportiona*l Zahl.

Zum andern addier $\frac{2}{4}$ der rothen Differenz zu der kleinen

rothen Zahl alß . . . . . . . . $\mathbf{5040\overset{\circ}{0}}$

**50400 die** $\mathbf{\frac{2}{4}}$

und die kleiner rothe Zahl . . . . <u>**17400.**</u>

Gibt die Rote Zall d[140] ander *proprotional* **118200** Zahl

welches ist ihre gebürende schwarze Zahl 32606976.[141]

Die ander begehrdte.

Zum driten *addir* $\frac{3}{4}$ der Roten *differenz* **50400.**

und der kleinere Rothe Zahl _ _ _ _ // **50400 .**

**50400 .**

**17400 .**

---

Diß ist die Rote Zahl d[142] driten *proprotional* **168600** Zahl.

welches ist ~~die Rote~~ Ihre gebührende schwarze Zahl .539735109[143]

die drit begehrte.[144]

Zwüschen zwaÿen vier *Medio proprotional* Zahl zu finden.

Es seindt die 2 bekhandten Zahlen alß 119004521 und 893423483[145]

ihre gebürende Rote Zahl ist _ _ _ _ _ **17400** der andern **219000**

ihr *differenz* ist ______________________ **201600**

die theil in 5 gleiche theil, der ist einer ___ **40320**

die kleinere Rothe Zahl *addir* zu der $\frac{1}{5}$ **17400**

diß ist die Rote Zahl der______________ **57720**

Gebürende schwarzen erster *Medio proprotional* Zahl 17809931[2][146]

---

[140] The article is shortened to the specific "d" character.

[141] This value should be 326069676.

[142] The article is shortened to the specific "d" character.

[143] This value should be 539739109.

[144] Page 20 of the "Kurzer Bericht" ends here.

[145] The final digit ("3") is cut off from the page; however, the corresponding red number is given in the next line.

[146] The final digit is cut off from the end of the page. The calculation $10^8(1.0001)^{5772}$ was used to determine the final assumed digit.

gibt die rothe Zahl der Anderen *Proportional* **118200** Zahl

Welches ist ihre gebüerende Schwarze Zahl .326069676

die ander begehrte.

Zum dritten *addier* $\frac{3}{4}$ . der rothen Differenz **50400**

**50400**

**50400**

Und die kleiner rothe Zahl . . **<u>17400</u>**

diß ist die rothe Zahl der dritten *Proportiona*l **168600** Zahl.

Welche ist ihre gepüerende Schwarze Zahl . . 539738109.

die dritte begehrte.

Zwischen .2. Vier *Medio Proportiona*l Zahlen zu finden.

Es zeigen die 2 bekandte Zahlen alß 119004521. und 893423483.

ihre gebürende rothe Zahl ist . . . **17400** der ander **219000**

ihre differenz ist . . . . . . **201600**

die theil in 5 gleiche theil der ist Eins . . . **40320**

die kleiner rothe Zahl *addier* zu der $\frac{1}{5}$ . . . **<u>17400</u>**

diß ist die rothe Zahl der . . . . . . . **57720**

gebürender Schwarzen Ersten *Medio Proportional* Zahl .178099312.

Zum Andern addier $\frac{2}{5}$ zu der kleiner roth Zahl **40320**

**40320 die $\frac{2}{5}$**

| | |
|---|---|
| Zum Andern *addir* $\frac{2}{5}$ Zu der kleiner Roten Zahl | **40320̊ die** __ |
| | **40320** |
| die kleine Rote Zahl ______________________________ . | **17400** |
| thuet zusamen die gebuerende Rote Zahl _ _ _ _ _ | **98040** |
| D[147] ander *Medio Proprotional* Zahl welche ist _ _ _ | 2665658[13][148] |
| Zum driten *addir* $\frac{3}{5}$ zu d[149] kleinen Roten Zahl _ | **40320̊** |
| | **40320** } **die** |
| die kleine Rote Zahl ______________________________ | **40320** |
| thuet zusamen die gebürende Rote Zahl der — — — | **17400** |
| | **138360** |
| driten *Medio Proprotional* welche ist _ _ _ . | 39889611[1][150] |
| Zum vierten *addir* $\frac{4}{5}$ Zu d[151] kleinen Roten Zahl | **161280**[152] **die** |
| die kleine Roth Zahl ___________________________ _ _ | **17400** |
| thuet zusamen die gebürende Rothe Zall | **178680** |
| der vierten *Medio Proprotional* welche ist __ __ . | 5969783[52][153] |

[147] Here, the specific "d" character is used for the article.

[148] The final two digits are cut off from the end of the page. However, these are the first 7 digits of the corresponding black number for **98050** (which is also incorrectly given in the Gk manuscript); the number should be 266539159.

[149] The specific "d" character is used for the article.

[150] The final digit is cut off from the end of the page. The calculation $10^8(1.0001)^{13836}$ was used to determine the final assumed digit.

[151] The specific "d" character is used for the article.

[152] Here Bürgi (or the copyist) just inserts the final result for **4×40320** instead of writing the one-fifth part four times.

[153] The final two digits are cut off from the page. The calculation $10^8(1.0001)^{17868}$ was used to determine the final assumed digits.

die kleinere Rothe Zahl . . . . . **17400**

thut zusammen die gebüerende Rote Zahl der. **98040**

ander *Medio Proportiona*l Zahl welche ist .266565813.

Zum dritten addire $\frac{3}{5}$ zu der kleinen rothen Zal **40320**

**40320**

**40320 der $\frac{3}{5}$**

die kleiner rothe Zahl . . . . . . **17400**

thut zusammen die gebüerendt rothe Zahl der **138360**

dritten *Medio Proportiona*l Zahl welche ist 398896111.

Zum vierten *addier* $\frac{4}{5}$ . zu der kleiner rothen Zahl **161280. die $\frac{4}{5}$**

die kleiner rothe Zahl . . . . . . **17400**

thut zusammen die gebürende rothe Zahl der **178680.**

vierten *Medio Proportiona*l Zahl, welche ist 596978352.

# Translation and Commentary

## Introduction

The translation and commentary of this section of Chapter 3 are divided into seven subsections, according to the purpose of the text and the type of examples given by Bürgi. The subsections—each is a paired translation and commentary—are organized as follows:

I. Title page and two-page foreword
II. "Kurzer Bericht": Pages 1–4
III. "Kurzer Bericht": Pages 4–6
IV. "Kurzer Bericht": Pages 7–10
V. "Kurzer Bericht": Pages 10–14
VI. "Kurzer Bericht": Pages 14–18
VII. "Kurzer Bericht": Pages 19–21

## I. Title Page and Two-Page Foreword

### Translation (Title Page)

*Arithmetic and Geometric Progression*

**Tables/with thorough instruction/for how these can be usefully applied in various calculations/and how they are to be understood**

**Printed/In Old City Prague/by Paul Sess/the Praiseworthy University Book Printers/in/1620.**

## Translation (Foreword, Pages 1–2)

### Foreword to the Truehearted Reader

Dear friendly reader: though many excellent and various tables have been invented to remove the difficulties involved in calculating multiplications, divisions, and extractions of roots, these have always been only for particular [calculations]. So multiplication and division have their own tables, e.g., the Pythagorean table, the extraction of square roots has its table of squares, the cubical extraction has its table of cubes, and thus continuing, every quantity needs its special tables; the multitude of tables is not only annoying but also cumbersome and difficult. I therefore searched for all time and worked to invent general tables with which you would like to do all of the above[mentioned] things. Consider therefore the property and correspondence of two progressions. One is arithmetic, the other geometric; what is multiplication [in the geometric progression] is only addition [within the arithmetic progression], and what is division [in the geometric progression] is subtraction in that [arithmetic progression], and what is in the extraction of a square root [in the geometric progression] is only halving in that [arithmetic progression], extraction of a cube root is only dividing in 3, extraction of a fourth root to divide in 4, [of a] fifth root [to divide] in 5, and so on in other quantities. I have considered nothing more useful than to create these tables so it may happen that all the numbers may be found in the same way.[154]

The objective out of which these tables grow, through which you [are] not only [able to remove the] difficulties of multiplication, division, and all sorts of root extraction, in which the Algebra or [the] Coss has an admirable advantage and [for which the difficulties] can be prevented. But also as many mean proportionals between two given numbers as one desires, may be found, which is difficult without these tables, so they are able to be understood with a few official exercises. And although I began these tables several years ago [and] my career has included the edition of the same, I wish to please the good-hearted readers with the tables and that they will favorably understand the following instructions through the several examples.

As follows hereafter.[155]

## Commentary (Title Page and Foreword)

The circular representation of Bürgi's tables on the title page is similar to that of circular slide rule systems that appeared later, such as William Ougthred's (1574–1660) circular version of a slide rule (Roegel 2010a; Sampson 1915). Although the circles are intended to be concentric, they are not (due to copyist or printer error). There are several errors in the graphical representation, including the last three digits of the black number associated with **5000** (the value is given as 105126407; it should be 105126847). Also, the black number associated with **230270** is missing a terminal "0." In the often-used image of the title page from the Gk edition, these two values have been corrected by someone's hand.

---

[154] Page 1 of the Foreword ends here.

[155] Page 2 of the Foreword ends here.

Many scholars note that the "gründlichem Unterricht" was never completed or issued, since Bürgi grandly referred to this "thorough instruction" in the title page. Instead, what was delivered (as is found in both the Gz and Gk editions) is the "Kurzer Bericht" (or "Short Report"). Since the solution of each example posed in the "Kurzer Bericht" is provided, the distinction between "thorough instruction" and "short report" may be an issue of interpretation. That is, Bürgi may have intended to include (in the "thorough instruction") information that illuminated his reasons for certain aspects of the construction of the tables, such as why he selected the factor of $10^8$ (100000000) or why he did not include trigonometric values. However, no surviving edition contains any content that clarifies such issues, nor does any document exist that promises anything further; thus, the "gründlichem Unterricht" comprises only the "Kurzer Bericht" as is given in the following pages.

In the foreword to the reader, Bürgi establishes the need for his tables, which would allow for users to perform all manner of calculations, and without the need of multiple tables. It is here that Bürgi introduces the fundamental mathematical idea of the tables and the corresponding calculations: the relationship of two sequences (progressions) of numbers, one that is arithmetic and one that is geometric. The calculations that Bürgi will explain in the "Kurzer Bericht" include multiplication and division and extracting roots. Furthermore, he connects the desired calculations to their corresponding operations using the tables: addition and subtraction (for multiplication and division, respectively) and halving and dividing by 3, 4, or 5 (for extracting square and cube, fourth, and fifth roots, respectively).

Bürgi completes the foreword by reminding the reader that he has sought to remove the difficulties involved when carrying out multiplication, division, and extraction of roots and to call attention to another important calculation: determining geometric means (or, as he refers to them, mean proportionals). In the *Aritmetische und Geometrische Progreß Tabulen*, the last eight pages of the "Kurzer Bericht" are dedicated to calculating not just the geometric mean between two given numbers but to determining multiple geometric means between two given numbers. As Lutstorf (2005) observed, determining any number of geometric means was of particular importance (p. 102), particularly for Bürgi, who was engaged in the construction and application of proportional drawing instruments. One example of Bürgi's need for the calculation of geometric means is found in his brother-in-law's report on his geometric triangular instrument. After Bürgi's death, Benjamin Bramer published the *Bericht zu M. Jobsten Burgi seligen Geometrischen Triangular Instruments mit schönen Kupferstücken hierzu geschnitten* (1648)—which he dedicated to Bürgi and which contained beautiful copper plates (see Figure 3.1 for an example) to accompany the examples in the book and which provided applications of calculating distances in one system after knowing measures in another proportional system.

**Figure 3.1** Illustration from Bramer's 1648 text on Bürgi's geometric triangular instrument (image courtesy of Toggenburger Museum, Lichtensteig, Switzerland)

There are numerous examples of differences in the content of the Gz manuscript when compared to the transcription of the Gk manuscript. Such an example is found on page 2 of the foreword, where the final five words of the Gz edition, "…günstig Zunehmen. Wie hernach folgt," differ from the last two words of the Gk copy's foreword ("wie folgt"). Here, the Gz copy provides a gentle invitation for the reader to take Bürgi's work (i.e., the instruction and accompanying examples) favorably.

## II. "Kurzer Bericht": Pages 1–4

### Translation

**Short Report of the Progression Tables**

How they can be usefully applied

In various calculations.

In these tables two rows of progressive numbers can be found, one with red characters, which are as easy to everyone, [nothing] other than an arithmetic progression, but the other [with] black [characters nothing] other than a geometric progression. For being faster on our way to go through these [examples], we want to call the arithmetic progression the red [numbers], and the geometric progression the black numbers. To enable everybody to understand the basics of these tables and to allow a better use of them, we will present the properties of these two progressions and explain them with several examples.

| ***Arithmetic*** | **0** · | **1** · | **2** · | **3** · | **4** · | **5** · | **6** · | **7** · | **8** · | **9** · | **10** · | **11** · | **12** · |
|---|---|---|---|---|---|---|---|---|---|---|---|---|---|
| *Geometric* | 1. | 2. | 4. | 8. | 16. | 32. | 64. | 128. | 256. | 512. | 1024. | 2048. | 4096. |

We have suggested in the foreword, as well as have been touched upon by some[156] Arithmeticians [such as] Simon Jacob [and] Moritius Zonz and others, that what is multiplying in the geometric progression or in the black numbers is adding in the arithmetic progression or in the red numbers.

When, for example, one should multiply 8 by 64, the red number for 64 is **6** and for 8 [it] is **3**. The sum for **6** and **3** is **9**, whose black number is 512. This is what we get when 8 [is] multiplied by 64.

In the same manner, 32 is multiplied with 256. Their red numbers are 5 and 8; together [they are] 13, whose black number is 8192, and [this] is as much as 32 multiplied by 256.

In the same manner, one wants to divide 16384 by 512. Their red numbers are **14** and **9**. Therefore, subtract **9** from **14**; **5** remains; its black number is 32, and that much is 16384 divided by 512.

[156] Page 1 of the "Kurzer Bericht" ends here.

In the meantime, Regula detri is not unlike what multiplying and dividing require; it follows that Regula detri may also be conducive to being performed by these tables. As an example,[157]

8 giving 128 which gives 32 giving <…> their corresponding
red numbers **3** **7** **5** and adding together

**7**
**5**
**12** From this subtract the first red
**3** number, [which] is 3, [**9**] remains, its

black

**9** number is 512, which is the desired
number outcome called [512].

In the same manner, you want to extract the square root of 256; its red number [is] **8**, this is halved [which is] **4**, whose black number is 16 which is the square root of 256.

In the same manner, you want to extract the cube root from 512; its red number is [**9**] which divided into 3 is [**3**], its black number is 8 and [is] the cube root of 512.

In the same manner, you want to extract Radicem Zonsi Zonsicum from 4096. Its red number is [**12**]. This divided into 4 is [**3**]. Its black number is 8, which is also the fourth root of 4096.

In the same manner, you want to find the mean proportion between [**64**] and [**512**]. Their red numbers are [**6**] and [**9**]. Then you subtract one from the other and 3 remains. This is divided into 3, [which] is 1. I add this 1 to the first of the two mean proportionals and so in turn one adds 1 to the 7 and gets 8, [whose] black number is 256, and continues with the other mean proportionals, and so on, as it will be shown later. This characteristic[158] does not have the two progressions shown above alone, but all others are like them when the arithmetic [progression begins] at 0 and the geometric [progression begins] at 1.

Then, the following tables are nothing else but 2 such progressions, and I am speaking here only about the progression above, [but] now we want to proceed to how to use our progression tables and we will learn seriously.

## Commentary

### *General Details and the Relation Between Two Progressions of Numbers*

To begin, the "Kurzer Bericht" Bürgi sets the stage for his system of logarithms,[159] by calling for the use of simpler operations (addition, subtraction, halving) performed on numbers from the arithmetic progression in place of computing more complex operations (multiplication, division, extraction of the square root) with much larger numbers taken from the geometric progression.

At the end of the first page of the "Kurzer Bericht" (see Figure 3.2), Bürgi presents the reader with an underlying fundamental structure for the tables, the juxtaposition

[157] Page 2 of the "Kurzer Bericht" ends here.

[158] This corresponds (approximately) to the end of page 3 of the "Kurzer Bericht."

[159] Bürgi never used the terms "logarithm" or "logarithms."

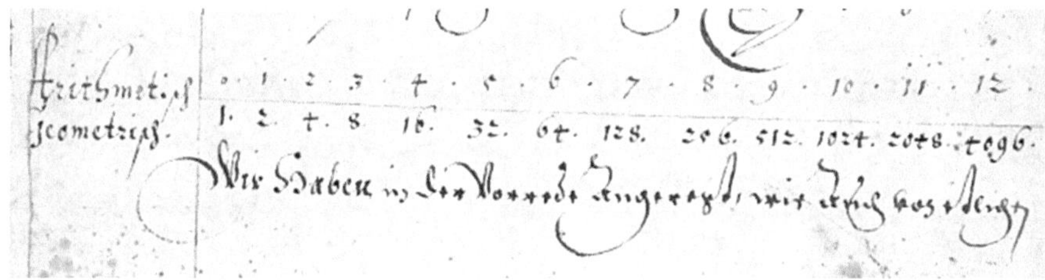


**Figure 3.2** The juxtaposition of an arithmetic and a geometric progression (page 1, "Kurzer Bericht")

of an arithmetic progression (the natural numbers, 0–12), and a geometric progression (the first 13 powers of 2). After establishing the relation of the two sequences of numbers, Bürgi presents eight examples of a variety of calculations (including multiplication, division, extracting square roots, and determination of mean proportionals) on the black numbers using corresponding yet simpler operations (e.g., addition, subtraction, halving) on the associated red numbers. This set of eight examples, using the two sequences given on page 1 of the "Kurzer Bericht," begins on page 2.

At the outset of the "Kurzer Bericht," Bürgi introduces the language he will use throughout the manuscript: the arithmetic progression (or sequence) will appear in red and the geometric progression will appear in black. Page 1 is the only page in which Bürgi used "*Charact*ern" (characters); "Zahl" or "Zahlen" (number or numbers) are used or implied for the remainder of the Gz edition.

Page 1 of the "Kurzer Bericht" also presents several examples of a change in script for either a Latin term or hybrid (Latin-German) term. These examples include *Fundamenta* as a Latin term and *Charact*ern, *Arithmet*ischer, and *Geomet*rischen for hybrid terms. Note, as well, that the copyist does not employ this strategy consistently, which is true for other attributes of the Gz manuscript.

The first use of the juxtaposition of red and black ink to emphasize the relationship of the two progressions is also found on page 1 of the "Kurzer Bericht." However, the hue of the red ink on this page is more faint than is found on subsequent pages.

Although it is almost certain that Bürgi learned of the idea to relate arithmetic and geometric progressions from some printed source, it is not certain that he learned of this from Michael Stifel (1487–1567), as many have claimed (e.g., Roegel 2010a), particularly since Stifel's famous work *Arithmetica Integra* was written in Latin, and Bürgi did not know Latin. It is also unknown whether Stifel's work was translated into German for Bürgi's use (as Copernicus' work was translated into German for Bürgi by Reimers).

Instead, Bürgi was more likely inspired to relate two progressions from the works of Simon Jacob (d. 1564) and Moritius Zonz (dates unknown), since these are the sixteenth-century German reckoning masters that Bürgi himself mentions in the *Aritmetische und Geometrische Progreß Tabulen.*[160] Since Bürgi could not navigate

[160] Indirectly, however, Stifel influenced Bürgi, since Jacob followed Stifel's work closely.

den / ſo ſtell erſtlich etliche ſovil dir geliebt / Als ich wil von 3. an-
fahen / ſo ſtehen darnach derſelbigen etliche in Ordnung folgend
alſo:

| 0. | 1. | 2. | 3. | 4. | 5. | 6. | 7. | 8. |
|---|---|---|---|---|---|---|---|---|
| 3. | 6. | 12. | 24. | 48. | 96. | 192. | 384. | 768. |

**Figure 3.3** Relationship between two sequences found in Jacob (1565, p. 14)[161]

publications in Latin, he relied solely on those in German to which he had access. Simon Jacob was a well-known reckoning master who published computation and arithmetic textbooks in 1560, 1565, and 1594 (Lutstorf 2005). Furthermore, Smith (1958) claimed that Simon Jacob's treatment of series and the nature of exponents was the more likely influence on Bürgi's work. This claim is justifiable when Simon Jacob's text, *Ein New und Wohlgegründt Rechenbuch* (1565), is examined. In Jacob's manuscript, examples such as the one given in Figure 3.3, which displays the arithmetic sequence for $n$ ranging from 0 to 8 and the geometric sequence equivalent to $g_n = 3 \cdot 2^n$, are similar to those found on page 1 of the "Kurzer Bericht." Note as well the punctuation on the cardinal numbers that Jacob employs and which Bürgi adopted in his own text (Figure 3.2). This was the practice of the time but which is now used for ordinal numbers instead of cardinal numbers.

### *Calculations with Powers of 2: Sequence of Eight Examples*

Bürgi begins with a series of eight examples on page 2 of the "Kurzer Bericht" in which a variety of calculations (multiplication, division, extracting square roots, rule of three, determining the mean proportional) are performed on the black numbers using corresponding yet simpler operations (e.g., addition, subtraction, halving) on the associated red numbers. In the first example, Bürgi wants to multiply 8 and 64 (2 numbers from the geometric sequence). The corresponding red numbers are **3** and **6**, respectively, since $8 = 2^{\mathbf{3}}$ and $64 = 2^{\mathbf{6}}$ and the numbers in the arithmetic sequence (the red numbers) are the exponents for the base used. Next, the corresponding, simpler operation in the arithmetic sequence is addition and the sum of **6** + **3**, or **9**, is the exponent of 2 for the product. Thus, the product of 8 and 64 is $2^9$, or 512.

The next example on page 2 of the "Kurzer Bericht" is also one of multiplication, with a slight increase in difficulty (two-digit number multiplied by a three-digit number): multiply 32 and 256. Again, the corresponding red numbers are identified (**5** and **8**, respectively), added together, and their sum (**13**) is associated with the product 8192, since $2^{13} = 8192$.

The third example calls for dividing 16384 by 512 and Bürgi first identifies the red numbers for each value ($2^{\mathbf{14}} = 16384$; $2^{\mathbf{9}} = 512$). Then, since division of the black numbers corresponds to subtraction within the sequence of red numbers, Bürgi

[161] Translation of the text: *Therefore put as first as many as you like/As I will start from 3./The next ones are following under the same order, like this: …*

subtracts the smaller from the larger (resulting in **5**) and, finally, reads the corresponding black number (32) to arrive at the division result.

The fourth example in the series of eight continues on page 3 of the "Kurzer Bericht" with an example of *Regula detri* or "rule of three." In this desired calculation, three quantities are known and the fourth is to be determined.

The problem posed by Bürgi ("Kurzer Bericht," page 3) reads

| | | | |
|---|---|---|---|
| | 8. geben. | 128. was geben. | 32 gib <…> der Zahl ihr gebürrende |
| Rote Zahl | **3.** | **7.** | **5** addir und zusammen. |

In this example, the black numbers 8, 128, and 32 are given, along with their red numbers (again, aligned just beneath the black numbers). Then, the product of 128 and 32 is determined, and then the result of that number divided by 8 is sought. However, using only the red numbers in a vertically oriented calculation, Bürgi adds the corresponding red numbers for 128 and 32 (**7** and **5**, respectively), subtracts the red number for 8 (which is **3**) from this sum, and uses the red number **9** from the subtraction result to find the corresponding black number (512) from the geometric progression, which is the desired result.

The remaining examples on page 3 of the "Kurzer Bericht" include one example each of extracting the square root, the cube root, and the fourth root and one example of calculating a mean proportional ("mitler *Proportional*," as the term first appears). In each root extraction calculation, Bürgi first determines the corresponding red number for the given radicand. Next, he divides by the integer of the root (e.g., 2 for square root). Finally, he reports the corresponding black number from the geometric sequence. In the first root extraction example, Bürgi seeks the square root of 256. The calculation entails:

1. Determine the corresponding red number for 256: **8**
2. Since the square root is desired, divide 8 by 2: **4**
3. Find the corresponding black number for 4: 16.

Bürgi concludes this example in the same way used for most of the other examples, by restating the example type with the result: or, in this case, that 16 is the square root of 256.

The second of the three examples is to extract the cube root of 512 ("Item man will *Radicem Cubicam* auß .512 *Extrahirn*"). In the same manner as the square root, the first step is to identify the red number associated with 512 (**9**). Then, this red number is divided by 3 (**3**); and finally, the black number associated with it is the result desired: 8 is the cube root of 512.

The final root extraction for values taken from the sequence of powers of 2 is to determine the fourth root of 4096. In this example, the red number for 4096, **12**, is divided by 4 to yield **3**, and the resulting associated black number, 8, is the fourth root of 4096.

Page 3 ends with providing the reader with instruction on how to determine a mean proportional between two given numbers from the geometric sequence used in these preliminary examples. This example also represents the first instance in the

*Aritmetische und Geometrische Progreß Tabulen* in which the example stated and the accompanying steps in its solution do not align. The goal of the example is to find the mean proportional between 4 and 64 ("Item man will Zwischen .4 und 64 die mitler *proportional* finden"), where 4 and 64 are two terms in the geometric progression. The solution provided, however, begins with stating that the corresponding red numbers are **6** and **9**, respectively. Unfortunately, from this point forward in the example, it appears that the copyist has actually transcribed the solution for the example, "Item man wil 2 *media proportionalia* zwischen 64 und 512 finden."[162] This example does not appear in the Gz copy. However, the error is most likely the result of copying the incorrect line from two similar lines of text. This is especially possible in the case of two examples, on two different lines, that share a common number (in this case, the number 64).

A bit of context is order regarding the presence of so many examples involving mean proportionals in the *Aritmetische und Geometrische Progreß Tabulen*. Bürgi dealt with large distances when measuring and calculating the positions of stars (see Chapter 1), and because he had to transfer them into small models with accurate gear systems in a way that they fulfilled and represented the same movements, the highest possible resolution and reproduction was a must for Bürgi. Indeed, such application of calculation was essential for remaining ahead of the competition. Thus, determining any number of mean proportionals is an important application in Bürgi's manuscript. Here, I present the solution for an example as it was originally stated on page 3 of the Gz manuscript: "Item man will Zwischen 4 und 64 die mitler *proportional* finden." First, in modern notation, solving for the mean proportional (or geometric mean) for this example requires the proportion

$$\frac{4}{x} = \frac{x}{64}.$$

Solving for $x$,

$$x^2 = 4 \times 64$$

$$x^2 = 256$$

$$x = 16.$$

Thus, 16 is the mean proportional between 4 and 64.

To equate the above process using simpler operations with numbers from the arithmetic sequence, Bürgi first uses addition (of the red numbers) to correspond to the product (in the black numbers) of 4 and 64. Using the associated red numbers (**2** and **6**, respectively) and the corresponding operation of addition, the sum is **8**. Next, halving a value of the arithmetic sequence would correspond to the square

[162] "One wants to find 2 mean proportionals between 64 and 512" (Gk edition; Gieswald 1856, p. 27).

root of an element in the geometric sequence, which is **4** for this example. Finally, converting back to an element of the geometric sequence yields the result sought as $2^4$, or 16.

I note here that the final example utilizing powers of 2 in the Gz manuscript is riddled with errors that most likely arose from copying lines from two different examples in different versions of the manuscript. The example as given in the Gz manuscript begins with 2 powers of 2 (4 and 64), but provides the red numbers for 64 and 512. The subtraction of the red numbers is associated correctly for 64 and 512; that is, **9 − 6** is **3**. However, the remainder of the example as written on page 3 does not make sense and in fact refers to determining two mean proportionals and not a single mean proportional originally called for in the example. When compared to the Gk edition (from Gieswald 1856), we find the source of both examples that have merged into one in the Gz manuscript. The two distinct examples given in the Gk copy are "Item mann wil zwischen 4. und 64. die mittler *Proportional* finden" and "Item mann wil 2. *media proportionalia* zwischen 64 und 512. finden."

Placing copyist errors aside, the process of determining a mean proportional is extended to two types of calculations that are found on pages 14 through 21 of the *Aritmetische und Geometrische Progreß Tabulen*: determining a single mean proportional number (i.e., geometric mean) between two given numbers of different magnitude and determining multiple mean proportionals between two given numbers of the same magnitude. These examples are treated in detail with the commentary that accompanies the translation of pages 14 through 21 of the "Kurzer Bericht."

### *Comments on Stylistic Elements*

The *Regula detri* example on page 3 of the "Kurzer Bericht" (Figure 3.4) shows a significant feature of Bürgi's manuscript with respect to his use of color within the text, that is, the careful alignment of the corresponding red numbers for each of the black numbers. Bürgi was deliberate in this alignment of the differently colored numbers within the layout of the examples used to illustrate the relation between the sequences of numbers.

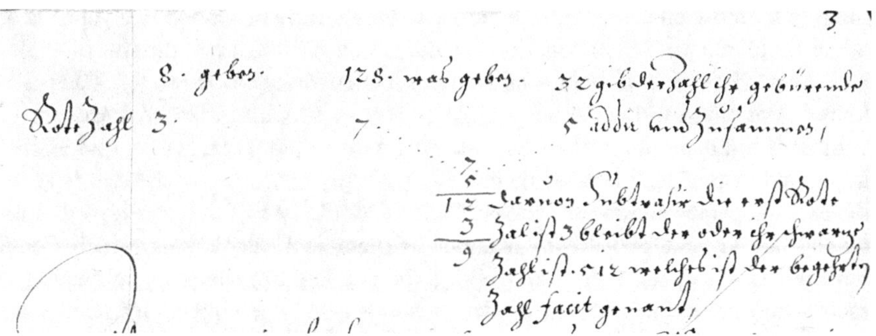

**Figure 3.4** An example of *Regula detri* (page 3, "Kurzer Bericht")

However, the use of red ink within the instructive examples also presents opportunity for copyist error. In four instances on page 3 alone, the copyist failed to return to spaces left in the text for which a change of ink color was required. This same error of omission occurs in subsequent pages of the "Kurzer Bericht," most significantly on page 19.

As previously mentioned, the Gz edition of the "Kurzer Bericht" presented here is not a corrected transcription, unlike the Gk edition, which Gieswald did correct when he published it in 1856. Thus, the transcription here retains the copyist errors. For example, the term used to indicate the fourth root of a given radicand in the Gz edition, given as *Radicem Zonsi Zonsicum*, appears correctly as *Radicem Zensi Zensicum* in Gieswald's Gk edition (and the Gk manuscript itself).

Bürgi gives no hint to why he chooses to begin with eight examples using the sequences he provided on page 1. It is likely that because Bürgi had access to the "Rechenbücher" of either Jacob or Zonz, or both, that he elected to present the basic notion of relating the two sequences as a way to situate the use of his tables within the known scholarship of his countrymen (who wrote in a language that Bürgi could read and apply). Also, beginning instruction on how to relate arithmetic and geometric sequences (or progressions of red and black numbers, respectively) using powers of 2—as opposed to beginning with the much longer nine-digit black numbers given in the tables—allowed Bürgi to focus on the procedures needed to use the tables. Though there is no evidence of explicit theoretical grounding (Lutstorf 2005), the first three pages of the "Kurzer Bericht" reveal much about the implied theoretical notions underlying Bürgi's conception of the logarithmic relationship. Furthermore, the presentation of the examples highlights important details of Bürgi's use of notation, his concern for the ease of use of the tables (particularly highlighted by the use of color), and his implied instructional techniques for the "Kurzer Bericht."

For whatever reason, Bürgi did not articulate any deeper conceptual foundations for what led to his choices for the construction of the tables. Nor did Bürgi provide any reason why he elected to construct the tables as he did, given that the community he worked within while in Prague was concerned with astronomical calculations that utilized trigonometric values. Thus, we can only speculate that the fact that Bürgi promised a "user's guide" (Gronau 1996, p. 1) may have meant that he intended users of his tables to include non-astronomers. That is, those concerned with calculations involving multiple operations and numbers of different magnitudes (e.g., those engaged with stereometry) would have benefitted from one set of tables for all manner of computation that did not require the use of the sine of angles and other trigonometric values as Napier's tables included. However, since it is known that Bürgi did construct a table of sines and with Folkerts' discovery of another Bürgi manuscript (the *Fundamentum Astronomiae*) that contains additional tables and examples, it is possible that Bürgi simply conceived of the two types of tables as separate entities. Whereas Napier saw the need for logarithms as intimately tied to matters of highly accurate astronomical calculations, which would have relied on trigonometric values, Bürgi did not. Unfortunately, a combination of factors may have contributed to modern scholars never knowing the intended order or audience for Bürgi's mathematical works, including his idiosyncrasies regarding publication agreements with peers and "over exaggerated modesty and dislike of literary activity" (Wolf 1872, p. 13), as well as disruptions caused by the Thirty Years' War affecting Prague after 1620.

## III. "Kurzer Bericht": Pages 4–6

### Translation

**I.** How to find for each black number in the tables, their corresponding red number.

As an example, you want to search for the red number for this number 133373810. This number can be found in the tables on the 8th page in the **28500** column and on the left side under **300**. The addition of **300** makes **28800**, which is the red number for 133373810, and in this way can, for every such number in the table, its red number be found.

**II.** How to find for each red number in the tables, [its] corresponding black number can be found. It would be desirable to know, for example, what black number is appropriate for the red number **28800**.[163]

To explore this, search for another red number, the number to be the same one listed above or the next smaller than that presented. This I find on the 8th page in the **28500** column, which still lacks **300** on the same page in the first column. Over in the **28500** column, [we] will find 133373810, which is the desired black number.

And so it is with any other, you can then find all the red numbers from **0** to **230270** and the desired black numbers in the same fashion.

But what is [there] to do when a number cannot be found in the tables and if one does not have pleasure with the approach of taking the red number, which is closest to the given number, and would prefer another approach in exploring its desired red number[?]

**III.** For an example, one wants to search for the true red number for 36, with seven 0s, which gives nine digits. All of the black numbers in our[164] tables have no less than 9 digits, so this black number is 360000000.
After this you look in the tables under the black number [for] the two [numbers], the next smaller and next larger [numbers] than 360000000. I find the black [numbers] for 360000000 on [the] 33rd page in the [**128000**] column on the left between

**90** this has the black [number] 359964763; this is too small
the difference of **10** 35996 the difference

[and]

**100** this has the black [number] 360000759; this is too large

The black [number] 359964763 of this smaller [red] number of **90** is subtracted from my given number 360000000
[what] remains [is] 000035237 [which is] the third difference
So to maintain the difference for the red [number], it is maintained from the third to the fourth

[163] Page 4 of the "Kurzer Bericht" ends here.

[164] Page 5 of the "Kurzer Bericht" ends here.

35996 **10000** 35237 to **9789**

This fourth [number] is added to the smaller red number.

The small red number **90**

The number from the column **128000**

This is the black number of 360000000 [and] its red [number] is **128099789**

It is to be understood that 36 has its equivalent red [number] as **128099** $\frac{789}{1000}$

And always it is completely understood that [what] follows below [after] the "o" is the fraction.

## Commentary

### *Introduction to Using the Tables and Determining Non-tabulated Values*

Page 4 of the "Kurzer Bericht" heralds the beginning of 26 examples that illustrate the numerous ways in which Bürgi's tables can be used. First, Bürgi explains how to associate red numbers with their black numbers, and vice versa. Thought of another way, Bürgi describes how to retrieve or determine tabulated values from his double-entry tables. Thus, to begin, given the black number, 133373810, the reader would page through to the eighth sheet of the tables and look for the nine-digit number within the body of the table. The instructions direct the reader to the **28500** column, and reading across to the left side of the table, note that the black number exists in the row labeled **300**. Thus, the desired red number that corresponds to 133373810 is **28500 + 300**, or **28800**.

Bürgi does not include the calculation since this would have been done to construct the tables in the first place; however, the correspondence between the red and black number is, using the relationship of Bürgi's tables, a modern calculator, and (2.1):

$$\begin{aligned}&10^8\left(1.0001\right)^{28800/10}\\&=10^8\left(1.0001\right)^{2880}\\&=10^8\left(1.33373809944\right)\\&=133373809.9 \text{ or } 133373810 \text{ (rounded to nine places)}.\end{aligned}$$

This calculation emphasizes the concept that Bürgi's tables are actually tables of antilogarithms; that is, the values that appear in the body of the table (the black numbers) are the results of the calculations performed with the logarithm values (the red numbers).

The first example on page 5 of the "Kurzer Bericht" is simply the converse of the final example on page 4: given the red number **28800**, determine its corresponding black number. This may be the more straightforward of the two examples in that the reader does not need to search through multiple pages of nine-digit numbers. Instead, the reader need only first to locate the correct page by reading the red-numbered column headings that are close to the given red number. Then, the reader can scan the left side of the page to locate the additional part of the red number (if needed). Finally, reading over (from the left) and down (from the top) of the table yields the desired black number.

Upon completion of this example, Bürgi introduces the idea that the red numbers can be determined from **0** to **230270** (or actually **230270.022**), and this important, fundamental notion of the tables will be used prominently in subsequent examples in the manuscript. This appears to be an inadequately addressed yet vital idea contained in Bürgi's tables. As previously mentioned (Chapter 2), Bürgi's tables are constructed with the logarithm of 1 equal to 0 ($\log 1 = 0$) and begin with the antilogarithm value of 100000000 (or $10^8$). Then, Bürgi tabulated all of the values until, essentially, he arrived at 1000000000 (or $10^9$), which he associated with the red number 230270.022. On page 5 of the "Kurzer Bericht," however, we see the truncated whole red number 230270.

### *Instruction on Linear Interpolation*

After the first two simple examples of reading (or locating) particular table values, the next example contains the more complex task of determining a red number for a non-tabulated black number. This example is introduced at the end of page 5 of the "Kurzer Bericht" and seeks the red number corresponding to 360000000. Here, Bürgi explains his process of linear interpolation for such non-tabulated values, and this is the only time he does so in the "Kurzer Bericht."

The whole of page 6 of the "Kurzer Bericht" is dedicated to an example of linear interpolation to determine the red number for a non-tabulated black number. A modern computation and explanation of the interpolation process is given below. However, for ease of notation in the explanation, all references to the "decimal zero" (Bürgi's decimal point) are omitted for a separate discussion.

First, the two black numbers closest to 360000000 (or, as Bürgi stated, "36 with seven 0s") are located in the table, along with their corresponding red numbers:

First black number, $B_1 = 359964763$; associated red number, $R_1 =$ **128090**[165]
Second black number, $B_2 = 360000759$; associated red number, $R_2 =$ **128100**

Next, the differences are calculated: between the black numbers, the difference is

$$360000759 - 359964763 = 35996;$$

[165] This is a correction from what is given on page 6 (**28000**) of the "Kurzer Bericht." Also, Bürgi uses just "**90**" and "**100**" in the first part of the calculation method.

and between the desired black number and the smallest black number, the difference is

$$360000000 - 359964763 = 35237.$$

Linear interpolation requires the following proportion so that the difference is maintained:

$$\frac{35996}{10000} = \frac{35237}{x}$$

$$x = 9789.143238,$$

where Bürgi reports only 9789 as the result and which he then adds to the smaller red number: **128090+9789=128099789**. Finally, he obtains the digit sequence for the associated red number for 360000000.

Bürgi begins the discussion of this example by asking for the red number (in our language, the logarithm) for the black number 360000000. Yet the next time we see the number, it appears as $3\overset{\circ}{6}0000000$ or, in modern notation, 36.0000000, since Bürgi's "decimal zero" marks the end of the whole number part of the digit sequence. Revisiting the calculations with this in mind, we have:

First black number, $B_1 = 35.9964763$; associated red number, $R_1 =$ **128090**
Second black number, $B_2 = 36.0000759$; associated red number, $R_2 =$ **128100**

Next, the differences are calculated: between the black numbers, the difference is

$$36.0000759 - 35.9964763 = 0.0035996;$$

and the difference between the desired black number and the smallest black number is

$$36.0000000 - 35.9964763 = 0.0035237.$$

The proportion used for the linear interpolation is

$$\frac{0.0035996}{10.000} = \frac{0.0035237}{x}$$

$$x = 9.789143238.$$

Finally, the desired red number is **128090+9.789** or **128099.789**. We can confirm the calculation using the modern relationship (2.1) given by

$$10^8 \left(1.0001\right)^{128099.789/10}$$
$$= 359999999.609.$$

When this value is rounded to the nearest whole number, we obtain the original black number digit sequence, 360000000. However, if we use the historical analysis provided by Wolf, all of Bürgi's table numbers are divided by the factor $10^8$ (Lutstorf 2005, p. 110).

### *Bürgi's "Decimal Zero"*

Pages 5 and 6 of the "Kurzer Bericht" are important pages of the manuscript because of the introduction of sophisticated notation used by Bürgi. His "decimal zero" is Stevin-like in its use and orientation. In 1585, Simon Stevin (1548–1620) published *De Thiende* (published in Flemish or Belgian Dutch; the work also appeared in French and later in English). In *De Thiende* (*The Art of Tenths*), Stevin introduced a notation for numbers based upon powers of tenths, and it represented "the first printed treatise on decimal fractions and the notation was instrumental in the development that followed" (Clark 2011, p. 2). Stevin's notation involved encircling whole numbers beginning with 0 and increasing according to the power of tenths associated with a given digit in a numeral. The placement of the "0"-tenths notation was to the right of the ones-place for the numeral, and the remaining notation in the decimal fraction appeared to the right of the number, each increasing power of one-tenth. For example, in the excerpt from *De Thiende* (Figure 3.5), Stevin represents two numbers, 32.57 and 89.46.

Although Stevin's notation did not gain popularity in the mainstream, it prompted further development toward representing values in an alternative format to conventional fraction notation of the sixteenth and seventeenth centuries. In the *Aritmetische und Geometrische Progreß Tabulen,* Bürgi's notation of placing a small "o" directly (or almost directly) above the final digit of the integer part of a decimal number represents one such development. Bürgi states as much in the last line of page 6 of the "Kurzer Bericht": "Und werden Alle Zeit biß under die o ganz verstanden und die folgen der bruch."[166] However, prior to the *Aritmetische und Geometrische Progreß Tabulen*, Bürgi employed a different "decimal zero" by placing a small "o" just beneath the final digit of the integer part of the number. Kepler stated in his *Wein-Visier-Büchlein* (1616) that he had seen the decimal fraction notation that Bürgi devised—as well as his calculations utilizing it—and ascribed credit to him and not to Stevin (Wolf 1872, p. 15).

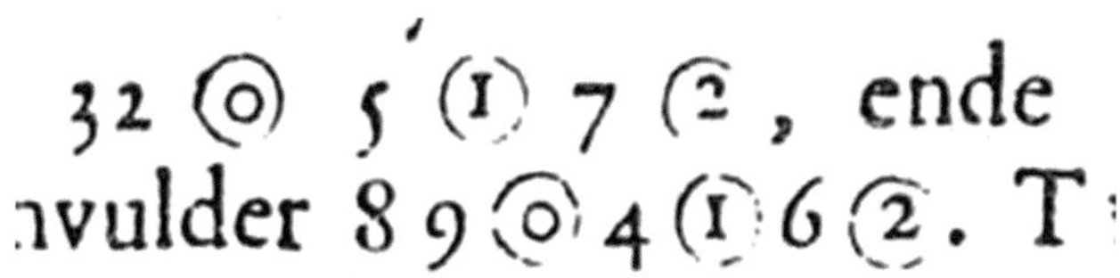


**Figure 3.5** Image from De Thiende, p. 16 (Available from http://www.maa.org/press/periodicals/convergence/in-these-numbers-we-use-no-fractions-a-classroom-module-on-stevins-decimal-fractions-element-iii)

[166] *And always it is completely understood that [what] follows below [after] the o is the fraction.*

## IV. "Kurzer Bericht": Pages 7–10

### Translation

How to multiply two numbers:

When one wants to multiply the number 154030185 by 205518112

Find their corresponding red numbers: **43200** and **72040**

Add the two red numbers together **43200**

**72040**

This red number is **115240**

The black [number] in 9 digits [is] [316559928], and these are the first nine digits of the product since our tables only have nine digits and the last or ninth [digit we] want to put only before [the] fraction for not getting many irrational numbers.

In the same manner, we want to multiply 551192902 by 709153668 .

Their red numbers are **170700** **195900**

Add the two red numbers together **170700**

**195900**

**366600**

[The] tables are not large enough to find this red number, thus subtracting:

**230270022** this is the whole red number

**136329978**

Look for its

black number [which] is 3908804680,

which are the first 9 digits of the sought-after product.

Here this is to note that I end this example with a digit more than the previous and that the tables have no more than 9 digits and should probably be 10. This is the reason that we need to subtract the whole red number 0, which is to be further expounded in the following.[167]

How to divide a number by another:

As one wants to divide .316559928 by .205518112 and [they have] the red number[s] **115240 und 72040**. Subtracting the red number of the divisor from the red number of the dividend, or **72040** from **115240**, [**43200**] remains, whose black number is 154030185, or $1\frac{54030185}{100000000}$.

In the same manner, one should divide 154030185 by 205518112.

---

[167] Page 7 of the "Kurzer Bericht" ends here.

Their red numbers are **4320$\mathring{0}$** and **7204$\mathring{0}$**. Subtracting the divisor's red number from the dividend's red number, or **7204$\mathring{0}$** from **4320$\mathring{0}$**, respectively, which [**43200**] is less, so add the whole red number **<u>230270022</u>**

[which] is **273470022**. [S]ubtracting the divisor's red number

**<u>72040000</u>**

**201430022** [and] search this red number[; our]

desired black number is 749472554, and as much as 154030185 divided by 205518112 is, it is not a whole [number], rather a fraction of the whole as $\mathring{0}749472554$ or $0\frac{749472554}{1000000000}$.

**V.** How to find the fourth proportional from three known numbers, which is usually called *Regula detri*.

As an example[168]:

The first the second the third the fourth

How 154030185 keeps [in proportion] to 205518112[,] so will 399854564 to the 4th number[.]

Their corresponding red numbers are

**4320$\mathring{0}$** **7204$\mathring{0}$** [**13860$\mathring{0}$**]

Adding the second and third red numbers together **138600**

**<u>72040</u>**

This is the red [number] of the black [number] **210640** from the multiplication that we have sought after

Subtracting the first red number **<u>43200</u>**

This is the red number of the fourth black [number] **167440**

[Which] gives 533514619[.] This is the fourth number.

The Second Example

The first the second the third the fourth

How 100160120 keeps [in proportion]

to [880122800] so will 945919848 to the 4th[.]

---

[168] Page 8 of the "Kurzer Bericht" ends here.

Their corresponding red numbers are

| | | |
|---|---|---|
| **160** | **[217500]** | **224710** |
| | | **[217500]** |

| | |
|---|---|
| Adding the second and third red numbers together is | |
| But because this number is larger than the whole red number | **452210** |
| Thus, you subtract the first whole red number | **160** |
| But because this number is greater than the whole red number | **452050** |
| So you subtract the whole red number from [this] | **230270022** |
| [What] remains is | **211779978** |
| Its black number is | 831194715 |

By placing an additional "0" at the end, we obtain the fourth desired proportional [from] being able to subtract the whole red number, which is as much as dividing by 10, and therefore [the final result] had to be multiplied by 10.

The Third Example[169]

| I | II | III | IV |
|---|---|---|---|

How 945919848 keeps [in proportion] to 100160120 so will 880122800 to the fourth.[170]

These are their red numbers

| | | |
|---|---|---|
| **224710** | **160** | **217500** |

| | |
|---|---|
| Adding the second and third red numbers together, | **160** |
| | **217660** |

and one should continue by subtracting the first of these [the red numbers].

But as this is less [than the other], add the whole red number

| | |
|---|---|
| | **230270022** |
| | **447930022** |
| After subtracting the first red number | **224710** |
| the red number is | **223220022** |

and its black number is 931931024, which one must cut off the last digit, as this happens when the whole red number is added to the sum, [when] searching for the fourth proportional.

[169] Page 9 of the "Kurzer Bericht" ends here.

[170] To clarify this particular passage, the example asks: "How 945919848 is in proportion to 100160120, so should 880122800 be to the fourth [number]."

## Commentary

### *Multiplication and Division of Tabulated and Non-tabulated Values*

There are two examples of multiplication of 2 nine-digit numbers on page 7 of the "Kurzer Bericht." To multiply two numbers (in Bürgi's terms: two black numbers), $a$ and $b$, we can use logarithmic multiplication:

$$\log(ab) = \log a + \log b,$$

which first requires locating the associated red numbers for $a = 154030185$ and $b = 205518112$, $R_a$ and $R_b$, respectively, which are:

$$R_a = \mathbf{43200}$$

$$R_b = \mathbf{72040}.$$

Again, the superscripted "°" symbol, or the "decimal zero," written above the final digit of each number in the Gz manuscript is represented as the decimal point here. Yet even in the Gz manuscript, the copyist does not carry the symbol throughout each instance of a given value. Instead the "decimal zero" is often written only at the first instance and sometimes at the top of a column of values that are being added.

Once the red numbers are located in the table, they are added together, and the sum (**115240**) is used to find the corresponding black number in the tables (which should be 316559928, but is written as 36559928 in the Gz manuscript[171]). Using a modern calculator and the algorithm underlying Bürgi's tables readily confirms this result. Calculating the result (using modern notation),

$$10^8\left(1.0001^{43200/10}\right)\cdot 10^8\left(1.0001^{72040/10}\right)$$
$$=(154030184.65)\cdot(205518112.428)$$

or

$$=10^8\left(1.0001^{4320+7204}\right)$$
$$=10^8\left(1.0001^{11524}\right)$$
$$=316559928.062.$$

The second example on page 7 is initially carried out in the same manner. However, Bürgi obtains the sum **366600** (where three trailing 0s are assumed, or **366600000**) and states that the tables are not large enough to accommodate this red number. Thus, the example yields a calculation result that is no longer in the interval

[171] Recall that I employ the convention of providing the corrected values (as editorial insertions, using square brackets) in this translation of the Gz manuscript, unless an exception is warranted, as in the case of the eighth example using powers of 2 in subsection II.

[100000000,1000000000]. To remedy this, Bürgi subtracts the "whole red number," **230270022**, from **366600000**, resulting in the red number **136329978**. Although the details of the linear interpolation are omitted in this example of the "Kurzer Bericht" (as they are for the remainder of the examples), the corresponding black number is given as 3908804680, which is a ten-digit number that Bürgi claims a nine-digit number. The additional digit of "0" is the result of knowing that the number obtained is 10 times too small (since the whole red number, **230270.022**, represents a factor of 10). Furthermore, the multiplication of the original two nine-digit numbers would result in an 18-digit number sequence. His comment, "that the tables have no more than 9 digits, and should probably be 10," is, in this case, an understatement!

Bürgi proceeds with examples on how to perform division via the use of his tables. However, the first example on page 8 of the "Kurzer Bericht" appears in the Gk manuscript, but is not completed; it is left out of Gieswald's edition altogether. The example (dividing 316559928 by 205518112) is a straightforward use of the tables and the application of an algorithm for logarithmic division, $\log\left(\frac{a}{b}\right) = \log a - \log b$. That is, Bürgi first locates the corresponding red numbers (**115240** and **72040**, respectively), subtracts the second from the first, and then searches for the black number corresponding to the result of the subtraction of the red numbers:

$$\mathbf{115240 - 72040 = 43200}.$$

The corresponding nine-digit number sequence for **43200** is 154030185. The result appears as 1.54030185 (with the use of the "decimal zero" above the left-most digit) in the Gz manuscript. Again, the details are not given; however, analyzing the division with modern notation yields

$$\begin{aligned}
\frac{316559928}{205518112} &\approx \frac{10^8\left(1.0001^{115240/10}\right)}{10^8\left(1.0001^{72040/10}\right)} \\
&\approx \frac{\left(1.0001^{115240/10}\right)}{\left(1.0001^{72040/10}\right)} \\
&\approx 1.0001^{11524-7204} \\
&\approx 1.0001^{4320} \\
&\approx 1.5403018465,
\end{aligned}$$

or, when rounded to the eighth decimal place, 1.54030185 (the original quotient is 1.540301849).

The second example on page 8 of the "Kurzer Bericht," dividing 154030185 by 205518112, is computed in the same fashion as the first division example of black numbers in the Gz copy and corresponds to the first completed division example (of nine digit numbers) that appears in the Gk copy. However, as Bürgi shows, the second example for the operation requires an additional step to complete. In this case, the dividend is less than the divisor; therefore, to subtract the red number of the latter from the former (or **43200 – 72040**) would result in a negative number, which is something that Bürgi could not deal with or wanted to avoid. He does, however, recognize that this is something negative: "dieweils aber weniger ist" ("which [**43200**] is less"); the translation of "less" in this case may be a result from Latin for "minus" or possibly from the phrase "less than nothing" (Lutstorf 2005, p. 115). Regardless, Bürgi first adds the whole red number to the dividend's red number:

**43200**
**230270022** [which] is
**273470022**.

Now Bürgi is able to subtract the divisor's red number, or **273470022 – 72040[000]**, which gives the resulting red number as **201430022**. Finally, Bürgi is able to linearly interpolate (again, the details are omitted in the manuscript) to arrive at the nine-digit black number sequence 749472554. To complete the example, Bürgi states, "and as much as 154030185 divided by 205518112 is, but it is not a whole [number], rather a fraction of the whole as $\overset{\circ}{0}749472554$ or $0\frac{749472554}{1000000000}$." Thought of another way, since the logarithm of 10 (**230270.022**) was added to the logarithmic calculation, it must be compensated for at the end of the calculation. Thus, the nine-digit number sequence corresponding to **201430022** would become 749472554 divided by 10 or 0.749472554 for the stated example.

### *Regula detri* Examples

The first two examples of *Regula detri* in which Bürgi searches for "the fourth proportional" are presented on page 9 of the "Kurzer Bericht." To solve a proportion for $x$ when given three of the four values, in modern notation and without the aid of logarithms, requires

$$\frac{a}{b} = \frac{c}{x}$$

$$x = \frac{bc}{a}.$$

Now, representing the proportion given in the first example on page 9 as

$$\frac{154030185}{205518112} = \frac{399854564}{x},$$

the equation

$$x = \frac{(205518112)\cdot(399854564)}{154030185}$$

would be used to solve without the aid of logarithms, which would be cumbersome with nine-digit numbers. However, the corresponding logarithmic operations involve (again, using modern notation)

$$\log x = \log b + \log c - \log a.$$

Using the red numbers, Bürgi performs the calculation (with the corrected values given here) **138600 + 72040 − 43200 = 167440** and determines the black number, 533514619, for this resulting red number.

The second example on page 9 is similar to other "second examples" in the Gz manuscript in that additional steps of some sort are required in order to complete the example using the tables appropriately. The first three values, or $a$, $b$, and $c$, for the proportion are 100160120, 889122800, and 945919848, respectively. Unfortunately, several copyist errors appear in the Gz manuscript for this example; namely, the red numbers and black numbers are not correctly associated. First, the value 889122800 does not correspond to the red number, **21750**. Examining the tables shows that 880122800 corresponds instead to the red number **217500**. If this corrected value were used, as opposed to the value **227500**, the remaining calculations would have appeared as follows:

Add the second and third red numbers together: **217500 + 224710.**
Subtract the first red number from the sum: **442210 − 160 = 442050.**
This red number exceeds the "whole red number," so the "whole red number" must be subtracted from the previous result: **442050[000] − 230270022 = 211779978.**

Although errors occur in each of the intermediary steps in the example due to the incorrectly recorded value of **227500** (instead of **217500**), the resulting red number is correctly recorded. Linear interpolation is used to determine the corresponding black number (for **211779978**), which is 831194715. This number, however, is too small by a factor 10, as Bürgi explains in the final paragraph of page 9. The completed proportion (with a tenfold fourth proportional) is

$$\frac{100160120}{880122800} = 0.11380243756 = \frac{945919848}{8311947150} = 0.11380243773.$$

The first example on page 10 of the "Kurzer Bericht" combines techniques from the preceding two *Regula detri* examples, as well as the final division example found on page 8. The example uses the same black number values from the second *Regula detri* example. However, because of the order of the given values, the proportion to be solved is

$$\frac{945919848}{100160120} = \frac{880122800}{x}.$$

The corresponding logarithmic calculation entails three steps:

Adding the red numbers for the second and third black numbers: **160 + 217500 = 217660**
Adding the whole red number to the sum: **217660 + 230270022 = 447930022**
Subtracting the red number of the first black number: **447930022 − 224710 = 223220022**

Next, linear interpolation is used to determine the nine-digit black number sequence associated with the fourth proportional, which is given as 931931024. However, this value must be adjusted to accommodate the factor of log10 (**230270022**) that was previously added. Thus, as Bürgi states, "one must cut off the last digit, as this happens when the whole red number is added to the sum," even though this is not shown in the resulting black number. If the magnitude of each of the original values was the same, then the fourth proportional would be 93193102.4. Consequently, the ratio for each term of the proportion (using the value 93193102.4) is

$$\frac{945919848}{100160120} = 9.44076624$$

$$\frac{880122800}{93193102.4} = 9.44076625.$$

### *Comments on Stylistic and Content Elements*

Page 8 of the "Kurzer Bericht" provides important examples of Bürgi's pedagogical inclinations of his instruction for using the tables. In both examples on page 8, he again elects to show both forms of the final black number: that is, the "decimal zero" form and the fractional form. There is also careful alignment of the operations carried out with the columns of red numbers (as they are referred to in Lutstorf (2005)). Despite several errors, what remains consistent throughout is the careful alignment of columns of red numbers, as well as the organization of the examples on a page, as found with the examples of *Regula detri*, which begin on page 9 of the "Kurzer Bericht." Finally, in the examples found on pages 7 through 10, we see Bürgi's repeated use of particular values in the division, *Regula detri*, and mean proportional examples. The decision to use values repeatedly was most likely one of convenience, for both the composer (or copyist) and the reader of the "Kurzer Bericht." For example, by calling upon the same black numbers in several instances, the reader can simply copy the corresponding red numbers rather than returning to the tables to search for the red number for a unique black number.

## V. "Kurzer Bericht": Pages 10–14

### Translation

To extract the square root from a given number. One wants to extract the square root of 4015374 for an example. Dots will be [placed] thus: first as in the extraction's use and so is $\dot{4}0\dot{1}5\dot{3}7\dot{4}$, and because of its four points here, there will also be four digits in the root. The red number of this is **$\mathbf{13902\mathring{0}}$** [and] halved this is **$\mathbf{6951\mathring{0}}$,** whose black number is $2003\mathring{8}3982$, or [which] should be understood as $20038\frac{3982}{10000}$.[172]

For a second example, one wants to extract the square root of 22033094. First, dots will be as used in the extraction and is thus $2\dot{2}0\dot{3}3\dot{0}9\dot{4}$, and because five dots are here, so also will there be five digits in the root. Those after the five [digits] are fractions; its red number is **$\mathbf{7900\mathring{0}}$.** But because the last point does not fall on the last digit in the black number as in the previous example but on the second digit, the whole red number must be added to and [the result] halved, so that the red number appears as follows:

| | |
|---|---|
| | **$\mathbf{7900\mathring{0}}$.** |
| Add the whole red number | **$\mathbf{230270\dot{0}22}$** |
| Halve this red number | **<u>309270022</u>** |
| Search the black number of this same red [number] | **154635011**[173] |

[which] is $4693\mathring{9}4227$ but as I have not found more than five points so my □ root is also $4693\mathring{9}4227$ or $46939\frac{4227}{10000}$.[174]

[172] This square root value is corrected in the commentary that follows.

[173] Page 10 of the "Kurzer Bericht" ends here.

[174] This square root value is corrected in the commentary that follows.

To extract the cube root from a given number:

As an example, one desires the cube root from 5612037.

This number and therefore its recorded points are [5̇612̇037̇]. [It] follows that the whole number of the [cube] root gets 3 digits; the others are in the fraction of a whole number. So I look for the red number, which is **172500**, to notice the point on the first digit [is] missing, so my root is also in the first whole number, and I divide my red number into three parts, as follows:

My red number is **172500**

A third is **[57500]**

Its due black number is 177707944 and 3 digits are known as given so I may reach this cube root 177707944 in my tables in 9 digits. But, reserving the rest of the 9 digits for the fraction, [it is] assumed that so many irrational numbers finish with [fractions], which may not be given enough satisfaction in 9 digits.

To extract the cube root from a given number, as one desires: As an example, to extract the cube root from 56120370. This number and therefore its recorded points [5̇612̇037̇0] [and it] follows that the whole number [integer] of the root has 3 known digits; the others are fractions of a whole number. So I look for the red number thereof, which is [**172500**], but because the point[175] is not placed on the first digit but on the second [digit], add to the red number a whole [red] number.

Bring also together **172500** and the whole [red] number **172500**

**230270022**

[175] Page 11 of the "Kurzer Bericht" ends here.

Cube of the three quantities is[176]

This is divided into three parts **402770022**

A third part in red [is] **134256674**

Search for the black number thereof 382°860159 [and this is] the cube root.

From a given number, one desires to extract the cube root. As an example, [extract the] cube root from 561203700. This number and therefore its recorded points here are [$56\dot{1}20\dot{3}70\dot{0}$] and stand with 3 points. But as the last point <appears> on the third digit, and although this is the same red number as in previous examples,

two whole [red] numbers need to be added to **172500**

[A]nd this is the reason that the first ["5"] together with the other **230270022**

digits, are the whole red number. Because the points did not **230270022**

fall on the first ["5"] and also not on the other [digit] like **633040044**

["6"] but did fall on the third, so has the[177] first ["5"] with the other digits **172500**

and the following ["6"] a whole red number **230270022**

Also for the third [digit], ~~so have the first ["5"]~~ showing a ["1"] the number lacks

**230270022**

So I have together the red number of the first 3 digits.

As the cube is the third quantity, take from the

same red number your third part [**211013348**]

[176] This phrase appears to the right of the column of numbers that begins centered above the end of a red dividing line.

[177] This begins a repeated passage of the three lines above.

This third part is the red number; its black number is the root [8̊24847208][178]

Extract the fourth quantity as *ZZ.R* from a given number. One desires as an example the *ZZ.* root from 56120370. This number is registered with its points 561̇2037̇0 [and] here falls two points so it is known that the root gets just 2 digits in the whole number. The other following digits [are of] the fraction, so seeking the corresponding red number for the given black number, which is **17250̊0**

But because the last point fell above the fourth digit, another three whole red numbers are added. **230270022**

**230270022**

**230270022**

The red number is divided into four equal parts **863310066**

This is the red number of the root **[215827516]**

Its due black number is [8̊655260259] or the root which we desire.

Extract a *Ss* [fifth] root from a given number. My given number as an example is the *Ss* [fifth] root of 671876768. Thus, this number is registered with its points [as] [671̇876768̇].

Consequently, the root's 2 digits are obtained without the fraction of a whole number.

| | |
|---|---|
| Seeking the given due red number | **19050̊0** __ 6 |
| Seeing then the last point[s] to the left | **230270022** – 7 |
| | **230270022** – 1 |
| Including the last digit [is the same but] | **230270022** 8 |

---

178 Page 12 of the "Kurzer Bericht" ends here.

The fourth is so due [for] the 4 digits

Its red number is **881310066**

These are the four digits as 6718 of its red number.[179]

The same in 5 equal parts is $\frac{1}{5}\left[\mathbf{17626\mathring{2}013\frac{1}{5}}\right]$, the red number of the due black number of the *Ss* root of 671876768 is [582717318] or $\left[58\frac{2717318}{10000000}\right]$.

### Commentary

## *Root Extraction*

The remaining two examples on page 10 launch a seven-example sequence of extracting various roots, from square roots to fifth roots. This sequence of examples extends the utility of the logarithmic equivalent of division; in this case, extracting a particular root is equivalent to dividing by the value of the index. Bürgi begins with two examples of extracting the square root, which is essentially equivalent to determining the mean proportional (or the geometric mean).

The first example of square root extraction presents the calculation for the square root of 4015374. This is the first example in the Gz copy of the *Aritmetische und Geometrische Progreß Tabulen* that does not begin with a nine-digit black number. The process for extracting a particular root requires the root to possess a number of digits that is proportional to the integer index. In order to provide for this, Bürgi used a process of "dotting" ("punctieren") the given radicand, which required the calculator (in this case, the human calculator) to begin by placing a dot above the right-most digit and then proceed to place dots over digits to the left according to the integer index. In the first root extraction example on page 10, Bürgi seeks to extract the square root, so the integer index is 2. The dotting procedure for the square root of the number sequence 4015374 would require the calculator to place a dot above the 4, 3, 1, and 4, which corresponds to beginning with the right-most digit (a "4") and moving from right to left while placing a dot above each second digit. In this case, the dotting procedure would yield $\dot{4}0\dot{1}5\dot{3}7\dot{4}$. The number of dots resulting from the dotting procedure indicates the number of digits of the integer part of the mixed number (i.e., the decimal fraction) that results.

Since the black number 4015374 represents the first seven digits of a number that does appear in the tables, no linear interpolation is needed. Bürgi identifies the corresponding red number as **139020**, which is the red number for the nine-digit number sequence 401537400. The remainder of the calculation in the Gz manuscript is correct until the final step. To begin, Bürgi divides the red number in half, which yields **69510**. Secondly, he retrieves the associated black number from the tables (200383982). Finally, the Gz manuscript gives a result with five digits of the integer component of the square root (20038.3982), and not four, as determined in the dotting procedure.

[179] Page 13 of the "Kurzer Bericht" ends here.

This is probably due to a copyist error, which may have resulted from omitting the final “00” of the original black number 401537400 in the example. If this black number was given in the Gz copy, the dotting procedure would have indeed included five dots, and the result is accurate to the thousandths place (in modern terms):

$$\sqrt{401537400} = 20038.3981396.$$

However, since the example as written asked for the square root of 4015374, the correct value should be reported as 2003.83982 (with rounding error accounted for in the tabulated value).

The second square root extraction example, to determine the square root of 22033094, also contains the same two types of errors that were found in the first square root extraction example. The first kind of error also entails the copyist omitting a “0” from the end of the number sequence for the radicand. The second error is again related to the mismatch between the number of digits that are determined to comprise the integer part of the square root and the radicand given in this example.

This square root extraction begins as in the first example on page 10: the dotting procedure concludes with four dots being placed in the progression described for the first example of this type on page 10. However, instead of concluding four digits for the integer part of the square root, Bürgi states that five digits are required. Again, if the number retrieved from the tables had been recorded correctly as 220330940, the dotting procedure would have indicated five dots: one placed above the 0, 9, 3, 0, and 2, moving from the right-most to left-most digit in the nine-digit number sequence, yielding a “dotted” radicand of $\dot{2}2\dot{0}3\dot{3}0\dot{9}4\dot{0}$. There are several errors or irregularities involving ending zeros of numbers used (or meant to be used) in the examples in the “Kurzer Bericht.” It is possible that the copyist may not have considered them meaningful, as Lutstorf (2005) reported: “zeros likely did not mean anything anyway and were omitted without prejudice” (pp. 122–123).

The process shown at the bottom of page 10 (and continued onto page 11) of the “Kurzer Bericht” for this square root extraction does not match the use of 220330940 as the radicand. Instead, the calculation would match 2203309400, if that were a tabulated value. Thus, a more reasonable square root extraction would match with the dotting procedure that was explained in the example as $2\dot{2}0\dot{3}3\dot{0}9\dot{4}$, meaning that one extra digit before the first dotted digit would call for adding the whole red number to the red number for 22033094 (i.e., adding **230270022** to **79000**), dividing the sum in half, and using that red number result (**154635011**) and linear interpolation to find the nine-digit sequence 469394227. Finally, since only four digits are needed for the whole number part of the square root, the desired root is 4693.94227. In the Gk manuscript, the solution for the square root extraction of 22033094 is not complete; the solution ends with reporting the red number of **154635011**.

The first complete example on page 11 of the “Kurzer Bericht” is a cube root extraction. In modern notation, we wish to solve for $x$:

$$x = \sqrt[3]{N}.$$

The affiliated logarithmic calculation becomes

$$\left(\log x\right)=\left(\log\left(\sqrt[3]{N}\right)\right)=\left(\log N^{\frac{1}{3}}\right)$$
$$=\frac{1}{3}\log N.$$

The example seeks to extract the cube root of the seven-digit number 5612037 and again, the first task is to complete the dotting procedure. For the cube root, this entails placing a dot above the right-most digit ("7" in this example) and continuing to "dot" every third digit to the left. Unlike the previous root extraction examples on page 10, no dotting appears in the examples on page 11. For reference, the dotting procedure for the present example would appear as $\dot{5}61\dot{2}03\dot{7}$ and would indicate that there are three digits in the integer part of the cube root.

Next, using the tables, the red number corresponding to the nine-digit number sequence 561203700 is **172500**. As the logarithmic calculation shows, extracting the cube root is equivalent to division by 3; thus, the red number **57500** is located in the tables (here, the copyist incorrectly recorded the value as **52500** in the manuscript) and the black number associated with it is 177707944. The dotting process requires the cube root to be 177.707944 (i.e., three dotted digits in the black number correspond to three digits in the integer part of the cube root). When cubed,

$$\left(177.707944\right)^{3}=5612037.017,$$

which is slightly too large. As Lutstorf (2005) pointed out, such errors are due to rounding that occurs from values in which more than nine digits are known. However, since the tables only include nine-digit numbers, such rounding from the tenth digit will produce errors. Bürgi himself recognized the nature of the values he tabulated: "I may reach this cube root 177707944 in my tables in 9 digits. But, reserving the rest of the 9 digits for the fraction, [it is] assumed that because so many irrational numbers finish with [fractions], which may not be given enough satisfaction in 9 digits" (p. 11).

The second cube root extraction is for the eight-digit number 56120370, and the calculation that begins at the bottom of page 11 proceeds as in the cube root extraction for 5612037. The dotting procedure results are again stated but not shown in the manuscript; however, Bürgi determines that the integer part of the cube root must have three digits. Moreover, the dotting procedure resulted in one digit of the cubed value $\left(5\dot{6}12\dot{0}37\dot{0}\right)$ to the left of the final "dotted" digit. Thus, to complete the logarithmic calculation, the whole red number is added to the corresponding red number (**172500[000] + 230270022**) and the result (**402770022**) is divided into 3. Linear interpolation is performed on this red number (**134256674**) to obtain the cube root 382.860159.

The third and final cube root extraction is for the now familiar radicand 561203700. The calculation unfolds in the same manner as in the previous example. Again, though not shown in the Gz manuscript, the dotting for this cube root process would appear as $56\dot{1}20\dot{3}70\dot{0}$. Then, it appears as if Bürgi twice explains why two whole red numbers must be added to the red number associated with the radicand. In a fairly broken manner, he states that the digits "5" and "6" remain after the final dot is placed in the dotting procedure; thus, a whole red number for each needs to be added. The first column of numbers is given as

**172500**

**230270022**

**230270022**

**633040044**.

Then, the second column is given:

**172500**

**230270022**

**230270022**

[2-line gap]

**211012246**.

Similar and repeated textual references (e.g., "…die erste .5 sambt den anderen Ziffern…" is similar to "…die erste 5 mit den andn Zifern…"), as well as a strikethrough phrase (i.e., "~~so hat die erste~~ 5"), hint that this is a lengthy copyist error.

An analysis of the logarithmic calculation using modern notation clearly shows the reason for the addition of two whole red numbers (i.e., **230270.022** = log10):

$$\begin{aligned}&\log\left(\sqrt[3]{561203700}\right)\\&=\log\left(561203700\right)^{\frac{1}{3}}\\&=\frac{1}{3}\log\left(561203700\right).\end{aligned}$$

Now, we can use the first cube root extraction (page 11 of the "Kurzer Bericht") by rewriting 561203700 as $5612037 \times 10^2$. Then,

$$\begin{aligned}&\frac{1}{3}\log\left(561203700\right)\\&=\frac{1}{3}\log\left(5612037\cdot 10^2\right)\\&=\frac{1}{3}\left[\log 5612037+\log 10^2\right]\\&=\frac{1}{3}\left[\log 5612037+2\log 10\right].\end{aligned}$$

Finally,

$$\begin{aligned}&\log\left(\sqrt[3]{561203700}\right)\\&=\frac{1}{3}\left[\mathbf{172500}[\mathbf{000}]\right]+2(\mathbf{230270.022})\\&=\frac{1}{3}[\mathbf{633040.044}]=\mathbf{211013.348}.\end{aligned}$$

Thus, Bürgi's calculation is correct except for the final digit of the red number obtained when the total (**633040.044**) is divided by 3; the Gz manuscript contains the incorrect value **211013.346** rather than the correct value **211013.348**. Consequently, linear interpolation for the former red number value would result in the value 824.847192 for the cube root of 561203700. This differs from the value obtained using the correct value of **211013.348**, which yields the cube root value of 824.847208.

The examples of root extraction conclude on page 13 of the "Kurzer Bericht" with two final examples: one a fourth root extraction ("ZZ.R" or "radicem zensi zensicam") and the other a fifth root extraction ("Ss" or "sursolidam"). The examples follow the same process as the previous root extraction examples, and, in particular, the fourth root extraction utilizes a version of the number 5612037xx. The fourth root extraction is a carefully maintained example in the sense that both the dotting procedure and the "decimal zero" are shown in the Gz manuscript.

To determine the fourth root of 56120370, the dotting procedure yields $561\dot{2}037\dot{0}$; thus, indicating the fourth root has a two-digit integer part. Moreover, since the three leading digits of $561\dot{2}037\dot{0}$ are without dots, three whole red numbers must be added to the red number for 56120370, resulting in **863310.066**, which is then divided by 4 to determine the red number for the fourth root. This red number is given in the Gz manuscript as **190827.516**; however, this value should be **215827.516**. Consequently, the fourth root is also written incorrectly; the fourth root is 86.5526026 instead of the value 67.080769. The fourth root given in the Gz manuscript is given to eight digits only (even though it is incorrect), whereas the Gk manuscript gives 67.4080769. The incorrect sum in the Gk manuscript (**763310.066**) would have resulted in the final red number, **190827.516**, and, correspondingly, the fourth root value that was too low.

Next, Bürgi selects a different black number for the fifth root extraction example. The dotting procedure is discussed but not shown for number 671876768. However, the process yields that the desired fifth root will have a two-digit integer part. The dotting would yield $671\dot{8}7676\dot{8}$ , and thus, with the leading three digits of the radicand after the final dot is placed, three whole red numbers would be added to the radicand's red number.

In this fifth root extraction example, the calculation is carried out as the others, with one small exception. In this example, to the right of the column of numbers being added, the copyist has included another column of numbers (in black):

**190500** __ 6

**230270022** – 7

**230270022** – 1

**230270022** 8

Here, the 6, 7, 1, and 8 serve as a guide for finalizing the root extraction process, as the text states, "Diß ist der vier Ziffern Alß 6718 ihr rote Zall" (or "These are the four digits as 6718 of its red number"). The calculation is without error until the final red number (which actually appears on page 14 of the "Kurzer Bericht"), in which the copyist records $\mathbf{176262015\frac{1}{2}}$ instead of $\mathbf{176262013\frac{1}{5}}$. As with other errors of this type, the corresponding black number is also incorrect, but not until the millionths digit. The correct fifth root should be 58.2717318.

### *Comments on Stylistic Elements*

It is not accidental that the first two cube root extractions, as well as the third and subsequent fourth root extraction, utilize the same black number, 5612037xx. In a similar way that a teacher would focus on a procedure as opposed to creating unique numerical content for their instruction, Bürgi retains the numerical value (or some magnitude of it) for four consecutive examples. Consequently, the focus remains on the procedure needed to complete the calculation rather than the need to search the 58 pages of tables for unique values in each instance. As previous examples have shown, and which will be seen in subsequent examples, this pedagogical technique is used throughout the "Kurzer Bericht."

Page 12 of the "Kurzer Bericht" is an interesting specimen of various types of copyist irregularities and omissions. As already mentioned, this one page includes a mostly repeated explanation for why two whole red numbers must be added (i.e., to represent the multiplication by 100) to compute the cube root of 561203700, a conspicuous strikethrough (related to the repeated text), and the incorrect values provided at the conclusion of the calculation. Additionally, the final cube root value is missing the important diacritical mark (Bürgi's "decimal zero") to mark the division between the integer and decimal parts of the number. Two of these irregularities, the strikethrough and the omitted "decimal zero," do not appear in the Gk edition or manuscript.

Finally, there are several interesting inclusions and irregularities represented in the examples on page 13 of the "Kurzer Bericht." The fact that Bürgi continues to emphasize the column calculation format with the red numbers—and with careful attention to alignment—is a significant component of the pedagogical tools of

Bürgi's "thorough instruction." Thus, emphasizing the simpler operations using the red numbers is still a key concern.

### VI. "Kurzer Bericht": Pages 14–18

#### Translation

**I.**

In the first place, to find a mean proportional number between two known numbers, the two numbers being 119004521 and 893423483. Their due red numbers

are **1740**$\overset{\circ}{\mathbf{0}}$ and **21900**$\overset{\circ}{\mathbf{0}}$

The difference between the red numbers is **201600**, the [whole]

In two equal parts or halved is **100800** the half

Added to the small red number is **<u>17400</u>**

Then the red number of mean proportional number is **11820.**$\overset{\circ}{\mathbf{0}}$

And its black [number] is the 326069676

Mean proportional number that we desire.

**II.**

Secondly, to find two mean proportional numbers

Divide the reported red difference in [3] equal parts.

Add the part to the small red number, so we have the first red number. The same mean proportional number, or adding the same second part to the small red number, so we have the other red number, of the same black mean proportional number.[180]

---

[180] Page 14 of the "Kurzer Bericht" ends here.

## III.

Thirdly, to find 3 mean proportionals, divide the reported difference into 4 equal parts, and add one part to the small red number, so we have the first red number of the same black mean proportional number. Or add the same [second] part to the same small red number so we have the second red number for [its] black mean proportional number; or add the same [third] to the small red number, [and] we have the third red number of its mean proportional number.

In this way can find all mean proportional numbers. Thus, this can be seen all the more in the following example for two given numbers with the same total [number of] digits.[181]

Find the mean proportional between 2 given numbers. It is, however, that the 2 given numbers do not have an equal number of digits. So the first has 7 digits and the other has 8; the first [one] is 2447471 and the other 33033604.

Search [for] their due red number[s] **89510** and **119500**.

Adding together **89510**

But this is the red number **209010**. Since [one] red [number] is

one digit longer **230270022**

Then the whole red number is added there

**439280022** This red number

is halved **219640011** the due

black [number] of this proportional number is 899159541.[182]

[181] However, the very next example does *not* use two given numbers with the same number of digits.

[182] Page 15 of the "Kurzer Bericht" ends here.

To find a mean proportional number between 2 numbers:

[B]ut the numbers do not have the same number of digits; the first has 7 digits, and the other has [9] and is thus

[The first] 2447471 and the other 330336040

| | |
|---|---|
| Their due red number[s are] $\mathbf{8951\overset{\circ}{0}}$ the other | $\mathbf{11950\overset{\circ}{0}}$**.** |
| Adding together | <u>**89510**</u> |
| Do together | **209010** to this |
| Add 2 whole red numbers because the larger | **230270022** |
| exceeds the smaller by 2 digits | **230270022** |
| | <u>**669550044**</u> |
| This red number is halved [and] is the red number | **334775022** |
| of the due black number. | |
| But it is greater than the whole red number. | |
| The whole red number is subtracted, so this is the red | **230270022** |
| Number of the mean proportional number | $\mathbf{10450\overset{\circ}{5}000}$ |
| which is | $28433921\overset{\circ}{3}$ |

Because I am able to subtract the whole red number from the halved red number, so can I also have more digits [in the second number] than the first, i.e., [9].

To find the mean proportional number between 2 numbers:

But the two numbers happen to be as follows:

The first with 6 digits, the other with 9 digits

| | | | |
|---|---|---|---|
| **I** | 303419 | **II** | 304939818 |
| Its due red number | **111000** | | **11150$\overset{\circ}{0}$0** |
| | | | **1110000**[183] |
| Adding together becomes this much | | | **22250$\overset{\circ}{0}$** |
| [To] this add 3 whole red numbers | | | **230270022** |
| Because one of the two numbers | | | **230270022** |
| Exceeds the other by [3] digits | | | **230270022** |
| Thus, the red number is the halved | | | **913310066** |
| From this halved number, subtract the whole red number | | | **456655033** |
| | | | **230270022** |
| So this is the red number of the due mean | | | [**226385011**] |
| Proportional number which is | | | [961896$\overset{\circ}{7}$44] |

And [this] is only one digit more than the first, and this is the proof that I no longer take all the red numbers once I have the halved red number.

[183] Page 16 of the “Kurzer Bericht” ends here.

To find a mean proportional number between 2 numbers:

But the two numbers happen to be as follows.

The first has 5 digits, the other 9, and the first [number] is

32891, and the other [number] is 454907654,

Its due red number is $\mathbf{11906\overset{\circ}{7}351}$ [and] $\mathbf{15\dot{1}500000}$

Adding together **119067351**

Doing this [addition], the red number [is] **270567351**

Add 4 whole red numbers because **230270022**

One [number] exceeds the other by four digits. **230270022**

**230270022**

**230270022**

So is this red number, halved **1191647439**

From the half, subtract the whole red number $\mathbf{595823719\frac{1}{2}}$ [184]

And as often as I like the same digit[s], in the mean proportional number, [I] have 2 more [digits] than the first [black number], and then I need [to subtract] the whole red number 2 times and it remains for me the red number of the mean proportional [number]

which is $\mathbf{13528\overset{\circ}{3}675.}$

Thus, the mean proportional number is $[386812\overset{\circ}{1}98]$

that we have desired.

[184] Page 17 of the "Kurzer Bericht" ends here.

To find a mean proportional number between 2 numbers:

But the two numbers occur to me as

the first with 4 digits, the other with 9 digits, and

so [one] is 5764 [and] the other [387649833]

| | | | |
|---|---|---|---|
| Its due red number is | **175170640** | [and] | **135500000** |
| Adding together | | | **175170640** |
| Making this red number | | | **310670640** |
| This constitutes 5 whole red numbers because | | | **230270022** |
| one exceeds the other by 5 digits | | | **230270022** |
| | | | **230270022** |
| | | | **230270022** |
| | | | **230270022** |
| This [is] added, halve the red number | | | **1462020750** |
| [The result] is this red number | | | **731010375** |

From this subtract the number of whole red numbers as many as I like; in this example, 3 times, wherefore the mean proportional number has 3 digits more than the first [black number], and its red number that remains [is]

**40200309**

This due black number is the mean proportional number [149479552].[185]

[185] Page 18 of the "Kurzer Bericht" ends here.

**Commentary**

## *Determining a Mean Proportional*

The first of several examples for how to determine the mean proportional, or any number of mean proportionals, between two given numbers is presented on page 14 of the Gz manuscript's "Kurzer Bericht." As previously mentioned, the remainder of the manuscript (pages 14–21) is dedicated to providing instruction on how to calculate mean proportionals. Interestingly, Bürgi begins with one example (section "I") and then presents brief descriptions of procedures (section "II" and the beginning of section "III") necessary for calculating more than one mean proportional between two given numbers. However, the descriptions provided are actually for the next set of examples, which begin on page 19 of the Gz manuscript, leading the reader to believe that this represents another copyist error.

The first example (appearing as the second example on page 14 of the Gz manuscript) calculates the mean proportional between 119004521 and 893423483. In theory (though of course not explicitly discussed in the Gz manuscript), the calculation involves two concepts. First, if $x$ is the mean proportional between $a$ and $b$, then the value $x$ between $a$ and $b$ differ by a factor $r$, such that $x=ar$ and $b=xr$ or $b=ar^2$. Solving for the square root,

$$r = \sqrt{\frac{b}{a}}$$

or

$$r = \left(\frac{b}{a}\right)^{1/2}.$$

Secondly, since $x$ is the mean proportional, the final calculation is

$$x = a\left(\frac{b}{a}\right)^{1/2}.$$

Logarithmically, the calculation is now equivalent to

$$\log x = \log a + \frac{1}{2}(\log b - \log a).$$

Now, keeping this logarithmic equation in mind, to calculate the geometric mean between 119004521 and 893423483, Bürgi simply locates the red numbers (**17400** and **219000**, respectively). Then, the first red number is subtracted from the second (**219000 – 17400 = 201600**); half the difference is recorded (**100800**); and the result is added to the first red number (**100800 + 17400 = 118200**). Finally, the resulting red number (**118200**) is located in the tables, and the number 326069676 is the mean proportional between 119004521 and 893423483.

### *Determining Mean Proportionals Between Two Numbers of Different Magnitude*

The next five examples in the Gz manuscript explain how to determine the mean proportional between two numbers of different magnitudes. In the first example on page 15, Bürgi must extend the calculation used in the mean proportional example given on page 14. That is, if the values for $a = 2447471$ and $b = 33033604$, the calculation is extended to include an additional log10 (the whole red number, **230270.022**). Using $\log x = \log a + \frac{1}{2}(\log b - \log a)$ as before:

$$\log x = \frac{1}{2}\left(\log\left(2.447471 \times 10^{6}\right) + \log\left(3.3033604 \times 10^{7}\right)\right)$$

$$= \frac{1}{2}(\log 2.447471 + 6\log 10 + \log 3.3033604 + 7\log 10).$$

Since the calculation requires the square root, the magnitude of the result of the logarithmic calculation must be considered. Thus, the calculation is rewritten as:

$$= \frac{1}{2}(\log 2.447471 + \log 3.3033604 + \log 10) + 6\log 10.$$

Bürgi considers the size of the result of the square root and employs only the truncated calculation, that is, the logarithmic sum that is increased only by log10. Bürgi's calculation procedure first entails (as expected) locating the red numbers for each of the black numbers. Next, the red numbers are added together (**89510 + 119500 = 209010**), and the whole red number is added to this (**230270022**) to yield **439280.022**. Here, Bürgi introduces the "decimal zero" into the column of numbers. Then, the final sum is divided in half and the resulting red number is **219640.011**. As expected, linear interpolation yields the corresponding black number 8991595.41.

The first example on page 16 of the "Kurzer Bericht" computes the mean proportional number between 2447471 (a seven-digit number) and 330336040 (a nine-digit number). The calculation is essentially the same as the example on page 15, although the nine-digit number is incorrectly identified as having eight digits.[186] To determine the mean proportional, the red numbers are repeated from the previous

[186] This error also exists in the Gk manuscript.

example and then added together. Then, two whole red numbers are added to the sum for the reason that "the larger exceeds the smaller by two digits." The sum is then divided by 2; however, the value, **334775022**, exceeds the whole red number, and thus, **230270022** is subtracted and the resulting red number (**104505.000**) is used to find the mean proportional (via linear interpolation), which is 28433921.3.

To clarify the result of the calculation, and in particular, the correct magnitude of the result, we have (in modern notation)

$$\begin{aligned}&\sqrt{\left(2.447471\times10^{6}\right)\cdot\left(3.30336040\times10^{8}\right)}\\&=\sqrt{8.084878782\times10^{14}}\\&=\sqrt{8.084878782}\times10^{7}\\&=2.8433992125\times10^{7}\\&=28433021.3.\end{aligned}$$

Next, the example stated at the bottom of page 16 seeks to determine the mean proportional number between the six-digit number 303419 and the nine-digit number 304939818. Similar to the previous examples, the two red numbers (**111000** and **111500**) are added together. Then, three whole red numbers are added to the sum, which is then divided by 2. However, the result, **456655033**, exceeds the largest red number value in the tables, and when subtracted the resulting red number (the actual value is **226385011**; the manuscript value is **226335011**) is used to find the mean proportional. Finally, the interpolated value in the manuscript which is 9614159.42 is correct (to the tenths place) for the incorrectly printed red number. However, the correct mean proportional number for the original two numbers is 9618967.44.

The second example presented on page 17 of the "Kurzer Bericht" determines the mean proportional number between a five-digit number (32891) and a nine-digit number (454907654). This example differs from the others of this type in that the red number corresponding to 32891 must be determined by linear interpolation (and, again, this is left to the user to compute). The computations are carried out as in the previous mean proportional examples: the red numbers are added together; four whole red numbers must be added to compensate for the difference in the number of digits of the two black numbers; and the total sum of the red numbers is halved. Then, as Bürgi explains, the whole red number must be subtracted twice in order to yield a red number that can be located in the tables.

Only the first nine digits of the red number $\mathbf{135283675\frac{1}{2}}$ are used to determine the associated black number through linear interpolation. The resulting black number, 386812198, appears on page 18 with a faint "decimal zero" above the "2," suggesting that the number 386812.198 is the mean proportional between 32891 and 454907654. However, the "decimal zero" should appear above the second "1," a number which actually looks like a thick column of ink that appears to have been changed from a "9" and which may have contributed to the incorrectly placed "decimal zero." Thus, the mean proportional between 32891 and 454907654 is 386121.98.

One interesting difference between the Gz manuscript and the Gk manuscript is the instance of the final red number in the first example on pages 16–17, **226335011**. This number is found in both manuscripts; however, as Lutstorf (2005) observed, this is a result of the incorrect column addition of the original two red numbers and three whole red numbers. In the Gk manuscript, the sum is given as **913210066**, and the remaining calculations are carried out using this value. However, in the Gz manuscript, the sum is correctly recorded as **913310066**, and all subsequent values are correctly saved for the final red number, which is given as **226335011** in both manuscript copies.

A second notable difference between the two manuscript copies with regard to the second example on page 17 (and completed on page 18) is that the final solution is not included in the Gk edition. Instead, the example simply ends with the halved red number found at the bottom of page 17 of the Gz copy $\left(\mathbf{595823719}\frac{\mathbf{1}}{\mathbf{2}}\right)$, along with and the statement:

So kombt dieße rothe Zahl (**1191647439**) die halbier $\left(\mathbf{595823719}\frac{\mathbf{1}}{\mathbf{2}}\right)$
von der halben rothen *Sub*: die ganze rothe Zahl
und so offt isch derselbigen mag so viel Zifffern wird die Zahl mehr haben dann die die erste[187]

Lutstorf (2005) explained that the remainder of this calculation is left to the user of the tables. However, a more likely explanation is copyist error (on the part of the copyist of the Gk manuscript), since the only evidence of "left to the user" aspects of the examples is in the case of linear interpolation.

The final example of finding a single mean proportional number between two given numbers of different magnitude appears on page 18 of the "Kurzer Bericht." Determining the mean proportional between the two numbers, 5764 and 387649833 (this black number is given as 287649833 in the Gz manuscript, but the red number (**135500.000**) corresponds to the corrected value given here), is conducted in the same way as the preceding examples. Notably, this example is another instance of the use of a conveniently selected nine-digit black number that appears at the top of a final column on a page in the tables (the number 387649833 appears on page 34 of the tables).

Similar to the just previous example, the four-digit given number in this example requires linear interpolation to determine its corresponding red number. The two red numbers are then added together (and the sum is given), along with five whole red numbers to reflect the difference in the number of digits for the given black numbers. This sum (**1462020750**) is halved (**731010375**), and since the resulting value exceeds largest value in the tables, Bürgi again states that as many whole red numbers that are needed must be subtracted in order to yield a red number that exists in

[187] *So when this red number (**1191647439**) is halved* $\left(595823719\frac{\mathbf{1}}{\mathbf{2}}\right)$, *and from the halved red [number] sub(tract) the whole red number and whenever I have digits, the number will have more than the first.*

the tables. For this example, he determines that three whole red numbers must be subtracted, since "…the mean proportional number has three digits more than the first [black number]." Thus, Bürgi knows that the mean proportional for this calculation must have seven digits. Furthermore, in this example, he does not show the physical subtraction of each of whole red numbers (similar to the previous example); at this point in the progression of examples, this particular aspect of the calculation is omitted. Instead, the final red number value in the calculation reflects the difference: **731010375 – 3(230270022) = 40200309**. Linear interpolation is left to the reader, and the resulting mean proportional number is 1494785.91.

Unfortunately, the recorded value for the linear interpolation to determine the red number for 5764 (given as **175170.640**) is inaccurate, and the error impacts the final mean proportional value. However, as Lutstorf (2005) noted, the error in the first linear interpolation in this example is consistently carried through the computation; thus, the procedure results in only a minor inaccuracy. In this instance, the corresponding proportion, using the corrected values, for this mean proportional example would yield

$$\frac{5764}{1494795.52} \approx \frac{1494795.52}{387649833}$$
$$0.00385605 \approx 0.00385605.$$

### *Comments on Stylistic Elements*

The mean proportional content found on page 14 is less onerous in terms of calculation when compared with previous pages. Consequently, fewer irregularities or errors exist in this one example than are found on other pages. Page 14 of the "Kurzer Bericht" also exhibits helpful organizational effects. The red lines dividing the example from the procedural description for determining multiple mean proportionals are used throughout the manuscript. (The second red line on page 14 is probably extraneous; dividing within an example is not particularly helpful.) However, on page 14, this organizational tool is also emphasized with a sort of Roman numeral outlining or section numbering in the left margin of the page. The notation "I" (with a hat-shaped emblem above it) appears with the sample calculation, and the notations "II" and "III" (also with a hat-shaped emblem above) accompany the procedural descriptions for calculating two and three mean proportional numbers between two given black numbers. Such section numbering can be found elsewhere in the Gz manuscript, beginning on page 4 with the examples associated with using the tables.[188]

[188] Some of the section numbers are difficult to read or appear to be missing due to their placement in the left margin of the pages.

Lutstorf (2005) observed that the "decimal zero" is used only sparingly in the example on page 16. Bürgi introduces the "decimal zero" late in the calculation; first at the point of the result of the subtraction (**104505.000**) and then with the corresponding result (28433921.3). The placement of the "decimal zero" in the mean proportional number, which is the square root of the product of the two given numbers, is again an easily understood calculation for a mathematician such as Bürgi. In the first example on page 17, both the Gz and the Gk manuscript copies include the "decimal zero" only at the first instances of red numbers and then only again at the reporting of the final result (in the black number determined). In contrast, the example on page 18 of the "Kurzer Bericht" includes a more careful incorporation of the "decimal zero." In particular, for the complete example on page 18, the "decimal zero" is given at the top of the column of red numbers to be added (and then halved), and the "decimal zero" is again reported with the resulting black number.

Bürgi's treatment of calculating a mean proportional number between 2447471 and 330336040 is reminiscent of previous examples in the "Kurzer Bericht" in which numbers with ending zeros are carefully selected for their use in examples. In particular, using the same values (with or without some number of trailing zeros) enables Bürgi to produce familiarity in his instructive examples. Additionally, the mean proportional calculation on pages 16 and 17 is another important example of Bürgi's use of carefully selected numbers for use in examples. For the calculation of the mean proportional number between 303419 and 304939818, he has selected two numbers that appear at the top of the final two columns on the 28th page of the tables, the first of which is one of only a half-dozen or so numbers with three zeros at the end. Indeed, the selection of these two numbers is both purposeful and efficient (Lutstorf 2005).

### *Errors in Content Organization*

In the sections labeled "II" and "III" on pages 14 and 15 of the Gz manuscript, Bürgi introduces how to find two and three mean proportionals, respectively, between two given numbers. As a method for preparing for the increased complexity of the calculation of mean proportionals, Bürgi provides a synopsis of the procedure needed. For determining two mean proportionals, first divide the difference between the two given numbers into three equal parts (although this is incorrectly written as four equal parts in the Gz manuscript). Next, add one of the equal parts just determined to the smaller red number (and this will yield the corresponding black number). And finally, add two of the equal parts to the smaller red number (and, again, the corresponding black number for the desired mean proportional results). Determining three mean proportionals between two given black numbers is similarly described. However, both of these descriptions appear before an inappropriate collection of examples. The descriptions should instead preface the final examples of the Gz manuscript, in which the reader is instructed on how to determine two, three, and four mean proportionals between two black numbers of the same magnitude.

## "Kurzer Bericht": Pages 19–21

### Translation

Between 2 numbers, find the mean proportional number.

It is in our opinion a slight alteration [is needed] to find 2, 3, 4, or more mean proportional numbers between 2 known numbers. Wherefore we want to make the change known, through an example, which is given as before through known numbers,

and the 2 numbers are 119004521 and 893423483.

Their due red number[s] [are] **17400** and **21900**

The difference of the red number(s) is $\mathbf{20160\overset{\circ}{0}}$

The [whole] part in 3 parts is [**67200**]

Add a third to the small[er] red number [**17400**]

Thus, the red number is the first proportional [**84600**] number

Its due black number is [233020839]

Two third [parts] of the difference of the red number[s] is [**67200**]

Add the small red number to it. [**67200**]

[**17400**]

This is the red number of the second proportional [**151800**] number.

Its due black number is the [233020839].

| *A*: | *B*: | *C*: | *D*: |
|---|---|---|---|
| 119004521 | [233020839] | [456274358] | 893423483 |
| **1740$\mathring{0}$** | **8460$\mathring{0}$** | **15180$\mathring{0}$** | **21900$\mathring{0}$** |

[The proportion that] holds for *A* to *B*: so it holds for *B* to *C*: and *C*: to *D*:[189]

Find three mean proportionals between 2 numbers.

[T]here are two known numbers as 119004521 and 893423483

Their due red number is **1740$\mathring{0}$** the other **21900$\mathring{0}$**

Their difference is **20160$\mathring{0}$**

If the [whole] part is divided into four equal parts, each part is **5040$\mathring{0}$**

**17400**

The one part is added to the smaller red number **6780$\mathring{0}$** is

the due red number [for the] black [number] 196986715[;]

this is the first mean proportional number.

Secondly, add $\frac{2}{4}$ (part) of the red difference to the small red number

As[:] **5040$\mathring{0}$** – the $\frac{2}{4}$

**50400**

And the smaller red number **17400**

[189] Page 19 of the "Kurzer Bericht" ends here.

| | |
|---|---|
| Gives the red number of the second proportional number | **118200** |
| which is its due black number | [326069676] |
| [of] the desired second [proportional number]. | |
| Thirdly, add $\frac{3}{4}$ of the red difference | **50400** |
| and the small red number | **50400** |
| | **50400** |
| | **17400** |
| This is the red number of the third proportional | **168600** number |
| which is its due black number [539739109] | |
| [of] the third desired.[190] | |
| Find four mean proportionals between two [numbers]. | |
| The 2 known numbers [are] 119004521 and 893423483 | |
| Their due red number[s] [are] | **1740$\overset{\circ}{0}$** the other **21900$\overset{\circ}{0}$** |
| Their difference is | **20160$\overset{\circ}{0}$** |
| If the [whole] part is divided in 5 equal parts, each part is | <u>**40320**</u> |
| Add the smaller red number to the $\frac{1}{5}$ (part) | <u>**17400**</u> |
| this is the red number | **57720** |
| of the first due mean proportional black number | 17809931[2] |
| Secondly, add $\frac{2}{5}$ [part] to the smaller red number | **4032$\overset{\circ}{0}$** |
| | **40320** |
| The smaller red number | <u>**17400**</u> |

[190] Page 20 of the “Kurzer Bericht” ends here.

| | |
|---|---|
| together making the due red number | **98040** |
| which is the second mean proportional number | [266539159] |
| Thirdly, add $\frac{3}{5}$ to the smaller red number | **40320°** |
| | **40320** |
| The smaller red number | **40320** |
| together making the due red number | **17400** |
| | **138360** |
| which is the third mean proportional | 39889611[1] |
| Fourthly, add $\frac{4}{5}$ [part] to the small red number | **161280 the** |
| The smaller red number | **17400** |
| together making the due red number | **178680** |
| which is the fourth mean proportional | 5969783[52] |

## Commentary

### *Determining 2, 3, and 4 Mean Proportionals*

Page 19, with its single example, is a curiously flawed page. The series of examples for determining multiple mean proportional numbers begins with an incorrect title: "Zwischen .2 Zahlen, die *Medio proprotional* Zahl zu finden."[191] This titling error is related to the incorrectly placed instructions for determining multiple mean proportionals between two black numbers (found on pages 14 and 15 of the Gz manuscript); indeed, the title for the example on page 19 and the title given at the beginning of section II of page 14 should be switched.

In the introduction to the example, Bürgi states that a "slight alteration" is needed when 2, 3, 4, or more mean proportional numbers between two given numbers are found. And, beginning with the example on page 19, he proceeds to give one example each of how to find 2, then 3, and, finally, 4 mean proportionals between two given numbers. In the same instructional manner found in other strings of related examples, Bürgi employs the same two given numbers (the boundary numbers) in each of the final three examples that appear in the Gz manuscript.

The first example of determining multiple mean proportional numbers between two given numbers is to find two mean proportionals between the boundary num-

[191] "Between 2 numbers, find the mean proportional number."

bers 119004521 and 893423483. After locating the corresponding red numbers (**17400** and **219000**, respectively), the difference is divided by 3 (**201600 ÷ 3 = 67200**). Next, one of the "third parts" is added to smaller of the red numbers (**17400 + 67200 = 84600**), and then the associated black number (233020839) yields the first mean proportional number. To determine the second mean proportional, two "third parts" are added to the smaller of the red numbers (**17400 + 67200 + 67200 = 151800**),[192] and the associated black number (456274358) gives the second mean proportional number.

Finally, Bürgi presents the results of the calculation (albeit with incorrectly recorded mean proportionals) using the letters *A*, *B*, *C*, and *D*, along with the colon notation for proportions as in *A* : *B* [: :] *C* : *D*. Furthermore, Bürgi concludes the example with "[The proportion that] holds for *A* to *B*: so it holds for *B* to *C*: and *C*: to *D*:" to emphasize the desired result of determining mean proportional numbers.

The proportion is confirmed using modern notation, and the calculations using the corrected mean proportional number values for the ratios (rounded to nine decimal places) are

$$\frac{A}{B} = \frac{119004521}{233020839} = 0.510703341,$$
$$\frac{B}{C} = \frac{233020839}{456274358} = 0.510703341,$$
$$\frac{C}{D} = \frac{456274358}{893423483} = 0.510703341.$$

If, however, the eight-digit numbers for the two mean proportional numbers are used in the ratios, all equivalency is lost:[193]

$$\frac{A}{B_M} = \frac{119004521}{23020839} = 5.16942588,$$
$$\frac{B_M}{C_M} = \frac{23020839}{45932698} = 0.501186300,$$
$$\frac{C_M}{D} = \frac{45932698}{893423483} = 0.051412011.$$

Moreover, it would not make sense that values less than 100 million (eight-digit numbers) are viable choices for given boundary numbers, each greater than 100 million (nine-digit numbers). Thus, an explanation for the other notable flaw found on page 19 of the "Kurzer Bericht" is only partially possible. For example, since the Gz

[192] Here, Bürgi would have shown each third part (**67200 + 67200**) to emphasize the additive nature of the logarithmic (in modern terms) calculation. This aspect of the calculation is seen in both examples found on pages 20 and 21 (except for the final addition on page 21).

[193] $B_M$ and $C_M$ correspond to the values for the mean proportional numbers given in the Gz manuscript.

manuscript value for the first mean proportional, 23020839, differs from the actual value, 233020839, by a single digit (i.e., an extra digit has been inserted), it is entirely possible that this is a simple copying error. However, the second mean proportional number in the manuscript, 45932698, is almost entirely incorrect (except for the first two digits (45) and the final digit (8)). One possibility offered by Lutstorf (2005) is that the third digit (9) is "set heads down"; that is, a "6" was written upside down. Beyond this conjecture, however, it is too difficult to explain the corruption of the actual value (456274358) into what is given in the manuscript (45932698).

When the Gieswald's Gk edition is examined, it is understood that the intermediary red and black numbers for this same example do appear in the copy from which Gieswald worked to set his transcription—which they in fact do. The "decimal zeros" appear in both manuscripts above the final digit of each red number (except for the first instance of **17400** in the Gz manuscript). In the Gk copy, however, the addition of two "third parts" is simple shown as

**134400**

<u>**17400**</u>

**151800**,

where **134400** = 2(**67200**).

The values for this step of the calculation in the Gz manuscript are missing; however, based on the second example in this series (where three mean proportional numbers are determined), it is more consistent to assume that Bürgi would have emphasized the additive nature of the red numbers for the two "third parts," particularly in this first example.

To determine three mean proportional numbers between the given numbers 119004521 and 893423483, the difference of the corresponding red numbers is first divided by 4 (**201600** ÷ **4** = **50400**). Then, in successive steps, one of these "gleiche teil" ("equal part[s]"), then two, and finally three are added to the smaller red number, and the resulting red number for the sum is used to locate the corresponding black number of the mean proportional. As expected from Bürgi's instructive style, each instance of the equal fourth part is shown in the addition, so that for the final calculation (to determine the third mean proportional number), we find

**50400**

**50400**

**50400**

<u>**17400**</u>

**168600**.

The two errors in reporting the second and third mean proportional numbers are minor and differ from those found in the Gk edition. According to the tables, the second mean proportional number for **118200** is 326069676. This is the same value as given in the Gk manuscript; however, in the Gz manuscript, this value appears as 32606976, where the second-to-last "6" has been omitted. In recording the corresponding black number for the third mean proportional's red number, **168600**, one digit is miscopied. The actual value from the tables is 539739109; however, the value in the Gz manuscript is 539735109. In the Gk manuscript, the third mean proportional number is reported as 539738109.

The final example of the "Kurzer Bericht" of the Gz manuscript presents the calculation for finding four mean proportional numbers between the familiar boundary numbers 119004521 and 893423483. In the example, the one-fifth part of **201600** (the red number corresponding to the difference between the red numbers of the two boundary values) is **40320**, and the calculation of each mean proportional number is carried out in the same manner as the two previous examples. The only difference in the accompanying explanation of this example is found at the determination of the fourth mean proportional. Here Bürgi does not show the addition of each of the four one-fifth parts; instead, only the total (**161280**) is added to the smaller red number. The computations are carried out with minimal copyist error. The most notable error is the correct identification of the red number corresponding to the second mean proportional. In the Gz manuscript, the black number given actually corresponds to **98050** and not the correct red number, which is **98040**.

The right-most edge of the final page of the "Kurzer Bericht" of the Gz manuscript is cut off. That is, the final digits of numbers that are written along the right-most edge of the page are unreadable. Each instance of a reconstructed mean proportional number (all four that were sought) is equivalent to those reported in the Gk manuscript. One reconstruction, the third proportional number 398896111, may not be the value that originally appeared in the Gz manuscript. When inspecting the extreme right edge of manuscript copy, this final digit appears more like a curved digit, which could possibly be a "0."

### *Comments on Stylistic Elements*

The "decimal zero" is only used for the first half of the example on page 20. This is seen in other examples; that is, once the calculation procedure and the magnitude of the numbers are established, the "decimal zero" appears less frequently. Unlike other examples, however, the "decimal zero" does not reappear on the red number at the conclusion of the calculation.

Notably, Bürgi elects to use the fractional notation for number of equal parts to be added in terms of the red numbers. After the first instance in which "one [fourth] part" is added to the smaller red number to determine the first mean proportional, he employs the directions of "*addir* $\frac{2}{4}$" and "*addir* $\frac{3}{4}$" for the next two mean proportional calculations. This fraction notation is also seen in the final example of the Gz manuscript.

The careful selection of the two boundary numbers used in the examples on pages 19, 20, and 21 is certainly not by accident. The difference of the corresponding red numbers, **201600**, is divisible by 3, 4, and 5 without remainder, which facilitates easy computation of 2, 3, and 4 mean proportional numbers. Moreover, when the one-third, one-fourth, or one-fifth parts (or their multiples) are added to the smaller red number (**17400**), each total will end in one or two zeros. At this point and because of the careful selection of boundary values, each mean proportional numbers is determined directly from reading the tables.

With this final example and no particular fanfare, the text of the *Aritmetische und Geometrische Progreß Tabulen*'s "Kurzer Bericht" ends. For all of its inconsistencies, copyist errors, and grammatical imperfections, the Gz manuscript—and the accompanying transcription, translation, and narrative provided in this book—may serve as an important addition to the existing English-language scholarship on Jost Bürgi. The intent of this chapter was to remove obstacles for non-German speakers (or those who have difficulty navigating handwritten sixteenth-century German texts) and to provide access to Bürgi's explanations and techniques in the same way that other resources have done to make Napier's conception of logarithms accessible. The task of focusing on this relatively short mathematical text (and in comparing it the Gk manuscript) has raised interesting questions regarding its own history, and these questions are discussed in Chapter 4.

# Chapter 4
# Final Perspectives

In the Preface, I established two aims for writing this book. First and foremost, I wanted to contribute to the existing English-language scholarship on Jost Bürgi's mathematical work by producing an edition, transcription, and English translation of his *Aritmetische und Geometrische Progreß Tabulen* (1620). My second goal was to offer readers an opportunity to explore Bürgi's contribution to the early development of logarithms. I felt that the best way to accomplish this was to provide a book-length treatment of the Gz copy of his manuscript and to include biographical and contextual information in order to situate Bürgi's mathematical contributions for readers interested in a figure often associated with the invention of logarithms.

Bürgi was a master clock- and watchmaker, skillful instrument designer, and capable mathematician in his lifetime and at a time when a great deal of computational activity occurred. Such activity of the sixteenth and seventeenth centuries required sophisticated and accurate methods in order to aid in understanding and explaining the physical universe. Although the world had already and favorably accepted Napier's logarithms by the time that Bürgi produced his tables in printed form (albeit on a very small scale), it is clear that he influenced a select few by providing access to his conception (e.g., Reimers, Kepler). For example, had Kepler not been familiar with Bürgi's methods—including the computational methods for using his tables of logarithms, the algebraic techniques of Bürgi's *Coss*, and the tables of his *Canon Sinuum*—it is unknown if or when he would have been able to adopt Napier's more difficult methods to construct his own tables of logarithms.

In very recent years, several scholars have sought to publish Bürgi's work in a variety of venues, including technical reports, conference papers, journal articles, monographs, and books. Among them, Fritz Staudacher (2014), Jörg Waldvogel (2012, 2014), and Heinz Theo Lutstorf (2005) have approached their study of Bürgi in a way dedicated to making sense of the existing resources and especially to highlight his mathematical methods. In a similar way, Denis Roegel (2010a) has focused on Bürgi's methods for the construction of tables of logarithms. This book provides what other recent or previous resources do not: a dedicated examination of the Gz

© Springer Science+Business Media New York 2015

K. Clark, *Jost Bürgi's Aritmetische und Geometrische Progreß Tabulen (1620)*, Science Networks. Historical Studies 53, DOI 10.1007/978-1-4939-3161-3_4

manuscript of the *Aritmetische und Geometrische Progreß Tabulen*. And providing an English translation of the manuscript, along with explanations for each example in the "Kurzer Bericht" in the form of a narrative commentary, enables those who are interested to compare Bürgi's conception and treatment of logarithms to those of Napier.[1]

There are, however, remaining unanswered questions regarding the identification of the copyist of the Gz manuscript of the *Aritmetische und Geometrische Progreß Tabulen* and the relationship between the Gz and Gk manuscripts—for example, whether one is a copy of the other, and if so, which came first.

A definitive response to the first question will perhaps never be known unless a manuscript is discovered which contains the name of the copyist. This is unlikely. At least one source (Wolf 1858) indicated that the Gk version of the manuscript was Bürgi's personal copy from Benjamin Bramer's estate. Lutstorf (2005) argued that there are certain egregious errors that occur in the Gk manuscript that Bürgi certainly would not have made.[2] Lutstorf also stated that such "gross errors" as those found in the computation of the fourth root of 56120370 ("Kurzer Bericht," p. 13) would indicate that not only did the Gk manuscript not come from Bürgi's own hand but that he certainly had not even viewed it. Such examples would contradict Wolf in his claim that the Gk manuscript—the copy discovered by Gieswald in 1855 and then published in 1856—was not the personal copy of Jost Bürgi.

The numerous copyist errors, including simple single-digit miscopying to entire numbers being omitted, miscopying of entire passages, and computational errors that would not have escaped an arithmetician of Bürgi's caliber, provide evidence that the Gz manuscript was not written in Bürgi's hand. And, consequently, we may never know the identity of the copyist of the Gz manuscript. Thus, a final question regarding the two manuscript copies remains: what is the relationship between them?

From the analysis in Chapter 3, I posit that the Gz manuscript is the most comprehensive version of the two versions available for study. In this regard, perhaps the Gz manuscript is a derivative of the Gk copy because of the additional examples that appear in the Gz manuscript that do not appear in the Gk copy. For example, it is possible that an owner or user of the Gk manuscript was dictating the Gz copy (e.g., using it as a guide to teach others or possibly just reading the manuscript aloud in order for a copy to be made) and expanded it by adding examples.

An example of this can be found on page 9 of the "Kurzer Bericht," in which the second example found in the Gz manuscript does not appear in the Gk copy. The inclusion of an additional example of *Regula detri* in the Gz manuscript and of other examples on pages 8, 9, 10, 16, and 18 of the "Kurzer Bericht" also indicates that

[1] To be fair, a full treatment of Napier's invention of logarithms is not included here, in much the same way that resources describing Napier's conception do not include a full treatment of Bürgi's. See Havil (2014) or Whiteside (2014), for example.

[2] An interesting example can be found on page 7 of the "Kurzer Bericht" with the phrase "…*dieweil viel ihr* rational *Zahl vorfalle.*" Here, the copyist has heard "their rational" ("*ihr* rational") and not the word, "irrational." The Gz manuscript contains the correct word.

further instruction was appropriate or warranted. For the case of *Regula detri*, on page 10, progressing immediately from the first to the third example (as they appear in the Gk edition) may have been too abrupt, with the first as a straightforward example and the third requiring the need to deal with a difference that is negative.

Another instance of altering a subsequent copy (in this case, the Gk copy) to accommodate modification of instruction or learning is found in the final square root extraction ("Kurzer Bericht," bottom of page 10). In this example, the completed solution for extracting the square root of 22033094 is not provided in the Gk manuscript. Instead, this example is left incomplete and ends with reporting the red number of **154635011**. This could represent error on the part of that manuscript's copyist, or perhaps the copyist (or instructor) did not deem it necessary to complete another interpolation.

Still other examples when comparing the two versions lead the reader to have more questions than answers. The discrepancies that appear within the two examples on pages 17 and 18 of the "Kurzer Bericht" keep the origination of the Gz manuscript copy in question. In particular, if the Gz version is a copy of the Gk manuscript (or more correctly: a derivative of the Gk copy), why would all of the computations with the red numbers in the first example on page 17 differ until the final result? Additionally, if the Gk manuscript does not contain the final solution steps for the second example of page 17 (which is completed on page 18 of the Gz copy), does this indicate that the Gk manuscript is a copy of the Gz version? Such questions, along with the numerous minor errors and discrepancies between the two copies, provide only the answer that the two manuscript copies are indeed different versions, written by two different persons. As Staudacher (2014) stated, the two versions of the manuscript were only proof copies (e.g., "trial copies") that were issued prior to Bürgi finally publishing the *Aritmetische und Geometrische Progreß Tabulen*, which unfortunately never happened.

An important conclusion to the speculation regarding the relationship between the two copies is that because of the examples that exist in the Gz version and not in the Gk copy, the focus on the Gz version in this book represents an important addition to previously known scholarship on the methods and examples of Bürgi's "thorough instruction."

I close with proposing a modification of the question that many raise when the value of Bürgi's contribution on the development of logarithms is considered. Indeed, many question whether he deserves a more prominent place than he received in the history of mathematics for constructing his tables, which, by any account, never reached the mainstream in the manner that Napier's version of the logarithmic relation and his tables did. Instead, I propose that a shift from concern over the magnitude of prominence to that of recognition of Bürgi's parallel insight is in order. To do so places the reader in a position to view Bürgi as his contemporaries did: almost an Archimedes, an ingenious mathematician, and inventor of logarithms.

# Appendix A: Bürgi Biography at a Glance

Biographical timeline for Jost Bürgi (specific to Bürgi the individual, in **boldface**), including relevant events related to the history of logarithms (Faustmann 1997; List and Bialas 1973; Staudacher 2014; Waldvogel 2012, 2014)

| Date | Event |
|---|---|
| 1546 | Birth of Tycho Brahe |
| 1550 | Birth of John Napier |
| **28 February 1552** | **Bürgi born in Lichtensteig (Switzerland)** |
| **1558** | **Bürgi enters public school at Lichtensteig for 6 years (until 1564)** |
| 1567 | Duke Philipp I of Hessen dies (father of Wilhelm IV); as the eldest of the four sons, Wilhelm gets the central Kassel area |
| 1571 | Birth of Johannes Kepler |
| 1575 | Tycho Brahe visits Landgrave (Duke) Wilhelm IV of Hessen in Kassel and meets Paul Wittich in Wittenberg |
| 1576 | Brahe builds his observatory on Hven (originally part of Denmark but is now in Sweden) |
| **1576** | **Bürgi in Nürnberg and finalizes Christoph Heiden's celestial sphere under construction in Heiden's workshop (after Heiden dies)** |
| **25 July 1579** | **Bürgi arrives in Kassel for Landgrave (Duke) Wilhelm IV of Hessen (as Watchmaker of the Duke)** |
| **1580** | **Bürgi builds his first Kassel celestial sphere** |
| Summer 1580 | Wittich goes to Hven and remains until November. First use of *prosthaphaeresis* |
| **1582** | **Bürgi begins development of new sextants in metal (brass, steel, copper)** |
| **1583** | **Invention of proportional compass by Bürgi (of his own type)** |
| **1584** | **Bürgi begins search to improve prosthaphaeresis formula** |
| 1584 | Emperor Rudolf II moves residence from Vienna to Prague |

(continued)

© Springer Science+Business Media New York 2015

K. Clark, *Jost Bürgi's Aritmetische und Geometrische Progreß Tabulen (1620)*,
Science Networks. Historical Studies 53, DOI 10.1007/978-1-4939-3161-3

(continued)

| Date | Event |
|---|---|
| 1584 | Christoph Rothmann is employed in Kassel |
| | Wittich stays several months in Kassel |
| September 1584 | Reimers visits Brahe at Uraniborg (Hven; then part of Denmark) for 8 days |
| **1584–1585** | **Bürgi creates the world's first clock precise enough to measure seconds and to indicate seconds visually and auditorially** |
| | **Kassel astronomers (including Bürgi, Rothmann, and Wilhelm IV) begin a new measurement program of the stars to obtain better data for navigation, science, and astrology** |
| Spring 1586 | Reimers goes to Kassel; leaves in June 1588 |
| **1586–1587** | **Bürgi designs and manufactures a 3D Astrarium model of his hybrid "Tychonian" world model for Reimers** |
| 6 May 1588 | Rothmann receives author privilege for 12 mathematical treatises |
| **About 1588** | **Bürgi invents logarithms** |
| **1588** | **Bürgi's new mathematical methods published in part by Reimers in *Fundamentum Astronicum*** |
| **1588** | **Bürgi introduces decimal fractions (and a symbol for the decimal point) and is one of earliest to do so** |
| 1589 | Christen Sørensen Longomontanus is Brahe's assistant (until 1597) |
| 1 August 1590 | Rothmann visits Brahe (until 1 September) |
| About July 1591 | Reimers hired by the Emperor as a mathematician |
| **1591** | **"Jost the watchmaker" becomes citizen (naturalized) of Kassel** |
| **1592** | **Bürgi gives the engraver A. Eisenhaut the order for the illustrations (21 total) for instructions on how to use the triangulation instrument** |
| | **Bürgi shows example of his *Canon Sinuum* tables to Brahe** |
| **1592** | **Bürgi's first trip to Prague** |
| **10 June** | **Arrives** |
| **4 July** | **Audience with and at the request of Emperor Rudolf II, where Bürgi delivers a silver Planetary Globe clock and a proportional compass to him** |
| **27 July** | **Receives payment/gift of $300** |
| 25 August 1592 | Landgrave Wilhelm's death; Moritz (his son) is his successor |
| **1 January 1593** | **Bürgi's contract renewal with Landgrave Moritz** |
| **1594** | **Bürgi finalizes his small Celestial Sphere** |
| 22 August 1596 | Reimers receives author privilege for astronomical journals |
| **End of 1596** | **Bürgi's second short trip to Prague** |
| Spring 1597 | Brahe leaves Hven (forced out by King Christian V) |
| 1597 | Reimers' *De astronomicis hypothesibus* ("The astronomical hypotheses") appears in Prague |
| **1597** | **Bürgi is described in a letter from Reimers to Kepler as "my teacher and master, combining the properties of Archimedes and Euclid"** (Staudacher 2014) |
| **1598** | **Bürgi finalizes the *Canon Sinuum* ("Canon of Sines")** |
| **June 1599** | **Brahe's arrival in Prague** |
| October 1599–January 1600 | Kepler's first trip to Prague |

(continued)

(continued)

| Date | Event |
|---|---|
| 11 January | Arrival in Prague; several secret meetings with Reimers |
| 4 February | Kepler meets Brahe for the first time at Benatek Castle |
| 1600 | Brahe becomes Imperial Court astronomer for Rudolf II |
| 1600 | Valentin Otto in Prague |
| 15 August 1600 | Reimers dies |
| 24 October 1601 | Brahe dies |
| 1601 | Kepler becomes Imperial Court astronomer for Rudolf II |
| **1603** | **Bürgi receives a patent protection privilege for his triangulation instrument from Rudolf II** |
| **Autumn 1603** | **Bürgi ordered to Prague by the Landgrave** |
| **1603** | **Bürgi's "proportional compasses"** |
| **23 December 1604** | **Bürgi is promoted to Emperor's watchmaker and paid 60 florins (guilders) monthly** |
| **15 May 1605** | **Bürgi receives his first salary, house, and workshop in the Imperial Castle** |
| 1608 | In another edition of *Trigonometria* Pitiscus publishes Bürgi's new algebraic methods |
| 1609 | Kepler publishes his revolutionary work *Astronomia Nova*, together with his two first laws |
| **1609** | **Bürgi's first wife (Benjamin Bramer's sister) dies in Prague** |
| **1610** | **Naturalization of Bürgi in Prague** |
| **3 February 1611** | **Bürgi ennobled; coat of arms issued** |
| **17 June 1611** | **Bürgi's second marriage (to Catharina Braun) takes place in Kassel** |
| 1612 | Rudolf II dies; new emperor is his brother, Matthias I |
| 1614 | Napier publishes his tables of logarithms (the *Descriptio*) in Edinburgh |
| 1617 | Henry Briggs publishes the first 1000 of his logarithms in London |
| **1617** | **Bürgi is briefly in Kassel; instructs Prince Hermann in Astronomy** |
| 1617 | Napier dies |
| 1618 | Thirty Years' War begins |
| 1619 | Napier publishes his *Constructio* (posthumously) |
| **28 February 1619** | **Sadeler draws Bürgi on his birthday; later an engraving is made** |
| **1620** | **Proof copies of Bürgi's *Aritmetische und Geometrische Progreß Tabulen* printed in Prague (a few are distributed)** |
| **29 October 1621** | **Bürgi receives a printing privilege for his book of logarithms and instruction** |
| 1624 | Kepler's *Chilias Logarithmorum* is printed in Marburg |
| 1624 | Henry Briggs publishes his *Arithmetica* in London |
| 1630 | Kepler dies |
| **1630/1631** | **Bürgi's final return to Kassel** |
| **31 January 1632** | **Bürgi dies** |
| 1648 | Manual for triangular instrument, began by Bürgi in 1592, published by Benjamin Bramer in Marburg |
| | End of Thirty Years' War |

# Appendix B: Napier's Argument and Construction of Logarithms

John Napier's (1550–1617) conception of the logarithmic relation was based upon a kinematic argument, as opposed to Bürgi's conception, which was algebraic. In his kinematic model, Napier described the movement of two particles (Figure B.1), one (point $P$) moving along a line segment of fixed distance ($AZ$) and another (point $Q$) moving along a line (or ray) of indefinite length ($A'Z'$). Also, Napier defined the line segment and the ray to be parallel to each other:

To define the movement of points $P$ and $Q$, Napier established three rules. First, the points $P$ and $Q$ begin movement along their paths with the same initial velocity. Second, point $Q$ keeps this velocity along its entire path. And lastly, point $P$'s velocity slows down in such a way that its velocity is proportional to the distance it has left to travel along segment $AZ$. Napier also defined the initial length of segment $AZ$ to be equivalent to $10^7$ units, since this was the value of the radius of the circles used to construct his tables of sines. By defining such a length, however, this meant that the initial velocities of points $P$ and $Q$ were also $10^7$, as well as point $Q$'s constant velocity.

Using the initial conditions Napier established, we can begin to describe subsequent movement along the segment and the ray, which will in turn provide the pair of sequences alluded to in comment 26[1] of Napier's *Mirifici Logarithmorum Canonis Constructio* (1619). First, we can consider $P$ and $Q$ moving along their respective paths to the next position:

Particle $P$'s velocity is diminishing at each point (Figure B.2, at point $B$) in such a way that the velocity is "proportional to the distance remaining in the line's terminus point of $Z$" (Calinger 1999, p. 488). A series of calculations will help to create the necessary sequences. (Units are omitted for convenience.)

[1] Comment 26 is: The logarithm of a given sine is that number which has increased arithmetically with the same velocity throughout as that with which radius began to decrease geometrically and in the same time as radius has decreased to the given sine.

© Springer Science+Business Media New York 2015

K. Clark, *Jost Bürgi's Aritmetische und Geometrische Progreß Tabulen (1620)*, Science Networks. Historical Studies 53, DOI 10.1007/978-1-4939-3161-3

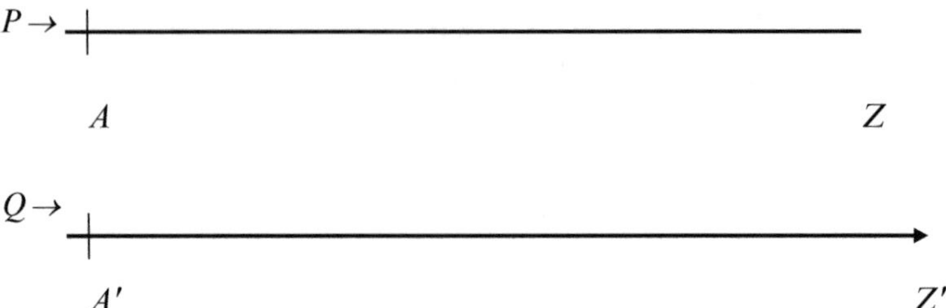


**Figure B.1** Napier's two-particle model

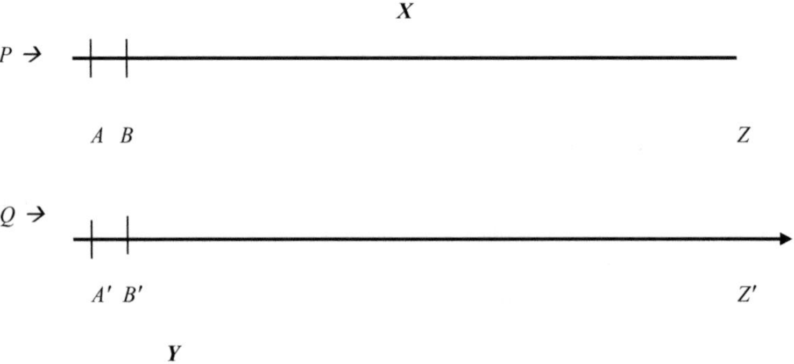


**Figure B.2** First movement

First, we can calculate the increment of time used for each movement:

The initial velocity ($v_1$) of $Q$ is $10^7$ and let the distance ($d_1$) $A'B'$ be defined as 1. Thus, the time it takes to travel from $A'$ to $B'$ is found by $t_1 = d_1/v_1$ or $10^{-7}$. Since this is a relatively short increment of time, the distance from $A$ to $B$ is also very close to 1. If these initial calculations are used, along with the same interval of time ($10^{-7}$), then the geometric sequence corresponding to the remaining distance left to travel along segment $AZ$ is found as follows:

$$BZ = 10^7 - 1 \text{ or } 10^7\left(1 - 10^{-7}\right)$$

$$BC = (\text{velocity at } B) \times (\text{time}) \left[\begin{array}{l}\text{since the velocity at each point is} \\ \text{proportional to the remaining distance along } AZ\end{array}\right]$$

$$BC = \left(10^7\left(1 - 10^{-7}\right)\right)\left(10^{-7}\right)$$

$$BC = \left(1 - 10^{-7}\right)$$

Now, $CZ$ will equal $AZ - AB - BC$ or $10^7 - 1 - (1 - 10^{-7})$. Simplifying,

$$CZ = 10^7\left(1 - 10^{-7}\right)^2.$$

Continuing this process yields the following geometric sequence corresponding to the remaining distance for particle $P$ to travel along $AZ$:

$$10^7\left(1-10^{-7}\right)^0, 10^7\left(1-10^{-7}\right)^1, 10^7\left(1-10^{-7}\right)^2, 10^7\left(1-10^{-7}\right)^3, \ldots$$
$$\text{which corresponds to } AZ, BZ, CZ, DZ, \ldots.$$

Alternatively, the arithmetic sequence corresponding to how far $Q$ has traveled on $A'Z'$ is increasing, or $A'A'$($Q$ has not moved yet), $A'B'$, $A'C'$, $A'D'$, … is given by:

$$0, 1, 2, 3, \ldots.$$

Finally, and numerically, Napier described his logarithms as the common ratio of the two sequences of numbers (Calinger 1999, p. 487). Thus, in Fig. B.2, $\boldsymbol{Y}$ is the logarithm of $\boldsymbol{X}$, or the logarithm of $10^7(1-10^{-7})^1$ would equal 1. Using the notation in the figure would yield $A'B' = \log(BZ)$.

# Appendix C

**The tables of the** Aritmeti*sche und* Geometri*sche Progreß Tabulen/Sambt gründlichem unterricht/wie solche nützlich in allerley Rechnungen zu gebrauchen/und verstanden werden sol*

(Bürgi 1620)

## Introduction

The 58 pages of tables that accompany the Gz version of the Artimetische und Geometrische Progreß Tabulen are published here (these tables may also be downloaded from www.springer.com/us/book/9781493931606). Several accounts are available (in German and English) that discuss the calculation of the table values, the errors that exist in the tables, and the potential causes for the errors, including Lutstorf (2005), Lutstorf and Walter (1992), Roegel (2010a), and Waldvogel (2012, 2014). This book does not include a transcription of the tables, discussion about the tabulated values, or an error analysis of tabulated values, due to the availability of the previously mentioned accounts. It is important to note, however, that the values from Waldvogel's identification of "isolated large errors" (2012, p. 24) also appear to be repeated in the Gz manuscript.

© Springer Science+Business Media New York 2015

K. Clark, *Jost Bürgi's Aritmetische und Geometrische Progreß Tabulen (1620)*,
Science Networks. Historical Studies 53, DOI 10.1007/978-1-4939-3161-3

| | 0 | 500 | 1000 | 1500 | 2000 | 2500 | 3000 | 3500 |
|---|---|---|---|---|---|---|---|---|
| 0 | 100000000 | 100501227 | 101004966 | 101511230 | 102020032 | 102531384 | 103045299 | 103561790 |
| 10 | ....10000 | ....11277 | ....15067 | ....21381 | ....30234 | ....41637 | ....55603 | ....72146 |
| 20 | ....20001 | ....21328 | ....25168 | ....31524 | ....40437 | ....51891 | ....65909 | ....82503 |
| 30 | ....30003 | ....31380 | ....35271 | ....41687 | ....50641 | ....62146 | ....76216 | ....92861 |
| 40 | ....40006 | ....41433 | ....45374 | ....51841 | ....60846 | ....72402 | ....86523 | 103603221 |
| 50 | ....50010 | ....51487 | ....55479 | ....61996 | ....71052 | ....82660 | ....96832 | ....13581 |
| 60 | ....60015 | ....61543 | ....65584 | ....72153 | ....81259 | ....92918 | 103107142 | ....23942 |
| 70 | ....70021 | ....71599 | ....75691 | ....82309 | ....91467 | 102603177 | ....17452 | ....34305 |
| 80 | ....80028 | ....81656 | ....85799 | ....92468 | 102101676 | ....13438 | ....27764 | ....44668 |
| 90 | ....90036 | ....91714 | ....95907 | 101602627 | ....11887 | ....23699 | ....38077 | ....55033 |
| 100 | 100100045 | 100601773 | 101106017 | ....12787 | ....22098 | ....33961 | ....48391 | ....65398 |
| 110 | ....10055 | ....11834 | ....16127 | ....22949 | ....32310 | ....44225 | ....58705 | ....75765 |
| 120 | ....20066 | ....21895 | ....26239 | ....33111 | ....42523 | ....54489 | ....69021 | ....86132 |
| 130 | ....30078 | ....31957 | ....36352 | ....43274 | ....52738 | ....64755 | ....79338 | ....96501 |
| 140 | ....40091 | ....42020 | ....46465 | ....53438 | ....62953 | ....75021 | ....89656 | 103706871 |
| 150 | ....50105 | ....52084 | ....56580 | ....63604 | ....73169 | ....85289 | ....99975 | ....17241 |
| 160 | ....60120 | ....62150 | ....66696 | ....73770 | ....83386 | ....95557 | 103210295 | ....27613 |
| 170 | ....70136 | ....72216 | ....76812 | ....83938 | ....93605 | 102705827 | ....20616 | ....37986 |
| 180 | ....80153 | ....82283 | ....86930 | ....94106 | 102203824 | ....16097 | ....30932 | ....48360 |
| 190 | ....90171 | ....92351 | ....97049 | 101704275 | ....14045 | ....26369 | ....41261 | ....58734 |
| 200 | 100200190 | 100702420 | 101207168 | ....14446 | ....24266 | ....36642 | ....51585 | ....69110 |
| 210 | ....10210 | ....12491 | ....17289 | ....24617 | ....34488 | ....46915 | ....61910 | ....7[illegible]4[illegible]7 |
| 220 | ....20231 | ....22562 | ....27411 | ....34790 | ....44712 | ....57190 | ....72237 | ....89865 |
| 230 | ....30253 | ....32634 | ....37533 | ....44963 | ....54936 | ....67466 | ....82564 | 103800244 |
| 240 | ....40276 | ....42707 | ....47657 | ....55138 | ....65162 | ....77742 | ....92892 | ....10624 |
| 250 | ....50300 | ....52782 | ....57782 | ....65313 | ....75388 | ....88020 | 103303221 | ....21005 |
| 260 | ....60325 | ....62857 | ....67908 | ....75490 | ....85616 | ....98299 | ....13552 | ....31387 |
| 270 | ....70351 | ....72933 | ....78035 | ....85667 | ....95845 | 102808579 | ....23883 | ....41770 |
| 280 | ....80378 | ....83011 | ....88162 | ....95846 | 102306074 | ....18860 | ....34216 | ....52155 |
| 290 | ....90406 | ....93189 | ....98291 | 101806025 | ....16305 | ....29142 | ....44549 | ....62540 |
| 300 | 100300435 | 100803168 | 101308421 | ....16206 | ....26536 | ....39425 | ....54883 | ....72926 |
| 310 | ....10465 | ....13248 | ....18552 | ....26387 | ....36769 | ....49708 | ....65219 | ....83313 |
| 320 | ....20496 | ....23330 | ....28684 | ....36570 | ....47003 | ....59993 | ....75555 | ....93702 |
| 330 | ....30528 | ....33412 | ....38817 | ....46754 | ....57237 | ....70279 | ....85893 | 103904091 |
| 340 | ....40562 | ....43496 | ....48950 | ....56939 | ....67473 | ....80566 | ....96232 | ....14481 |
| 350 | ....50596 | ....53580 | ....59085 | ....67124 | ....77710 | ....90855 | 103406571 | ....24873 |
| 360 | ....60631 | ....63665 | ....69221 | ....77311 | ....87947 | 102901144 | ....16912 | ....35265 |
| 370 | ....70667 | ....73752 | ....79358 | ....87499 | ....98186 | ....11434 | ....27254 | ....45659 |
| 380 | ....80704 | ....83839 | ....89496 | ....97687 | 102408426 | ....21725 | ....37596 | ....56053 |
| 390 | ....90742 | ....93927 | ....99635 | 101907877 | ....18667 | ....32017 | ....47940 | ....66449 |
| 400 | 100400781 | 100904017 | 101409775 | ....18068 | ....28909 | ....42310 | ....58285 | ....76846 |
| 410 | ....10821 | ....14107 | ....19916 | ....28260 | ....39152 | ....52604 | ....68631 | ....87243 |
| 420 | ....20862 | ....24199 | ....30058 | ....38453 | ....49396 | ....62900 | ....78978 | ....9[illegible]642 |
| 430 | ....30904 | ....34291 | ....40201 | ....48646 | ....59641 | ....73196 | ....89326 | 104008042 |
| 440 | ....40948 | ....44384 | ....50345 | ....58841 | ....69887 | ....83493 | ....99674 | ....18443 |
| 450 | ....50991 | ....54479 | ....60489 | ....69037 | ....80133 | ....93792 | 103510024 | ....28844 |
| 460 | ....61037 | ....64574 | ....70636 | ....79234 | ....90381 | 103004091 | ....20375 | ....39247 |
| 470 | ....71083 | ....74671 | ....80783 | ....89432 | 102500630 | ....14391 | ....30727 | ....49651 |
| 480 | ....[illegible]113[illegible] | ....84768 | ....90931 | ....99631 | ....10880 | ....24693 | ....41080 | ....60056 |
| 490 | ....91178 | ....94867 | 101501080 | 102009831 | ....21132 | ....34995 | ....51435 | ....70462 |
| 500 | 100501227 | 101004966 | ....11230 | ....20032 | ....31384 | ....45299 | ....61790 | ....80816 |

**Figure C.1**

| | 4000 | 4500 | 5000 | 5500 | 6000 | 6500 | 7000 | 7500 |
|---|---|---|---|---|---|---|---|---|
| 0 | 104080869 | 104602551 | 105126847 | 105653791 | 106183336 | 106715556 | 107250443 | 107788011 |
| 10 | ....91277 | ....13011 | ....37359 | ....64336 | ....93954 | ....26227 | ....61168 | ....98790 |
| 20 | 104101686 | ....23472 | ....47873 | ....74903 | 106204574 | ....36900 | ....71894 | 107809570 |
| 30 | ....12097 | ....33935 | ....58388 | ....85470 | ....15194 | ....47574 | ....82621 | ....20351 |
| 40 | ....225[illegible]8 | ....44398 | ....68904 | ....96039 | ....25816 | ....58248 | ....93349 | ....31133 |
| 50 | ....[illegible] | ....54[illegible]62 | ....79421 | 105706608 | ....36438 | ....68924 | 107304079 | ....41916 |
| 60 | ....43333 | ....65328 | ....89939 | ....17179 | ....47062 | ....79601 | ....14809 | ....52700 |
| 70 | ....5374[illegible] | ....75794 | 105200458 | ....27751 | ....57686 | ....90279 | ....25541 | ....63485 |
| 80 | ....64163 | ....86262 | ....10978 | ....38324 | ....68312 | 106800958 | ....36273 | ....74271 |
| 90 | ....74580 | ....96731 | ....21499 | ....48897 | ....78939 | ....11638 | ....47007 | ....85059 |
| 100 | ....84997 | 104707200 | ....32021 | ....59472 | ....89567 | ....22319 | ....57742 | ....95847 |
| 110 | ....95416 | ....17671 | ....42544 | ....70048 | 106300196 | ....33001 | ....68477 | 603906637 |
| 120 | 104205835 | ....28143 | ....53069 | ....80625 | ....10826 | ....43685 | ....79214 | ....17428 |
| 130 | ....16256 | ....38616 | ....63594 | ....91203 | ....21457 | ....54369 | ....89952 | ....28219 |
| 140 | ....26677 | ....49090 | ....74120 | 105801782 | ....32089 | ....65055 | 107400691 | ....39012 |
| 150 | ....37100 | ....59564 | ....84648 | ....12363 | ....42723 | ....75741 | ....11431 | ....49806 |
| 160 | ....47524 | ....70040 | ....95176 | ....22944 | ....53357 | ....86429 | ....22172 | ....60601 |
| 170 | ....57948 | ....80517 | 105305706 | ....33526 | ....63992 | ....97117 | ....32914 | ....71397 |
| 180 | ....68374 | ....90995 | ....16236 | ....44010 | ....74629 | 106907807 | ....43658 | ....82194 |
| 190 | ....78801 | 104801474 | ....26768 | ....54694 | ....85266 | ....18498 | ....54402 | ....92992 |
| 200 | ....89229 | ....1105[illegible] | ....37301 | ....65279 | ....95905 | ....29189 | ....65148 | 108003792 |
| 210 | ....99658 | ....22436 | ....47834 | ....75866 | 106406544 | ....39882 | ....75894 | ....14592 |
| 220 | 104310088 | ....32918 | ....58369 | ....86453 | ....17185 | ....50576 | ....86642 | ....25394 |
| 230 | ....20519 | ....43401 | ....68905 | ....97042 | ....27827 | ....61271 | ....97390 | ....36196 |
| 240 | ....30951 | ....53886 | ....79442 | 105907632 | ....38470 | ....71968 | 107508140 | ....47000 |
| 250 | ....41394 | ....64371 | ....89980 | ....18223 | ....49113 | ....82665 | ....18891 | ....57804 |
| 260 | ....51818 | ....74857 | 105400519 | ....28814 | ....59758 | ....93363 | ....29643 | ....68610 |
| 270 | ....62253 | ....85345 | ....11059 | ....39407 | ....70404 | 107004062 | ....40396 | ....79417 |
| 280 | ....72689 | ....95833 | ....21600 | ....50001 | ....81051 | ....14763 | ....51150 | ....90225 |
| 290 | ....83127 | 104906223 | ....32142 | ....60596 | ....91699 | ....25464 | ....61905 | 108101034 |
| 300 | ....93565 | ....16814 | ....42685 | ....71192 | 106502348 | ....36167 | ....72661 | ....11844 |
| 310 | 104404004 | ....27305 | ....53229 | ....81789 | ....12999 | ....46870 | ....83418 | ....22656 |
| 320 | ....14445 | ....37798 | ....63775 | ....92388 | ....23650 | ....57575 | ....94177 | ....33468 |
| 330 | ....24886 | ....48292 | ....74321 | 106002987 | ....34302 | ....68281 | 107604936 | ....44281 |
| 340 | ....35329 | ....58787 | ....84869 | ....13587 | ....44956 | ....78988 | ....15696 | ....55096 |
| 350 | ....4577[illegible] | ....60[illegible]82 | ....95417 | ....24189 | ....55610 | ....89696 | ....26458 | ....65911 |
| 360 | ....56217 | ....79720 | 105505967 | ....34791 | ....66266 | 107100405 | ....37221 | ....76728 |
| 370 | ....66663 | ....90277 | ....16517 | ....45394 | ....76923 | ....11115 | ....47985 | ....87545 |
| 380 | ....77109 | 105000776 | ....27069 | ....55999 | ....87580 | ....21826 | ....58749 | ....98364 |
| 390 | ....87557 | ....11277 | ....37622 | ....66605 | ....98239 | ....32538 | ....69515 | 108209184 |
| 400 | ....98006 | ....21778 | ....48175 | ....77211 | 106608899 | ....43251 | ....80282 | ....20005 |
| 410 | 104508456 | ....32280 | ....58730 | ....87819 | ....19560 | ....53966 | ....91050 | ....30827 |
| 420 | ....18906 | ....42783 | ....69286 | ....98428 | ....30222 | ....64681 | 107701819 | ....41650 |
| 430 | ....29358 | ....53287 | ....79843 | 106109038 | ....40885 | ....75398 | ....12589 | ....52474 |
| 440 | ....39811 | ....63793 | ....90401 | ....19648 | ....51549 | ....86115 | ....23361 | ....63299 |
| 450 | ....50265 | ....74299 | 105600960 | ....30260 | ....62214 | ....96834 | ....34133 | ....74126 |
| 460 | ....60720 | ....84807 | ....11520 | ....40873 | ....72880 | 107207553 | ....44906 | ....84953 |
| 470 | ....71176 | ....95315 | ....22081 | ....51487 | ....83547 | ....18274 | ....55681 | ....95782 |
| 480 | ....81633 | 105105825 | ....32643 | ....62103 | ....94216 | ....28996 | ....66456 | 108306611 |
| 490 | ....92091 | ....16335 | ....43207 | ....72719 | 106704885 | ....39719 | ....77233 | ....17442 |
| 500 | 104602551 | ....26847 | ....53791 | ....83336 | ....15556 | ....50443 | ....88011 | ....28274 |

A 2

Figure C.2

| | 8000 | 8500 | 9000 | 9500 | 10000 | 10500 | 11000 | 11500 |
|---|---|---|---|---|---|---|---|---|
| 0 | 108328274 | 108871244 | 109416936 | 109965363 | 110516539 | 111070478 | 111627193 | 112186699 |
| 10 | ····39106 | ····82131 | ····27878 | ····76360 | ····27591 | ····81585 | ····38356 | ····97917 |
| 20 | ····49940 | ····93019 | ····38821 | ····87357 | ····38644 | ····92693 | ····49120 | 112209137 |
| 30 | ····60775 | 108903909 | ····49764 | ····98356 | ····49697 | 111103802 | ····60685 | ····20358 |
| 40 | ····71611 | ····14799 | ····60709 | 110009356 | ····60752 | ····14913 | ····71851 | ····31580 |
| 50 | ····82448 | ····25691 | ····71655 | ····20356 | ····71809 | ····26024 | ····83018 | ····42803 |
| 60 | ····93287 | ····36583 | ····82603 | ····31359 | ····82866 | ····37137 | ····94186 | ····54028 |
| 70 | 108404126 | ····47477 | ····93551 | ····42362 | ····93924 | ····48251 | 111705356 | ····65253 |
| 80 | ····14966 | ····58372 | 109504500 | ····53366 | 110604984 | ····59365 | ····16526 | ····76480 |
| 90 | ····25808 | ····69267 | ····15451 | ····64371 | ····16044 | ····70481 | ····27698 | ····87707 |
| 100 | ····36651 | ····80164 | ····26402 | ····75378 | ····27106 | ····81598 | ····38871 | ····98936 |
| 110 | ····47494 | ····91062 | ····37355 | ····86386 | ····38168 | ····92717 | ····50045 | 112310166 |
| 120 | ····58339 | 109001961 | ····48309 | ····97394 | ····49232 | 111203836 | ····61220 | ····21397 |
| 130 | ····69185 | ····12862 | ····59263 | 110108404 | ····60297 | ····14956 | ····72396 | ····32629 |
| 140 | ····80032 | ····23763 | ····70219 | ····19415 | ····71363 | ····26078 | ····83573 | ····43862 |
| 150 | ····90880 | ····34665 | ····81176 | ····30427 | ····82430 | ····37200 | ····94751 | ····55097 |
| 160 | 108501729 | ····45569 | ····92135 | ····41440 | ····93498 | ····48324 | 111805931 | ····66332 |
| 170 | ····12579 | ····56473 | 109603094 | ····52454 | 110704568 | ····59449 | ····17111 | ····77569 |
| 180 | ····23430 | ····67379 | ····14054 | ····63469 | ····15638 | ····70575 | ····28293 | ····88806 |
| 190 | ····34283 | ····78286 | ····25015 | ····74485 | ····26710 | ····81702 | ····39476 | 112400045 |
| 200 | ····45136 | ····89194 | ····35978 | ····85503 | ····37782 | ····92830 | ····50660 | ····11285 |
| 210 | ····55990 | 109100102 | ····46942 | ····96521 | ····48856 | 111303959 | ····61845 | ····22526 |
| 220 | ····66846 | ····11012 | ····57906 | 110207541 | ····59931 | ····15090 | ····73031 | ····33769 |
| 230 | ····77703 | ····21924 | ····68872 | ····18562 | ····71007 | ····26221 | ····84218 | ····45012 |
| 240 | ····88561 | ····32836 | ····79839 | ····29584 | ····82084 | ····37354 | ····95407 | ····56257 |
| 250 | ····99419 | ····43749 | ····90807 | ····40607 | ····93162 | ····48488 | 111906596 | ····67502 |
| 260 | 108610279 | ····54603 | 109701776 | ····51631 | 110804242 | ····59622 | ····17787 | ····78749 |
| 270 | ····21140 | ····65579 | ····12746 | ····62656 | ····15322 | ····70758 | ····28979 | ····89997 |
| 280 | ····32002 | ····76496 | ····23717 | ····73682 | ····26404 | ····81895 | ····40172 | 112501246 |
| 290 | ····42866 | ····87413 | ····34690 | ····84710 | ····37486 | ····93034 | ····51366 | ····124[illegible]6 |
| 300 | ····53730 | ····98332 | ····45663 | ····95738 | ····48570 | 111404173 | ····62561 | ····23747 |
| 310 | ····64595 | 109209252 | ····56638 | 110306768 | ····59655 | ····15313 | ····73757 | ····35000 |
| 320 | ····75462 | ····20173 | ····67614 | ····17799 | ····70741 | ····26455 | ····84955 | ····46253 |
| 330 | ····86329 | ····31095 | ····78590 | ····28830 | ····81828 | ····37597 | ····96153 | ····57508 |
| 340 | ····97198 | ····42018 | ····89568 | ····39863 | ····92916 | ····48741 | 112007353 | ····68764 |
| 350 | 108708068 | ····52942 | 109800547 | ····50897 | 110904005 | ····59886 | ····18553 | ····80020 |
| 360 | ····18938 | ····63867 | ····11527 | ····61932 | ····15096 | ····71032 | ····29755 | ····91278 |
| 370 | ····29810 | ····74794 | ····22508 | ····72969 | ····26187 | ····82179 | ····40958 | 112602538 |
| 380 | ····40683 | ····85721 | ····33491 | ····84006 | ····37280 | ····93327 | ····52162 | ····13798 |
| 390 | ····51557 | ····96650 | ····44474 | ····95044 | ····48374 | 111504477 | ····63367 | ····25059 |
| 400 | ····62133 | 109307579 | ····55458 | 110406084 | ····59468 | ····15627 | ····74574 | ····36322 |
| 410 | ·0··73309 | ····18510 | ····66444 | ····17124 | ····70564 | ····26779 | ····85781 | ····47585 |
| 420 | ····84186 | ····29442 | ····77431 | ····28166 | ····81601 | ····37931 | ····96990 | ····5885[illegible] |
| 430 | ····95065 | ····40375 | ····88418 | ····39209 | ····92760 | ····49085 | 112108200 | ····70116 |
| 440 | 108805944 | ····51309 | ····99407 | ····50253 | 111003859 | ····60240 | ····19410 | ····8138[illegible] |
| 450 | ····16825 | ····62244 | 109910397 | ····61298 | ····14959 | ····71396 | ····30622 | ····9265[illegible] |
| 460 | ····27706 | ····73180 | ····21388 | ····72344 | ····26061 | ····82553 | ····41835 | 11270392[illegible] |
| 470 | ····38589 | ····84118 | ····32380 | ····83391 | ····37163 | ····9371[illegible] | ····53050 | ····[illegible] |
| 480 | ····49473 | ····95056 | ····43373 | ····94439 | ····48267 | 111604871 | ····64265 | ····[illegible]2646 |
| 490 | ····60358 | 109405995 | ····54363 | 110505489 | ····59372 | ····16032 | ····75481 | ····37735 |
| 500 | ····71244 | ····16936 | ····65363 | ····16539 | ····70478 | ····27193 | ····86699 | ····49009 |

**Figure C.3**

| | 12000 | 12500 | 13000 | 13500 | 14000 | 14500 | 15000 | 15500 |
|---|---|---|---|---|---|---|---|---|
| 0 | 112749009 | 113314137 | 113882098 | 114452906 | 115026575 | 115603119 | 116182553 | 116764891 |
| 10 | ····60284 | ····25469 | ····93486 | ····64351 | ····38078 | ····14679 | ····94171 | ····76568 |
| 20 | ····71560 | ····36801 | 113904876 | ····75798 | ····49581 | ····26241 | 116205791 | ····88246 |
| 30 | ····82837 | ····48135 | ····16266 | ····87245 | ····61086 | ····37803 | ····17411 | ····99924 |
| 40 | ····94115 | ····59470 | ····27658 | ····98694 | ····72592 | ····49367 | ····29033 | 116811604 |
| 50 | 112805395 | ····70805 | ····39051 | 114510144 | ····84099 | ····60932 | ····40656 | ····23286 |
| 60 | ····16675 | ····82143 | ····50445 | ····21595 | ····95608 | ····72498 | ····52280 | ····34968 |
| 70 | ····27957 | ····93481 | ····61840 | ····33047 | 115107118 | ····84065 | ····63905 | ····46651 |
| 80 | ····39240 | 113404820 | ····73236 | ····44500 | ····18628 | ····95634 | ····75532 | ····58336 |
| 90 | ····50523 | ····16161 | ····84633 | ····55955 | ····30140 | 115707203 | ····87159 | ····70022 |
| 100 | ····61809 | ····27502 | ····96032 | ····67410 | ····41653 | ····18774 | ····98788 | ····81709 |
| 110 | ····73095 | ····38845 | 114007431 | ····78867 | ····53167 | ····30346 | 116310418 | ····93397 |
| 120 | ····84382 | ····50189 | ····18832 | ····90325 | ····64683 | ····41919 | ····22049 | 116905086 |
| 130 | ····95670 | ····61534 | ····30234 | 114601784 | ····76199 | ····53493 | ····33681 | ····16777 |
| 140 | 112906960 | ····72880 | ····41637 | ····13244 | ····87717 | ····65069 | ····45314 | ····28468 |
| 150 | ····18251 | ····84227 | ····53041 | ····24706 | ····99236 | ····76645 | ····56949 | ····40161 |
| 160 | ····29541 | ····95576 | ····64446 | ····3616[illegible] | 115210755 | ····88223 | ····68585 | ····51855 |
| 170 | ····40835 | 113506925 | ····75853 | ····47632 | ····22277 | ····99802 | ····80221 | ····63550 |
| 180 | ····52130 | ····18276 | ····87260 | ····59096 | ····33799 | 115811382 | ····91859 | ····75247 |
| 190 | ····63425 | ····29628 | ····98669 | ····70562 | ····45322 | ····22963 | 116403499 | ····86944 |
| 200 | ····74721 | ····40981 | 114110079 | ····82029 | ····56847 | ····34545 | ····15139 | ····98643 |
| 210 | ····86019 | ····52335 | ····21490 | ····93498 | ····68372 | ····46129 | ····26780 | 117010343 |
| 220 | ····97317 | ····63690 | ····32902 | 114704967 | ····79899 | ····57713 | ····38423 | ····22044 |
| 230 | 113008617 | ····75047 | ····44315 | ····16437 | ····91427 | ····69299 | ····50067 | ····33746 |
| 240 | ····19918 | ····86404 | ····55730 | ····27909 | 115302956 | ····80886 | ····61712 | ····45450 |
| 250 | ····31220 | ····97763 | ····67145 | ····39382 | ····14487 | ····92474 | ····73358 | ····57154 |
| 260 | ····42523 | 113609122 | ····78562 | ····50856 | ····26018 | 115904063 | ····85005 | ····68860 |
| 270 | ····53827 | ····20483 | ····89980 | ····62331 | ····37551 | ····15654 | ····96654 | ····80567 |
| 280 | ····65132 | ····31845 | 114201399 | ····73807 | ····49084 | ····27245 | 116508304 | ····92275 |
| 290 | ····76439 | ····43209 | ····12819 | ····85284 | ····60619 | ····38838 | ····19954 | 117103984 |
| 300 | ····87746 | ····54563 | ····24240 | ····96763 | ····72155 | ····50432 | ····31607 | ····15694 |
| 310 | ····99055 | ····65928 | ····35663 | 114808243 | ····83693 | ····62027 | ····43460 | ····27406 |
| 320 | 113110365 | ····77295 | ····47086 | ····19723 | ····95231 | ····73623 | ····54914 | ····39119 |
| 330 | ····21676 | ····88663 | ····58511 | ····31205 | 115406770 | ····85220 | ····66570 | ····50833 |
| 340 | ····32989 | 113700032 | ····69937 | ····42689 | ····18311 | ····96819 | ····78226 | ····62548 |
| 350 | ····44302 | ····11402 | ····81364 | ····54173 | ····29853 | 116008419 | ····89884 | ····74264 |
| 360 | ····55616 | ····22773 | ····92792 | ····65658 | ····41396 | ····20019 | 116601543 | ····85981 |
| 370 | ····66932 | ····34145 | 114304221 | ····77145 | ····52940 | ····31621 | ····13203 | ····97700 |
| 380 | ····78249 | ····45518 | ····15652 | ····88633 | ····64485 | ····43224 | ····24865 | 117209420 |
| 390 | ····89566 | ····56893 | ····27083 | 114900121 | ····76032 | ····54829 | ····36527 | ····21141 |
| 400 | 113200885 | ····68269 | ····38516 | ····11611 | ····87579 | ····66434 | ····48191 | ····32863 |
| 410 | ····12205 | ····79656 | ····49950 | ····23103 | ····99128 | ····78041 | ····59856 | ····44586 |
| 420 | ····23527 | ····91034 | ····61385 | ····34595 | 115510678 | ····89649 | ····71522 | ····56311 |
| 430 | ····34849 | 113802413 | ····72821 | ····46088 | ····22229 | 116101258 | ····83189 | ····68036 |
| 440 | ····46172 | ····1379[illegible] | ····84258 | ····57583 | ····33781 | ····12868 | ····94857 | ····79763 |
| 450 | ····57497 | ····25174 | ····95697 | ····69079 | ····45335 | ····24479 | 116706526 | ····91421 |
| 460 | ····68823 | ····36557 | 111407136 | ····80576 | ····56889 | ····36092 | ····18197 | 117303220 |
| 470 | ····80150 | ····47940 | ····18577 | ····92074 | ····68445 | ····47705 | ····29869 | ····14950 |
| 480 | ····91478 | ····59325 | ····30019 | 115003573 | ····80002 | ····59320 | ····41542 | ····26682 |
| 490 | 113302807 | ····70711 | ····41462 | ····15073 | ····91560 | ····70936 | ····53216 | ····38415 |
| 500 | ····14137 | ····82098 | ····52906 | ····26575 | 115603119 | ····82553 | ····64891 | ····5014 |

**Figure C.4**

| | 16000 | 16500 | 17000 | 17500 | 18000 | 18500 | 19000 | 19500 |
|---|---|---|---|---|---|---|---|---|
| 0 | 117350148 | 117933339 | 118529478 | 119123579 | 119720659 | 120320731 | 120923811 | 121529514 |
| 10 | ....61883 | ....50133 | ....41331 | ....354[illegible] | ....32631 | ....32763 | ....35903 | ....42067 |
| 20 | ....73620 | ....61926 | ....53155 | ....4740[illegible] | ....44614 | ....44797 | ....47997 | ....54221 |
| 30 | ....85357 | ....73724 | ....65047 | ....59320 | ....56579 | ....56831 | ....60092 | ....66376 |
| 40 | ....97095 | ....85521 | ....76897 | ....71236 | ....68555 | ....68867 | ....72188 | ....78533 |
| 50 | 117408835 | ....97320 | ....88754 | ....83153 | ....80531 | ....80904 | ....84285 | ....90691 |
| 60 | ....20576 | 118009120 | 118600613 | ....95071 | ....92510 | ....92942 | ....96384 | 121602850 |
| 70 | ....32318 | ....20921 | ....12473 | 119206991 | 119804489 | 120404981 | 121008483 | ....15010 |
| 80 | ....44061 | ....32723 | ....24334 | ....18912 | ....16469 | ....17022 | ....20584 | ....27172 |
| 90 | ....55806 | ....44526 | ....36197 | ....30834 | ....28451 | ....29063 | ....32686 | ....39335 |
| 100 | ....67551 | ....56330 | ....48061 | ....42757 | ....40434 | ....41106 | ....44789 | ....51498 |
| 110 | ....79298 | ....68136 | ....59925 | ....54631 | ....52418 | ....53150 | ....56894 | ....63664 |
| 120 | ....91046 | ....79943 | ....71791 | ....66606 | ....64403 | ....65195 | ....69000 | ....75830 |
| 130 | 117502795 | ....91751 | ....83659 | ....78533 | ....76389 | ....77242 | ....81106 | ....87998 |
| 140 | ....14546 | 118103560 | ....95527 | ....90461 | ....88377 | ....89290 | ....93215 | 121700166 |
| 150 | ....26297 | ....15370 | 118707397 | 119302390 | 119900366 | 120501339 | 121105324 | ....12336 |
| 160 | ....38050 | ....27182 | ....19267 | ....14320 | ....12356 | ....13389 | ....17434 | ....24508 |
| 170 | ....49803 | ....38995 | ....31139 | ....26252 | ....24347 | ....25440 | ....29146 | ....36680 |
| 180 | ....61558 | ....50808 | ....43012 | ....38184 | ....36339 | ....37492 | ....41659 | ....48854 |
| 190 | ....73315 | ....62623 | ....54887 | ....50118 | ....48333 | ....49546 | ....53773 | ....61029 |
| 200 | ....85072 | ....74440 | ....66762 | ....62053 | ....60328 | ....61601 | ....65889 | ....73205 |
| 210 | ....96830 | ....86257 | ....78639 | ....73989 | ....72324 | ....73657 | ....78005 | ....85382 |
| 220 | 117608590 | ....98076 | ....90517 | ....85927 | ....84321 | ....85715 | ....90123 | ....97561 |
| 230 | ....20351 | 118209896 | 118802396 | ....97865 | ....96319 | ....97773 | 121202242 | 121809740 |
| 240 | ....32113 | ....21717 | ....14276 | 119409805 | 120008319 | 120609833 | ....14362 | ....21921 |
| 250 | ....43876 | ....33539 | ....26157 | ....21746 | ....20320 | ....21894 | ....26484 | ....34104 |
| 260 | ....55641 | ....45362 | ....38040 | ....33688 | ....32322 | ....33956 | ....38606 | ....46287 |
| 270 | ....67406 | ....57187 | ....49924 | ....45632 | ....44325 | ....46020 | ....50730 | ....58472 |
| 280 | ....79173 | ....69012 | ....61809 | ....57576 | ....56329 | ....58084 | ....62855 | ....70657 |
| 290 | ....90941 | ....80839 | ....73695 | ....69522 | ....68335 | ....70150 | ....74982 | ....82844 |
| 300 | 117702710 | ....92668 | ....85582 | ....81469 | ....80342 | ....82217 | ....87109 | ....95033 |
| 310 | ....14480 | 118304497 | ....97471 | ....93417 | ....92350 | ....94285 | ....99238 | 121907222 |
| 320 | ....26252 | ....16327 | 118909361 | 119505366 | 120104359 | 120706355 | 121311368 | ....19417 |
| 330 | ....38024 | ....28159 | ....21252 | ....17317 | ....16370 | ....18425 | ....23499 | ....31605 |
| 340 | ....49798 | ....39992 | ....33144 | ....29269 | ....28381 | ....30497 | ....35631 | ....43798 |
| 350 | ....61573 | ....51826 | ....45037 | ....41222 | ....40394 | ....42570 | ....47765 | ....55992 |
| 360 | ....73349 | ....63661 | ....56931 | ....53176 | ....52408 | ....54645 | ....59900 | ....68188 |
| 370 | ....85126 | ....75497 | ....68827 | ....65131 | ....64423 | ....66720 | ....72036 | ....80385 |
| 380 | ....96905 | ....87335 | ....80724 | ....77087 | ....76440 | ....78797 | ....84173 | ....92583 |
| 390 | 117808685 | ....99173 | ....92622 | ....89045 | ....88457 | ....90875 | ....96311 | 122004782 |
| 400 | ....20466 | 118411013 | 119004521 | 119601004 | 120300476 | 120802954 | 121408451 | ....16983 |
| 410 | ....32248 | ....22854 | ....16422 | ....12964 | ....12496 | ....15034 | ....20592 | ....29184 |
| 420 | ....44031 | ....34697 | ....28323 | ....24925 | ....24517 | ....27116 | ....32734 | ....41387 |
| 430 | ....55815 | ....46540 | ....40226 | ....36888 | ....36540 | ....39198 | ....44877 | ....53591 |
| 440 | ....67601 | ....58385 | ....52130 | ....48852 | ....48564 | ....51282 | ....57021 | ....65707 |
| 450 | ....79388 | ....70231 | ....64035 | ....60816 | ....60589 | ....63367 | ....69167 | ....78003 |
| 460 | ....91175 | ....82078 | ....75942 | ....72733 | ....72615 | ....75454 | ....81314 | ....90211 |
| 470 | 117902965 | ....93926 | ....87849 | ....84750 | ....84642 | ....87541 | ....93462 | 122102420 |
| 480 | ....14755 | 118505775 | ....99758 | ....96718 | ....96671 | ....99630 | 121505611 | ....14630 |
| 490 | ....26546 | ....17626 | 119111668 | 119708688 | 120308700 | 120911720 | ....17762 | ....26842 |
| 500 | ....38339 | ....29478 | .0..23579 | ....20659 | ....20731 | ....23811 | ....29914 | ....39055 |

**Figure C.5**

| | 20000 | 20500 | 21000 | 21500 | 22000 | 22500 | 23000 | 23500 |
|---|---|---|---|---|---|---|---|---|
| 0 | 122139055 | 122751248 | 123300511 | 123984857 | 124606303 | 125230863 | 125858553 | 126489391 |
| 10 | ····51268 | ····63524 | ····78847 | ····97255 | ····18763 | ····43386 | ····71139 | 126502040 |
| 20 | ····63484 | ····75800 | ····91185 | 124009655 | ····31225 | ····55910 | ····83726 | ····14690 |
| 30 | ····75700 | ····88078 | 123403524 | ····22056 | ····43688 | ····68436 | ····96315 | ····27341 |
| 40 | ····87917 | 122800356 | ····15865 | ····34458 | ····56153 | ····80963 | 125908904 | ····39994 |
| 50 | 122200136 | ····12636 | ····28206 | ····46862 | ····68618 | ····93491 | ····21495 | ····52648 |
| 60 | ····12356 | ····24918 | ····40549 | ····59266 | ····81085 | 125306020 | ····34087 | ····65303 |
| 70 | ····24577 | ····37200 | ····52893 | ····71672 | ····93553 | ····18551 | ····46681 | ····77960 |
| 80 | ····36800 | ····49484 | ····65235 | ····84080 | 124706023 | ····31083 | ····59275 | ····90618 |
| 90 | ····49024 | ····61769 | ····77585 | ····96488 | ····18493 | ····43616 | ····71871 | 126603277 |
| 100 | ····61249 | ····74055 | ····89933 | 124108898 | ····30965 | ····56150 | ····84469 | ····15937 |
| 110 | ····73475 | ····86242 | 123502282 | ····21309 | ····43438 | ····68686 | ····97067 | ····28599 |
| 120 | ····85702 | ····98031 | ····14632 | ····33721 | ····55912 | ····81223 | 126009667 | ····41262 |
| 130 | ····97930 | 122510921 | ····26984 | ····46134 | ····68388 | ····93761 | ····22268 | ····53929 |
| 140 | 122310166 | ····23212 | ····39336 | ····58549 | ····80865 | 125406300 | ····34870 | ····66591 |
| 150 | ····22391 | ····35504 | ····51690 | ····70965 | ····93343 | ····18841 | ····47474 | ····79258 |
| 160 | ····34624 | ····47798 | ····64045 | ····83382 | 124805822 | ····31383 | ····60078 | ····91926 |
| 170 | ····46857 | ····60093 | ····76402 | ····95800 | ····18303 | ····43926 | ····72684 | 126704595 |
| 180 | ····59092 | ····72385 | [illegible] | 124208220 | ····30785 | ····56470 | ····85292 | ····17265 |
| 190 | ····71328 | ····84686 | 123601115 | ····20640 | ····43268 | ····69016 | ····97900 | ····29937 |
| 200 | ····83565 | ····96984 | ····13478 | ····33062 | ····55752 | ····81563 | 126110510 | ····42610 |
| 210 | ····95803 | 123009284 | ····25840 | ····45486 | ····68238 | ····94111 | ····23121 | ····55284 |
| 220 | 122408043 | ····21585 | ····38202 | ····57910 | ····80724 | 125506660 | ····35733 | ····67960 |
| 230 | ····20284 | ····33887 | ····50566 | ····70336 | ····93213 | ····19211 | ····48347 | ····80637 |
| 240 | ····32526 | ····46190 | ····62931 | ····82763 | 124905702 | ····31763 | ····60962 | ····93315 |
| 250 | ····44769 | ····58495 | ····75297 | ····95191 | ····18192 | ····44316 | ····73578 | 126805994 |
| 260 | ····57013 | ····70801 | ····87665 | 124307621 | ····30684 | ····56870 | ····86195 | ····18675 |
| 270 | ····69259 | ····83108 | 123700034 | ····20052 | ····43177 | ····69426 | ····98814 | ····39357 |
| 280 | ····81506 | ····95416 | ····12404 | ····32484 | ····55672 | ····81983 | 126211434 | ····44040 |
| 290 | ····93754 | 123107726 | ····24775 | ····44917 | ····68167 | ····94541 | ····24055 | ····56724 |
| 300 | 122506004 | ····20037 | ····37147 | ····57351 | ····80664 | 125607101 | ····36677 | ····69410 |
| 310 | ··18254 | ····32349 | ····49521 | ····69787 | ····93162 | ····19661 | ····49301 | ····82097 |
| 320 | ····30506 | ····44662 | ····61896 | ····82224 | 125005661 | ····32223 | ····61926 | ····94785 |
| 330 | ····42759 | ····56976 | ····74272 | ····94662 | ····18162 | ····44787 | ····74552 | 126907474 |
| 340 | ····55013 | ····69292 | ····86650 | 124407102 | ····30664 | ····57351 | ····87180 | ····20165 |
| 350 | ····67269 | ····81609 | ····99028 | ····19542 | ····43167 | ····69917 | ····99808 | ····32857 |
| 360 | ····79520 | ····93927 | 123811408 | ····31984 | ····55671 | ····82484 | 126312438 | ····45550 |
| 370 | ····91784 | 123206247 | ····23789 | ····44428 | ····68177 | ····95052 | ····25070 | ····58245 |
| 380 | 122604043 | ····18567 | ····36172 | ····56872 | ····80684 | 125707622 | ····37702 | ····70941 |
| 390 | ····10303 | ····30889 | ····48555 | ····69318 | ····93192 | ····20193 | ····50336 | ····83638 |
| 400 | ····28565 | ····43212 | ····60940 | ····81765 | 125105701 | ····32764 | ····62971 | ····96336 |
| 410 | ····40828 | ····55536 | ····73326 | ····94213 | ····18212 | ····45338 | ····75607 | 127009036 |
| 420 | ····53092 | ····67862 | ····85714 | 124506662 | ····30723 | ····57912 | ····88245 | ····21737 |
| 430 | ····65357 | ····89189 | ····98102 | ····19113 | ····43236 | ····70488 | 126400884 | ····34439 |
| 440 | ····77623 | ····92517 | 123910492 | ····31565 | ····55751 | ····83065 | ····13524 | ····47142 |
| 450 | ····89891 | 123304846 | ····22883 | ····44018 | ····68266 | ····95643 | ····26165 | ····59847 |
| 460 | 122702160 | ····17177 | ····35275 | ····56473 | ····80783 | 125808222 | ····38808 | ····72553 |
| 470 | ····14430 | ····29508 | ····47669 | ····68928 | ····93301 | ····20803 | ····51452 | ····85260 |
| 480 | ····26702 | ····41841 | ····60064 | ····81385 | 125205821 | ····33385 | ····64097 | ····9756[illegible] |
| 490 | ····38975 | ····54176 | ····72460 | ····93843 | ····18341 | ····45969 | ····76743 | 12711067[illegible] |
| 500 | ····51248 | ····66511 | ····84857 | 124606303 | ····30863 | ····58553 | ····89391 | ····2335[illegible] |

**Figure C.6**

| | 24000 | 24500 | 25000 | 25500 | 26000 | 26500 | 27000 | 275000 |
|---|---|---|---|---|---|---|---|---|
| 0 | 127123390 | 127760566 | 128400937 | 129044517 | 129691323 | 130341370 | 130994677 | 131651257 |
| 10 | ....36102 | ....73343 | ....13777 | ....57421 | 129704292 | ....54405 | 131007776 | ....64423 |
| 20 | ....48816 | ....86120 | ....26618 | ....70322 | ....17262 | ....67440 | ....20877 | ....77589 |
| 30 | ....61530 | ....98898 | ....39461 | ....83234 | ....30234 | ....80477 | ....33979 | ....90757 |
| 40 | ....74247 | 127811678 | ....52305 | ....96142 | ....43207 | ....93515 | ....47083 | 131703926 |
| 50 | ....86964 | ....24459 | ....65150 | 129109052 | ....56181 | 130406554 | ....60187 | ....17096 |
| 60 | ....99683 | ....37242 | ....77997 | ....21963 | ....69157 | ....19595 | ....73293 | ....30268 |
| 70 | 127212403 | ....50026 | ....90845 | ....34875 | ....82134 | ....32637 | ....86401 | ....13441 |
| 80 | ....25124 | ....62811 | 128503691 | ....47789 | ....95112 | ....45680 | ....99509 | ....56615 |
| 90 | ....37846 | ....75597 | ....16544 | ....60703 | 129808091 | ....58725 | 131112619 | ....69791 |
| 100 | ....50570 | ....88384 | ....29396 | ....73620 | ....21073 | ....71771 | ....25731 | ....82968 |
| 110 | ....63295 | 127901173 | ....42249 | ....86537 | ....34055 | ....84818 | ....38843 | ....96146 |
| 120 | ....76022 | ....13963 | ....55103 | ....99456 | ....47038 | ....97867 | ....51957 | 131809326 |
| 130 | ....88749 | ....26755 | ....67958 | 129212376 | ....60023 | 130510916 | ....65072 | ....22507 |
| 140 | 127301478 | ....39547 | ....80815 | ....25297 | ....73009 | ....23963 | ....78189 | ....35689 |
| 150 | ....14208 | ....52341 | ....93673 | ....38219 | ....85996 | ....37020 | ....91306 | ....48873 |
| 160 | ....26940 | ....65137 | 128606533 | ....51143 | ....98985 | ....50074 | 131204426 | ....62058 |
| 170 | ....39672 | ....77933 | ....19393 | ....64068 | 129911975 | ....63129 | ....17546 | ....75244 |
| 180 | ....52406 | ....90731 | ....32255 | ....76995 | ....24966 | ....76185 | ....30668 | ....88431 |
| 190 | ....65142 | 128003530 | ....45118 | ....89922 | ....37958 | ....89243 | ....43791 | 131901620 |
| 200 | ....77878 | ....16330 | ....57983 | 129302851 | ....50952 | 130602301 | ....56915 | ....14810 |
| 210 | ....90616 | ....29132 | ....70849 | ....15782 | ....63947 | ....15361 | ....70041 | ....28002 |
| 220 | 127403355 | ....41935 | ....83716 | ....28713 | ....76944 | ....28423 | ....83168 | ....41194 |
| 230 | ....16095 | ....54739 | ....96584 | ....41646 | ....89941 | ....41436 | ....96296 | ....54389 |
| 240 | ....28837 | ....67545 | 128709454 | ....54580 | 130002940 | ....54550 | 131309426 | ....67584 |
| 250 | ....41580 | ....80351 | ....22325 | ....67516 | ....15941 | ....67615 | ....22557 | ....80781 |
| 260 | ....54324 | ....93159 | ....35197 | ....80453 | ....28942 | ....80682 | ....35689 | ....93979 |
| 270 | ....67069 | 128105969 | ....48070 | ....93391 | ....41945 | ....93750 | ....48823 | 132007178 |
| 280 | ....79816 | ....18779 | ....60945 | 129406330 | ....54949 | 130706820 | ....61958 | ....20379 |
| 290 | ....92564 | ....31591 | ....73821 | ....19270 | ....67955 | ....19890 | ....75094 | ....33581 |
| 300 | 127505313 | ....44404 | ....86699 | ....32212 | ....80962 | ....32962 | ....88231 | ....46784 |
| 310 | ....18064 | ....57219 | ....99577 | ....45156 | ....93970 | ....46036 | 131401370 | ....59989 |
| 320 | ....30816 | ....70034 | 128812457 | ....58110 | 130106979 | ....59110 | ....14510 | ....73195 |
| 330 | ....43569 | ....82851 | ....25339 | ....71046 | ....19990 | ....72186 | ....27652 | ....86402 |
| 340 | ....56323 | ....95670 | ....38221 | ....83993 | ....33002 | ....85263 | ....40794 | ....99611 |
| 350 | ....69079 | 128208439 | ....51105 | ....96941 | ....46015 | ....98342 | ....53938 | 132112821 |
| 360 | ....81836 | ....21310 | ....63990 | 129509891 | ....59030 | 130811422 | ....67084 | ....26032 |
| 370 | ....94594 | ....34132 | ....76876 | ....22842 | ....72046 | ....24503 | ....80231 | ....39245 |
| 380 | 127607353 | ....46956 | ....89764 | ....35794 | ....85063 | ....37585 | ....93379 | ....52459 |
| 390 | ....20114 | ....59730 | 128902653 | ....48748 | ....98031 | ....50667 | 131506528 | ....65674 |
| 400 | ....32876 | ....72606 | ....15543 | ....61703 | 130211101 | ....63754 | ....19679 | ....78891 |
| 410 | ....45639 | ....85434 | ....28435 | ....74659 | ....24122 | ....76841 | ....32831 | ....92109 |
| 420 | ....58404 | ....98262 | ....41328 | ....87617 | ....37144 | ....89928 | ....45984 | 132205328 |
| 430 | ....71170 | 128311092 | ....54222 | 129600575 | ....50163 | 130903017 | ....59138 | ....18548 |
| 440 | ....83937 | ....23923 | ....67117 | ....13535 | ....63193 | ....16108 | ....72294 | ....31770 |
| 450 | ....96705 | ....36756 | ....80014 | ....26497 | ....76219 | ....29199 | ....85452 | ....44993 |
| 460 | 127709475 | ....49589 | ....92912 | ....39459 | ....89247 | ....42292 | ....98610 | ....58218 |
| 470 | ....22246 | ....62424 | 129005811 | ....52423 | 130302276 | ....55386 | 131611770 | ....71444 |
| 480 | ....35018 | ....75260 | ....18712 | ....65388 | ....15306 | ....68482 | ....24931 | ....84671 |
| 490 | ....47792 | ....88098 | ....31614 | ....78355 | ....28338 | ....81579 | ....38094 | ....97899 |
| 500 | ....60566 | 128400937 | ....44517 | ....91323 | ....41370 | ....94677 | ....51257 | 132311129 |

Figure C.7

| | 28000 | 28500 | 29000 | 29500 | 30000 | 30500 | 31000 | 31500 |
|---|---|---|---|---|---|---|---|---|
| 0 | 132311129 | 132974308 | 133640811 | 134210655 | 134983856 | 135660432 | 136340398 | 137023773 |
| 10 | ....24360 | ....87605 | ....54175 | ....24086 | ....97355 | ....75998 | ....54032 | ....37476 |
| 20 | ....37593 | 133000904 | ....67541 | ....37518 | 135010854 | ....87565 | ....67668 | ....51179 |
| 30 | ....50826 | ....14204 | ....80907 | ....50952 | ....24355 | 135701134 | ....81305 | ....64884 |
| 40 | ....64061 | ....27506 | ....94276 | ....64387 | ....37858 | ....14704 | ....94943 | ....78591 |
| 50 | ....77295 | ....40809 | 133707645 | ....77824 | ....51362 | ....28275 | 136408582 | ....92299 |
| 60 | ....90536 | ....54113 | ....21016 | ....91262 | ....64867 | ....41848 | ....22223 | 137106008 |
| 70 | 132403775 | ....67418 | ....34388 | 134404701 | ....78373 | ....55422 | ....35865 | ....19719 |
| 80 | ....17015 | ....80725 | ....47761 | ....18141 | ....91881 | ....68998 | ....49509 | ....33431 |
| 90 | ....30257 | ....94033 | ....61136 | ....31583 | 135105390 | ....82575 | ....63154 | ....47144 |
| 100 | ....43500 | 133107342 | ....74512 | ....45026 | ....18501 | ....96153 | ....76800 | ....60859 |
| 110 | ....56744 | ....20653 | ....87890 | ....58471 | ....32413 | 135809733 | ....90448 | ....74575 |
| 120 | ....69990 | ....33965 | 133801268 | ....71916 | ....45926 | ....23314 | 136504097 | ....88292 |
| 130 | ....83237 | ....47272 | ....14649 | ....85364 | ....59440 | ....36896 | ....17747 | 137202011 |
| 140 | ....96485 | ....60593 | ....28030 | ....98812 | ....72956 | ....50480 | ....31399 | ....15731 |
| 150 | 132509735 | ....73909 | ....41413 | 134512262 | ....86474 | ....64065 | ....45052 | ....29453 |
| 160 | ....22986 | ....87227 | ....54797 | ....25713 | ....99992 | ....77651 | ....58707 | ....43176 |
| 170 | ....36238 | 133200545 | ....68182 | ....39166 | 135213512 | ....91239 | ....72363 | ....56900 |
| 180 | ....49492 | ....13865 | ....81569 | ....52620 | ....27034 | 135904828 | ....86020 | ....70626 |
| 190 | ....62746 | ....27187 | ....94957 | ....66075 | ....40556 | ....18419 | ....99678 | ....84353 |
| 200 | ....76003 | ....40510 | 133908347 | ....79532 | ....54081 | ....32010 | 136613338 | ....98081 |
| 210 | ....89260 | ....53834 | ....21738 | ....92990 | ....67606 | ....45604 | ....27000 | 137311811 |
| 220 | 132602519 | ....67159 | ....35130 | 134606449 | ....81133 | ....59198 | ....40662 | ....25542 |
| 230 | ....15780 | ....80486 | ....48523 | ....19910 | ....94661 | ....72794 | ....54326 | ....39275 |
| 240 | ....29041 | ....93814 | ....61918 | ....33372 | 135308190 | ....86391 | ....67992 | ....53009 |
| 250 | ....42304 | 133307143 | ....75314 | ....46835 | ....21721 | ....99990 | ....81659 | ....66744 |
| 260 | ....55568 | ....20474 | ....88712 | ....60300 | ....35253 | 136013590 | ....95327 | ....80481 |
| 270 | ....68834 | ....33806 | 134002111 | ....73766 | ....48787 | ....27191 | 136708996 | ....94219 |
| 280 | ....82101 | ....47139 | ....15511 | ....87233 | ....62322 | ....40794 | ....22667 | 137407958 |
| 290 | ....95369 | ....60474 | ....28913 | 134700702 | ....75858 | ....54398 | ....36340 | ....21699 |
| 300 | 132708639 | ....73810 | ....42316 | ....14172 | ....89395 | ....68004 | ....50013 | ....35441 |
| 310 | ....21909 | ....87147 | ....55720 | ....27643 | 153402934 | ....81610 | ....63688 | ....49184 |
| 320 | ....35182 | 133400486 | ....69125 | ....41116 | ....16475 | ....95219 | ....77365 | ....62929 |
| 330 | ....48455 | ....13826 | ....82532 | ....54590 | ....30016 | 136108828 | ....91042 | ....76676 |
| 340 | ....61730 | ....27167 | ....95941 | ....68066 | ....43559 | ....22439 | 136804721 | ....90423 |
| 350 | ....75006 | ....40510 | 134109350 | ....81542 | ....57104 | ....36051 | ....18402 | 137504172 |
| 360 | ....88284 | ....53854 | ....22761 | ....95020 | ....70649 | ....49665 | ....32084 | ....17923 |
| 370 | 132801562 | ....67200 | ....36173 | 134808500 | ....84196 | ....63280 | ....45767 | ....31675 |
| 380 | ....14843 | ....80546 | ....49587 | ....21981 | ....97745 | ....76896 | ....59451 | ....45428 |
| 390 | ....28124 | ....93894 | ....63002 | ....35463 | 135511295 | ....90514 | ....73137 | ....59182 |
| 400 | ....41407 | 133507244 | ....76418 | ....48946 | ....24846 | 136204133 | ....86825 | ....72938 |
| 410 | ....54691 | ....20594 | ....89836 | ....62431 | ....38398 | ....17753 | 136900513 | ....86695 |
| 420 | ....67977 | ....33947 | 134203255 | ....75918 | ....51952 | ....31375 | ....14203 | 137600454 |
| 430 | ....81263 | ....47300 | ....16675 | ....89405 | ....65507 | ....44998 | ....27895 | ....14214 |
| 440 | ....94551 | ....60655 | ....30097 | 134902[illegible]94 | ....79064 | ....58623 | ....41588 | ....27976 |
| 450 | 132907841 | ....74[illegible] | ....43520 | ....16384 | ....92622 | ....72249 | ....55282 | ....41739 |
| 460 | ....21132 | ....8736[illegible] | ....56944 | ....29876 | 135606181 | ....85876 | ....68977 | ....55503 |
| 470 | ....34424 | 133600727 | ....70370 | ....43369 | ....19742 | ....99504 | ....82674 | ....69268 |
| 480 | ....47717 | ....14087 | ....83797 | ....56863 | ....33304 | 136313134 | ....96373 | ....83035 |
| 490 | ....61012 | ....27448 | ....97225 | ....70359 | ....46867 | ....26766 | 137010072 | ....96804 |
| 500 | ....74308 | ....40811 | ....10655 | ....83856 | ....60432 | ....40398 | ....23773 | 137710573 |

B

**Figure C.8**

| | 32000 | 32500 | 33000 | 33500 | 34000 | 34500 | 35000 | 35500 |
|---|---|---|---|---|---|---|---|---|
| 0 | 137710573 | 138400816 | 139094518 | 139791697 | 140492371 | 141196557 | 141904272 | 142615534 |
| 10 | ....24344 | ....14656 | 139108428 | 139805677 | 140506420 | 141210676 | ....18463 | ....29796 |
| 20 | ....38117 | ....28497 | ....22338 | ....19657 | ....20471 | ....24797 | ....32655 | ....44059 |
| 30 | ....51891 | ....42340 | ....36250 | ....33639 | ....34523 | ....38920 | ....46848 | ....58323 |
| 40 | ....65666 | ....56184 | ....50164 | ....47623 | ....48576 | ....53044 | ....61043 | ....72589 |
| 50 | ....79442 | ....70030 | ....64079 | ....61607 | ....62631 | ....67169 | ....75239 | ....86856 |
| 60 | ....93220 | ....83877 | ....77995 | ....75594 | ....76687 | ....81296 | ....89437 | 142701125 |
| 70 | 137807000 | ....97725 | ....91913 | ....89581 | ....90745 | ....95424 | 142003635 | ....15395 |
| 80 | ....20780 | 138511575 | 139205832 | 139903570 | 140604804 | 141309553 | ....17836 | ....29667 |
| 90 | ....34562 | ....25426 | ....19753 | ....17561 | ....18865 | ....23684 | ....32038 | ....43940 |
| 100 | ....48346 | ....39279 | ....33675 | ....31552 | ....32927 | ....37817 | ....46240 | ....58214 |
| 110 | ....62131 | ....53133 | ....47598 | ....45545 | ....46990 | ....51951 | ....60444 | ....72490 |
| 120 | ....75917 | ....66988 | ....61523 | ....59540 | ....61055 | ....66086 | ....74650 | ....86767 |
| 130 | ....89704 | ....80845 | ....75449 | ....73536 | ....75121 | ....80222 | ....88858 | 142801046 |
| 140 | 137903493 | ....94703 | ....89377 | ....87533 | ....89188 | ....94360 | 142103067 | ....15326 |
| 150 | ....17284 | 138608562 | 139303306 | 140001532 | 140703257 | 141408499 | ....17277 | ....29607 |
| 160 | ....31075 | ....22423 | ....17236 | ....15532 | ....17327 | ....22640 | ....31489 | ....43890 |
| 170 | ....44869 | ....36286 | ....31168 | ....29534 | ....31399 | ....36783 | ....45702 | ....58175 |
| 180 | ....58663 | ....50149 | ....45101 | ....43537 | ....45472 | ....50926 | ....59917 | ....72460 |
| 190 | ....72459 | ....64014 | ....59035 | ....57541 | ....59547 | ....65072 | ....74133 | ....86748 |
| 200 | ....86256 | ....77881 | ....72971 | ....71547 | ....73623 | ....79218 | ....88350 | 142901036 |
| 210 | 138000055 | ....91748 | ....86909 | ....85554 | ....87700 | ....93366 | 142202569 | ....15326 |
| 220 | ....13855 | 138705618 | 139400847 | ....99563 | 140801779 | 141507515 | ....16789 | ....29618 |
| 230 | ....27656 | ....19488 | ....14787 | 140113573 | ....15859 | ....21666 | ....31011 | ....43911 |
| 240 | ....41459 | ....33360 | ....28729 | ....27584 | ....29941 | ....35818 | ....45234 | ....58205 |
| 250 | ....55263 | ....47233 | ....42672 | ....41596 | ....44024 | ....49972 | ....59458 | ....72501 |
| 260 | ....69069 | ....61108 | ....56616 | ....55610 | ....58108 | ....64127 | ....73684 | ....86798 |
| 270 | ....82876 | ....74984 | ....70562 | ....69626 | ....72194 | ....78283 | ....87912 | 143001097 |
| 280 | ....96684 | ....88862 | ....84509 | ....83643 | ....86281 | ....92441 | 142302141 | ....15397 |
| 290 | 138110494 | 138802741 | ....98457 | ....97661 | 140900370 | 141606600 | ....16371 | ....29699 |
| 300 | ....24305 | ....16621 | 139512407 | 140211681 | ....14460 | ....20761 | ....30602 | ....44002 |
| 310 | ....38117 | ....30503 | ....26358 | ....25702 | ....28551 | ....34923 | ....44835 | ....58306 |
| 320 | ....51931 | ....44386 | ....40311 | ....39725 | ....42644 | ....49087 | ....59070 | ....72612 |
| 330 | ....65746 | ....58270 | ....54265 | ....53749 | ....56738 | ....63251 | ....73306 | ....869[illegible]9 |
| 340 | ....79563 | ....72156 | ....68220 | ....67774 | ....70834 | ....77418 | ....87543 | 143101228 |
| 350 | ....93381 | ....86043 | ....82177 | ....81801 | ....84931 | ....91585 | 142401782 | ....15538 |
| 360 | 138207200 | ....99932 | ....96135 | ....95829 | ....99030 | 141705755 | ....16022 | ....29850 |
| 370 | ....21021 | 138913822 | 139610095 | 140309859 | 141013129 | ....19925 | ....30264 | ....44163 |
| 380 | ....34843 | ....27713 | ....24056 | ....23890 | ....27231 | ....34097 | ....44507 | ....58477 |
| 390 | ....48666 | ....41606 | ....38018 | ....37922 | ....41333 | ....48271 | ....58751 | ....72793 |
| 400 | ....62491 | ....55500 | ....51982 | ....51956 | ....55438 | ....62445 | ....72997 | ....87110 |
| 410 | ....76317 | ....69396 | ....65947 | ....65991 | ....69543 | ....76622 | ....87244 | 143201429 |
| 420 | ....90145 | ....83293 | ....79914 | ....80028 | ....83650 | ....90799 | 142501493 | ....15749 |
| 430 | 138303974 | ....97191 | ....93882 | ....94066 | ....97758 | 141804978 | ....15743 | ....30070 |
| 440 | ....17804 | 139011091 | 139707851 | 140408105 | 141111868 | ....19159 | ....29995 | ....44394 |
| 450 | ....31636 | ....24992 | ....21822 | ....22146 | ....25979 | ....33341 | ....44248 | ....58718 |
| 460 | ....45469 | ....38894 | ....35795 | ....36188 | ....40092 | ....47524 | ....58502 | ....73044 |
| 470 | ....59304 | ....52798 | ....49768 | ....50232 | ....54206 | ....61709 | ....72758 | ....87371 |
| 480 | ....73140 | ....66703 | ....63743 | ....64277 | ....68321 | ....75895 | ....87015 | 143301700 |
| 490 | ....86977 | ....80610 | ....77720 | ....78323 | ....82438 | ....90083 | 142601274 | ....16030 |
| 500 | 138400816 | ....94518 | ....91697 | ....92371 | ....96557 | 14190[illegible] | ....15534 | ....30362 |

**Figure C.9**

| | 36000 | 36500 | 37000 | 7500 | 8000 | 38500 | 39000 | 39500 |
|---|---|---|---|---|---|---|---|---|
| 0 | 143330362 | 144043772 | 144770783 | 145496414 | 146225681 | 146958603 | 147695199 | 148435488 |
| 10 | ....44695 | ....63177 | ....85260 | 145510963 | ....40304 | ....73299 | 147709969 | ....50331 |
| 20 | ....59025 | ....77583 | ....99739 | ....25514 | ....54927 | ....87997 | ....24740 | ....65176 |
| 30 | ....73365 | ....91991 | 144814219 | ....40067 | ....69553 | 147002695 | ....39513 | ....80023 |
| 40 | ....87702 | 144106400 | ....28700 | ....54621 | ....84180 | ....17396 | ....54287 | ....94871 |
| 50 | 143402041 | ....20811 | ....43183 | ....69176 | ....98808 | ....32097 | ....69062 | 148509720 |
| 60 | ....16381 | ....35223 | ....57668 | ....83733 | 146313438 | ....46801 | ....83839 | ....24571 |
| 70 | ....30723 | ....49637 | ....72153 | ....98292 | ....28070 | ....61505 | ....98617 | ....39424 |
| 80 | ....45066 | ....64052 | ....86641 | 145612852 | ....42702 | ....76212 | 147813397 | ....54278 |
| 90 | ....59411 | ....78468 | 144901129 | ....27413 | ....57337 | ....90919 | ....28178 | ....69133 |
| 100 | ....73757 | ....92886 | ....15619 | ....41976 | ....71972 | 147105628 | ....42961 | ....83990 |
| 110 | ....88104 | 144207305 | ....30111 | ....56540 | ....86610 | ....20339 | ....57746 | ....98848 |
| 120 | 143502453 | ....21726 | ....44604 | ....71105 | 146401248 | ....35051 | ....72531 | 148613708 |
| 130 | ....16803 | ....36148 | ....59098 | ....85673 | ....15888 | ....49764 | ....87319 | ....28570 |
| 140 | ....31155 | ....50572 | ....73504 | 145700241 | ....30530 | ....64479 | 147902107 | ....43433 |
| 150 | ....45508 | ....64997 | ....88092 | ....14811 | ....45173 | ....79196 | ....16897 | ....58297 |
| 160 | ....59862 | ....79423 | 145002590 | ....29383 | ....59818 | ....93914 | ....31689 | ....73163 |
| 170 | ....74218 | ....93851 | ....17091 | ....43955 | ....74464 | 147208633 | ....46482 | ....88030 |
| 180 | ....88576 | 144308280 | ....31593 | ....58530 | ....89111 | ....23354 | ....61277 | 148702899 |
| 190 | 143602935 | ....22711 | ....46096 | ....73106 | 146503760 | ....38076 | ....76073 | ....17769 |
| 200 | ....17295 | ....37144 | ....60600 | ....87683 | ....18410 | ....52800 | ....90871 | ....32641 |
| 210 | ....31657 | ....51577 | ....75106 | 145802262 | ....33062 | ....67525 | 148005669 | ....47514 |
| 220 | ....46020 | ....66012 | ....89614 | ....16842 | ....47715 | ....82252 | ....20470 | ....62389 |
| 230 | ....60384 | ....80449 | 145104123 | ....31424 | ....62370 | ....96980 | ....35272 | ....77265 |
| 240 | ....74750 | ....94887 | ....18633 | ....46007 | ....77026 | 147311710 | ....50076 | ....92143 |
| 250 | ....89118 | 144409327 | ....33145 | ....60591 | ....91684 | ....26441 | ....64881 | 148807022 |
| 260 | 143703487 | ....23767 | ....47658 | ....75178 | 146606343 | ....41174 | ....79687 | ....21903 |
| 270 | ....17857 | ....38210 | ....62173 | ....89765 | ....21004 | ....55908 | ....54495 | ....36785 |
| 280 | ....32229 | ....52654 | ....76689 | 145904354 | ....35666 | ....70643 | 148109305 | ....51669 |
| 290 | ....46602 | ....67099 | ....91207 | ....18944 | ....50330 | ....85381 | ....24116 | ....66554 |
| 300 | ....60977 | ....81546 | 145205726 | ....33536 | ....64995 | 147400119 | ....38928 | ....81441 |
| 310 | ....75353 | ....95994 | ....20247 | ....48130 | ....79661 | ....14859 | ....53742 | ....96329 |
| 320 | ....89730 | 144510443 | ....34769 | ....62725 | ....94329 | ....29601 | ....68558 | 148911218 |
| 330 | 143804109 | ....24894 | ....49292 | ....77321 | 146708998 | ....44344 | ....83374 | ....26109 |
| 340 | ....18490 | ....39347 | ....63817 | ....91919 | ....23669 | ....59088 | ....98193 | ....41002 |
| 350 | ....32872 | ....53801 | ....78343 | 146006518 | ....38342 | ....73834 | 148213013 | ....55896 |
| 360 | ....47255 | ....68256 | ....92871 | ....21118 | ....53016 | ....88581 | ....27834 | ....70792 |
| 370 | ....61640 | ....82713 | 145307401 | ....35721 | ....67691 | 147503330 | ....42657 | ....85689 |
| 380 | ....76026 | ....97171 | ....21931 | ....50324 | ....82368 | ....18080 | ....57481 | 149000587 |
| 390 | ....90413 | 144611631 | ....36463 | ....64929 | ....97046 | ....32832 | ....72307 | ....15487 |
| 400 | 143904802 | ....26092 | ....50997 | ....79536 | 146811726 | ....47586 | ....87134 | ....30389 |
| 410 | ....19193 | ....40555 | ....65532 | ....94144 | ....26407 | ....62840 | 148301963 | ....45292 |
| 420 | ....33585 | ....55019 | ....80069 | 146108753 | ....41089 | ....77097 | ....16793 | ....60196 |
| 430 | ....47978 | ....69484 | ....94607 | ....23364 | ....55773 | ....91854 | ....31624 | ....75102 |
| 440 | ....62373 | ....83951 | 145409146 | ....37976 | ....70459 | 147606614 | ....46458 | ....90010 |
| 450 | ....79769 | ....98420 | ....23687 | ....52590 | ....85146 | ....21374 | ....61292 | 149104919 |
| 460 | ....91167 | 144712890 | ....38230 | ....67205 | ....99835 | ....36136 | ....76128 | ....19829 |
| 470 | 144005566 | ....27761 | ....52773 | ....81822 | 146914625 | ....50900 | ....90966 | ....34741 |
| 480 | ....19967 | ....41834 | ....67319 | ....96440 | ....29216 | ....65665 | 148405805 | ....49655 |
| 490 | ....34369 | ....56308 | ....81865 | 146211060 | ....43909 | ....80432 | ....20646 | ....64570 |
| 500 | ....48772 | ....70783 | ....96414 | ....25681 | ....58603 | ....95199 | ....35488 | ....79486 |

B 2

Figure C.10

| | 40000 | 40500 | 41000 | 41500 | 42000 | 42500 | 43000 | 43500 |
|---|---|---|---|---|---|---|---|---|
| 0 | 149179486 | 149927214 | 150678689 | 151432932 | 152192960 | 152955782 | 153722448 | 154492946 |
| 10 | ····94404 | ····42207 | ····93758 | ····49076 | 152208179 | ····71087 | ····37820 | 154508395 |
| 20 | 149209324 | ····57201 | 150708827 | ····64220 | ····23400 | ····86385 | ····53194 | ····23846 |
| 30 | ····24245 | ····72197 | ····23898 | ····79367 | ····38622 | 153001683 | ····68569 | ····39298 |
| 40 | ····39167 | ····87194 | ····38970 | ····94515 | ····53846 | ····16983 | ····83946 | ····54753 |
| 50 | ····54091 | 150002193 | ····54041 | 151509664 | ····69071 | ····32285 | ····99324 | ····70208 |
| 60 | ····69016 | ····17193 | ····69120 | ····24815 | ····84298 | ····47588 | 153814704 | ····85665 |
| 70 | ····83943 | ····32195 | ····84197 | ····39968 | ····99527 | ····62893 | ····30086 | 154601123 |
| 80 | ····98872 | ····47198 | ····99275 | ····55122 | 152314757 | ····78199 | ····45469 | ····16584 |
| 90 | 149313802 | ····62203 | 150814355 | ····70277 | ····29988 | ····93507 | ····60853 | ····32045 |
| 100 | ····28733 | ····77209 | ····29436 | ····85434 | ····45221 | 153108817 | ····76239 | ····47508 |
| 110 | ····43666 | ····92216 | ····44519 | 151600593 | ····60456 | ····24127 | ····91627 | ····62973 |
| 120 | ····58600 | 150107226 | ····59604 | ····15753 | ····75692 | ····39440 | 153907016 | ····78440 |
| 130 | ····73536 | ····22236 | ····74689 | ····30914 | ····90929 | ····54754 | ····22407 | ····93507 |
| 140 | ····88474 | ····37249 | ····89777 | ····46077 | 152406168 | ····70069 | ····37799 | 154709377 |
| 150 | 149403412 | ····52263 | 150904866 | ····61242 | ····21409 | ····82386 | ····53193 | ····24848 |
| 160 | ····18353 | ····67278 | ····19957 | ····76408 | ····36651 | 153200705 | ····68588 | ····40320 |
| 170 | ····33295 | ····82294 | ····35049 | ····91576 | ····51895 | ····16025 | ····83985 | ····55794 |
| 180 | ····48238 | ····97313 | ····50142 | 151706745 | ····67140 | ····31346 | ····99383 | ····71270 |
| 190 | ····63183 | 150212332 | ····65237 | ····21916 | ····82387 | ····46670 | ····14783 | ····86747 |
| 200 | ····78129 | ····27354 | ····80334 | ····37088 | ····97635 | ····61994 | ····30185 | 154802226 |
| 210 | ····93077 | ····42376 | ····95432 | ····52262 | 152512885 | ····77320 | ····45588 | ····17706 |
| 220 | 149508026 | ····57400 | 151010531 | ····67437 | ····28136 | ····92648 | ····60992 | ····33188 |
| 230 | ····22977 | ····72426 | ····25632 | ····82613 | ····43389 | 153307977 | ····76398 | ····48671 |
| 240 | ····37929 | ····87453 | ····40735 | ····97792 | ····58643 | ····23308 | ····91806 | ····64156 |
| 250 | ····52883 | 150302482 | ····55839 | 151812972 | ····73899 | ····38641 | 154107215 | ····79642 |
| 260 | ····67838 | ····17512 | ····70945 | ····28153 | ····19156 | ····53974 | ····22626 | ····95130 |
| 270 | ····82795 | ····32544 | ····86052 | ····43336 | 152604415 | ····69310 | ····38038 | 154910620 |
| 280 | ····97753 | ····47577 | 151101160 | ····58520 | ····19676 | ····84647 | ····53452 | ····26111 |
| 290 | 149612713 | ····62612 | ····16270 | ····73706 | ····34938 | ····99985 | ····68867 | ····41603 |
| 300 | ····27674 | ····77649 | ····31382 | ····88893 | ····50201 | 153415325 | ····84284 | ····57097 |
| 310 | ····42637 | ····92686 | ····46495 | 151904082 | ····65466 | ····30667 | ····99703 | ····72593 |
| 320 | ····57602 | 150407726 | ····61610 | ····19273 | ····80733 | ····46010 | 154215123 | ····88090 |
| 330 | ····72567 | ····22767 | ····76726 | ····34464 | ····96001 | ····61354 | ····30544 | 155003589 |
| 340 | ····87535 | ····37809 | ····91844 | ····49658 | 152711270 | ····76701 | ····45967 | ····19090 |
| 350 | 149702503 | ····52853 | 151206963 | ····64853 | ····26542 | ····92048 | ····61392 | ····34592 |
| 360 | ····17473 | ····67898 | ····22083 | ····80049 | ····41814 | 153507397 | ····76818 | ····50095 |
| 370 | ····32445 | ····82945 | ····37206 | ····95247 | ····57088 | ····22748 | ····92246 | ····65600 |
| 380 | ····47418 | ····97993 | ····52329 | 152010447 | ····72364 | ····38100 | 154307675 | ····81107 |
| 390 | ····62393 | 150513043 | ····67455 | ····25648 | ····87641 | ····53454 | ····23106 | ····96615 |
| 400 | ····77369 | ····28094 | ····82581 | ····40850 | 152802920 | ····68810 | ····38538 | 155112124 |
| 410 | ····92347 | ····43147 | ····97710 | ····56055 | ····18200 | ····84167 | ····53972 | ····27636 |
| 420 | 149807326 | ····58201 | 151312839 | ····71260 | ····33482 | ····99525 | ····69407 | ····43148 |
| 430 | ····22307 | ····73257 | ····27971 | ····86467 | ····48766 | 153614885 | ····84844 | ····58663 |
| 440 | ····37289 | ····88314 | ····43104 | 152101676 | ····64050 | ····30246 | 154400283 | ····74179 |
| 450 | ····52273 | 150603373 | ····58238 | ····16886 | ····79337 | ····45609 | ····15723 | ····89696 |
| 460 | ····67258 | ····18433 | ····73374 | ····32098 | ····94625 | ····60974 | ····31164 | 155205215 |
| 470 | ····82245 | ····33495 | ····88511 | ····47311 | 152909914 | ····76340 | ····46607 | ····20735 |
| 480 | ····97233 | ····48558 | 151403650 | ····62526 | ····25205 | ····91708 | ····62052 | ····36[illegible]5 |
| 490 | 149912223 | ····63623 | ····18790 | ····77742 | ····40498 | 153707077 | ····77498 | ····51781 |
| 500 | ····27214 | ····78689 | ····33932 | ····92960 | ····55782 | ····22448 | ····92946 | ····67306 |

**Figure C.11**

| | 44000 | 44500 | 45000 | 45500 | 46000 | 46500 | 47000 | 47500 |
|---|---|---|---|---|---|---|---|---|
| 0 | 155267306 | 156045548 | 156827690 | 157613753 | 158403755 | 159197718 | 159995660 | 160797601 |
| 10 | ····82833 | ····61152 | ····43373 | ····29514 | ····19596 | 159213638 | 160011659 | 160813681 |
| 20 | ····98361 | ····76759 | ····59057 | ····45277 | ····35438 | ····29559 | ····27661 | ····29762 |
| 30 | 155313891 | ····92366 | ····74743 | ····61042 | ····51281 | ····45482 | ····43663 | ····45845 |
| 40 | ····29422 | 156107975 | ····90431 | ····76808 | ····67126 | ····61406 | ····59668 | ····61930 |
| 50 | ····44955 | ····23586 | 156906125 | ····92575 | ····82973 | ····77333 | ····75674 | ····78016 |
| 60 | ····60490 | ····39192 | ····21810 | 157708345 | ····98821 | ····93260 | ····91681 | ····94104 |
| 70 | ····76026 | ····54812 | ····37502 | ····24116 | 158514671 | 159309190 | 160107691 | 160910193 |
| 80 | ····91563 | ····70428 | ····53196 | ····39888 | ····30523 | ····25120 | ····23701 | ····26284 |
| 90 | 155407103 | ····86045 | ····68892 | ····55662 | ····46376 | ····41053 | ····39714 | ····42377 |
| 100 | ····22643 | 156201664 | ····84588 | ····71438 | ····62231 | ····56987 | ····55727 | ····58471 |
| 110 | ····38186 | ····17284 | 157000287 | ····87215 | ····78087 | ····72923 | ····71743 | ····74567 |
| 120 | ····53730 | ····32905 | ····15987 | 157802993 | ····93945 | ····88860 | ····87760 | ····90664 |
| 130 | ····69275 | ····48529 | ····31689 | ····18774 | 158609804 | 159404799 | 160203779 | 161006763 |
| 140 | ····84822 | ····64154 | ····47392 | ····34556 | ····25665 | ····20739 | ····19799 | ····22864 |
| 150 | 155500370 | ····79780 | ····63096 | ····50339 | ····41528 | ····36682 | ····35821 | ····38966 |
| 160 | ····15920 | ····95408 | ····78803 | ····66124 | ····57392 | ····52625 | ····51845 | ····55070 |
| 170 | ····31472 | 156311038 | ····94511 | ····81911 | ····73257 | ····68571 | ····67870 | ····71176 |
| 180 | ····47025 | ····26669 | 157110220 | ····97699 | ····89125 | ····84518 | ····83897 | ····87283 |
| 190 | ····62580 | ····42301 | ····25931 | 157913489 | 158704994 | 159500466 | ····99925 | 161103392 |
| 200 | ····78136 | ····57936 | ····41644 | ····29280 | ····20864 | ····16416 | 160315955 | ····19502 |
| 210 | ····93694 | ····73571 | ····57358 | ····45073 | ····36736 | ····32368 | ····31987 | ····35614 |
| 220 | 155609253 | ····89209 | ····73074 | ····60867 | ····52610 | ····48321 | ····48020 | ····51728 |
| 230 | ····24814 | 156404848 | ····88791 | ····76663 | ····68485 | ····64276 | ····64055 | ····67843 |
| 240 | ····40377 | ····20488 | 157204510 | ····92461 | ····84362 | ····80232 | ····80091 | ····83960 |
| 250 | ····55941 | ····36130 | ····20230 | 158008260 | 158800240 | ····96190 | ····96129 | 161200078 |
| 260 | ····71506 | ····51774 | ····35952 | ····24061 | ····16120 | 159612150 | 160412169 | ····16198 |
| 270 | ····87073 | ····67419 | ····51676 | ····39864 | ····32002 | ····28111 | ····28210 | ····32320 |
| 280 | 155702642 | ····83066 | ····67401 | ····55668 | ····47885 | ····44074 | ····44253 | ····48443 |
| 290 | ····18212 | ····98714 | ····83128 | ····71473 | ····63770 | ····60038 | ····60297 | ····64568 |
| 300 | ····33784 | 156514364 | ····98856 | ····87280 | ····79656 | ····76004 | ····76343 | ····80694 |
| 310 | ····49358 | ····30015 | 157314583 | 158103089 | ····95544 | ····91972 | ····92391 | ····96822 |
| 320 | ····64933 | ····45668 | ····30318 | ····18899 | 158911434 | 159707941 | 160508440 | 161312952 |
| 330 | ····80509 | ····61323 | ····46051 | ····34711 | ····27325 | ····23912 | ····24491 | ····29083 |
| 340 | ····96087 | ····76979 | ····61785 | ····50525 | ····43218 | ····39884 | ····40543 | ····45216 |
| 350 | 155811667 | ····92637 | ····77521 | ····66340 | ····59112 | ····55858 | ····56597 | ····61350 |
| 360 | ····27248 | 156608296 | ····93259 | ····82156 | ····75008 | ····71834 | ····72653 | ····77487 |
| 370 | ····42831 | ····23967 | 157408998 | ····97975 | ····90906 | ····87811 | ····88710 | ····93624 |
| 380 | ····58415 | ····39619 | ····24739 | 158213794 | 159006805 | 159803790 | 160604769 | 161409764 |
| 390 | ····74001 | ····55283 | ····40482 | ····29[illegible]16 | ····22705 | ····19770 | ····20830 | ····25905 |
| 400 | ····89588 | ····70949 | ····56226 | ····45439 | ····38608 | ····35752 | ····36892 | ····42047 |
| 410 | 155905177 | ····86616 | ····71971 | ····61263 | ····54512 | ····51736 | ····52956 | ····58191 |
| 420 | ····20768 | 156702285 | ····87719 | ····77089 | ····70417 | ····67721 | ····69021 | ····74337 |
| 430 | ····36360 | ····17955 | 157503467 | ····29217 | ····86324 | ····83708 | ····85088 | ····90485 |
| 440 | ····51953 | ····33627 | ····19218 | 158308746 | 159102233 | ····99696 | 160701156 | 161506634 |
| 450 | ····67548 | ····49300 | ····34969 | ····24577 | ····18143 | 159915686 | ····17226 | ····22784 |
| 460 | ····83145 | ····64975 | ····50723 | ····40410 | ····34055 | ····31677 | ····33298 | ····38937 |
| 470 | ····98744 | ····80651 | ····66478 | ····56244 | ····49968 | ····47671 | ····49371 | ····55091 |
| 480 | 156014343 | ····96329 | ····82235 | ····72079 | ····65883 | ····63665 | ····65446 | ····71246 |
| 490 | ····29945 | 156812009 | ····97993 | ····87917 | ····81800 | ····79662 | ····81523 | ····87403 |
| 500 | ····45548 | ····27690 | 157613753 | 158403755 | ····97718 | ····95660 | ····97601 | 161603562 |

Figure C.12

| | 48000 | 48500 | 49000 | 49500 | 50000 | 50500 | 51000 | 51500 |
|---|---|---|---|---|---|---|---|---|
| 0 | 161603562 | 162413563 | 163227623 | 164045764 | 164868006 | 165694369 | 166524873 | 167359541 |
| 10 | ....19722 | ....29804 | ....43946 | ....62169 | ....84493 | 165710938 | ....41526 | ....76277 |
| 20 | ....35884 | ....46047 | ....6027[illegible] | ....7[illegible] | 164900981 | ....27509 | ....58180 | ....93015 |
| 30 | ....52045 | ....6229[illegible] | ....[illegible] | ....94[illegible] | ....17471 | ....44082 | ....74836 | 167409754 |
| 40 | ....63213 | ....78538 | ....92924 | 164111392 | ....33963 | ....60656 | ....91493 | ....26495 |
| 50 | ....84380 | ....94786 | 163309253 | ....27803 | ....50456 | ....77232 | 166608152 | ....43237 |
| 60 | 161700543 | 162511035 | ....25584 | ....44216 | ....66951 | ....93810 | ....24813 | ....59982 |
| 70 | ....16718 | ....27286 | ....41917 | ....60631 | ....83448 | 163810389 | ....41476 | ....76728 |
| 80 | ....32890 | ....43539 | ....58251 | ....77047 | ....99946 | ....26970 | ....58140 | ....93475 |
| 90 | ....49063 | ....59793 | ....74587 | ....93464 | 165016446 | ....43553 | ....74806 | 167510225 |
| 100 | ....65238 | ....76049 | ....90924 | 164209884 | ....32948 | ....60137 | ....91473 | ....26976 |
| 110 | ....81415 | ....92307 | 163407263 | ....26305 | ....49451 | ....76723 | 166708142 | ....43729 |
| 120 | ....97593 | 162608566 | ....23604 | ....42727 | ....65956 | ....93311 | ....24813 | ....60483 |
| 130 | 161813773 | ....24827 | ....39947 | ....59152 | ....82463 | 165909900 | ....41486 | ....77239 |
| 140 | ....29954 | ....41089 | ....56291 | ....75577 | ....98971 | ....26491 | ....58160 | ....93997 |
| 150 | ....46137 | ....57354 | ....72636 | ....92005 | 165115481 | ....43084 | ....74836 | 167610756 |
| 160 | ....62322 | ....73619 | ....88983 | 164308434 | ....31992 | ....59678 | ....91513 | ....27517 |
| 170 | ....73508 | ....89887 | 163505332 | ....24365 | ....48505 | ....76274 | 166808192 | ....44280 |
| 180 | ....94696 | 162706156 | ....21683 | ....41298 | ....65020 | ....92872 | ....24873 | ....61044 |
| 190 | 161910885 | ....22426 | ....38035 | ....57732 | ....81537 | 166009471 | ....41556 | ....77810 |
| 200 | ....27076 | ....38698 | ....54389 | ....74167 | ....98055 | ....26072 | ....58240 | ....94578 |
| 210 | ....43269 | ....54972 | ....70744 | ....90605 | 165214575 | ....42675 | ....74926 | 167711348 |
| 220 | ....59463 | ....71243 | ....87101 | 164407044 | ....31036 | ....59279 | ....91613 | ....28119 |
| 230 | ....75659 | ....87525 | 163603460 | ....23485 | ....47619 | ....75885 | 166908302 | ....44892 |
| 240 | ....91857 | 162803804 | ....19820 | ....39927 | ....64144 | ....92493 | ....24993 | ....61666 |
| 250 | 162008056 | ....20084 | ....36182 | ....56371 | ....80671 | 166109102 | ....41686 | ....78442 |
| 260 | ....24257 | ....36366 | ....52546 | ....72817 | ....97199 | ....25713 | ....58380 | ....95220 |
| 270 | ....40459 | ....52650 | ....68911 | ....89261 | 165313728 | ....42325 | ....75076 | 167812000 |
| 280 | ....56663 | ....63935 | ....85278 | 164505713 | ....30260 | ....58940 | ....91773 | ....28781 |
| 290 | ....72869 | ....85222 | 163701647 | ....22163 | ....46793 | ....75555 | 167008472 | ....45564 |
| 300 | ....89076 | 162901510 | ....18017 | ....38616 | ....63328 | ....92173 | ....25173 | ....62348 |
| 310 | 162105285 | ....17801 | ....34389 | ....55070 | ....79864 | 166208792 | ....41876 | ....79134 |
| 320 | ....21496 | ....34092 | ....50762 | ....71525 | ....96402 | ....25413 | ....58580 | ....95922 |
| 330 | ....37708 | ....50386 | ....67137 | ....87982 | 165412941 | ....42036 | ....75286 | 167912712 |
| 340 | ....53922 | ....66681 | ....83514 | 164604441 | ....29483 | ....58660 | ....91993 | ....29503 |
| 350 | ....70137 | ....82978 | ....99892 | ....20901 | ....46026 | ....75286 | 167108702 | ....46296 |
| 360 | ....86354 | ....99276 | 163816272 | ....37364 | ....62570 | ....91913 | ....25413 | ....63091 |
| 370 | 162202573 | 163015576 | ....32654 | ....53827 | ....79117 | 166308542 | ....42126 | ....79887 |
| 380 | ....18793 | ....31877 | ....49037 | ....70293 | ....95664 | ....25173 | ....5[illegible]840 | ....96685 |
| 390 | ....35015 | ....48181 | ....65422 | ....86760 | 165512214 | ....41806 | ....75556 | 168013485 |
| 400 | ....51233 | ....64485 | ....81808 | 164703228 | ....28765 | ....58440 | ....92273 | ....30286 |
| 410 | ....67463 | ....80792 | ....98197 | ....19699 | ....45318 | ....75076 | 167208993 | ....47089 |
| 420 | ....83690 | ....97100 | 163914586 | ....36171 | ....61873 | ....91713 | ....25714 | ....63894 |
| 430 | ....99919 | ....13410 | ....30978 | ....52644 | ....78429 | 166408353 | ....42436 | ....80700 |
| 440 | 162316149 | ....29721 | ....47371 | ....69120 | ....94987 | ....24993 | ....59160 | ....97508 |
| 450 | ....32380 | ....46034 | ....63766 | ....85596 | 165611546 | ....41636 | ....75886 | 168114318 |
| 460 | ....48613 | ....62343 | ....80162 | 164802075 | ....28107 | ....58280 | ....92614 | ....31130 |
| 470 | ....64848 | ....78665 | ....96560 | ....18555 | ....44670 | ....74926 | 167309343 | ....47943 |
| 480 | ....81085 | ....94983 | 164012960 | ....35037 | ....61235 | ....91573 | ....26074 | ....64757 |
| 490 | ....97323 | 163211302 | ....29361 | ....51521 | ....77801 | 166508223 | ....42807 | ....81574 |
| 500 | 162413563 | ....27623 | ....45764 | ....68006 | ....94369 | ....24873 | ....59541 | ....98392 |

**Figure C.13**

| | 52000 | 52500 | 53000 | 53500 | 54000 | 4500 | 55000 | 55500 |
|---|---|---|---|---|---|---|---|---|
| 0 | 168198392 | 169041448 | 169888729 | 170740257 | 171596053 | 172456139 | 173320530 | 174189265 |
| 10 | 168215212 | ····58352 | 169905718 | ····57331 | 171613213 | ····73385 | ····37868 | 174206684 |
| 20 | ····32033 | ····75258 | ····22709 | ····74407 | ····30374 | ····90632 | ····55202 | ····24105 |
| 30 | ····48857 | ····92165 | ····39701 | ····91484 | ····47537 | 172507881 | ····72537 | ····41527 |
| 40 | ····65682 | 169109074 | ····56695 | 170808564 | ····64702 | ····25132 | ····89875 | ····58951 |
| 50 | ····82508 | ····25985 | ····73691 | ····25644 | ····81869 | ····42384 | 173407213 | ····76377 |
| 60 | ····99336 | ····42896 | ····90688 | ····42727 | ····99037 | ····59639 | ····24554 | ····93805 |
| 70 | 168316166 | ····59812 | 170007688 | ····59811 | 171716207 | ····76895 | ····41897 | 174311234 |
| 80 | ····32998 | ····76728 | ····24688 | ····76897 | ····33378 | ····94152 | ····59241 | ····28665 |
| 90 | ····49831 | ····93646 | ····41691 | ····93985 | ····50552 | 172611412 | ····70567 | ····46098 |
| 100 | ····66666 | 169210565 | ····58694 | 170911074 | ····67727 | ····28673 | ····93934 | ····63533 |
| 110 | ····83503 | ····27486 | ····75700 | ····28166 | ····84904 | ····45936 | 173511284 | ····80969 |
| 120 | 168400341 | ····44405 | ····92708 | ····45255 | 171802082 | ····63200 | ····28635 | ····98407 |
| 130 | ····17181 | ····61334 | 170109717 | ····62353 | ····19262 | ····80467 | ····45988 | 174415847 |
| 140 | ····34023 | ····78260 | ····26725 | ····79449 | ····36444 | ····97735 | ····63342 | ····33289 |
| 150 | ····50966 | ····95188 | ····43741 | ····96547 | ····53628 | 172715004 | ····80699 | ····50732 |
| 160 | ····67712 | 164312117 | ····60755 | 171013647 | ····70813 | ····32276 | ····98057 | ····68177 |
| 170 | ····84558 | ····29048 | ····77771 | ····30748 | ····88000 | ····49549 | 173615417 | ····85624 |
| 180 | 168501407 | ····45981 | ····94789 | ····47851 | 171905189 | ····66824 | ····32778 | 174503072 |
| 190 | ····18257 | ····62916 | 170211809 | ····64956 | ····22380 | ····84101 | ····50141 | ····20523 |
| 200 | ····35109 | ····79852 | ····28830 | ····82062 | ····39572 | 172801379 | ····67506 | ····37975 |
| 210 | ····51962 | ····96790 | ····45853 | ····99171 | ····56766 | ····18659 | ····84873 | ····55429 |
| 220 | ····68817 | 169413730 | ····62877 | 171116280 | ····73561 | ····35941 | 173702242 | ····72884 |
| 230 | ····85674 | ····30671 | ····79904 | ····33392 | ····91159 | ····53225 | ····19612 | ····90341 |
| 240 | 168602533 | ····47614 | ····96932 | ····50505 | 172008358 | ····70510 | ····36984 | 174607800 |
| 250 | ····19393 | ····64559 | 170313961 | ····67630 | ····25559 | ····87797 | ····54358 | ····25261 |
| 260 | ····36255 | ····81506 | ····30992 | ····84733 | ····42761 | 172905086 | ····71733 | ····42724 |
| 270 | ····53119 | ····98454 | ····48025 | 171201856 | ····59966 | ····22377 | ····89110 | ····60188 |
| 280 | ····69984 | ····15404 | ····65060 | ····18976 | ····77172 | ····39669 | 173806489 | ····77654 |
| 290 | ····86851 | ····32355 | ····82097 | ····36098 | ····94379 | ····56963 | ····72870 | ····95122 |
| 300 | 168703720 | ····49308 | ····99135 | ····53221 | 172111589 | ····74256 | ····41252 | 174712591 |
| 310 | ····20590 | ····66263 | 170416175 | ····70347 | ····28800 | ····91556 | ····58636 | ····30063 |
| 320 | ····37462 | ····83220 | ····33216 | ····87474 | ····46013 | 173008855 | ····76022 | ····47536 |
| 330 | ····54336 | 169600172 | ····50260 | 171304602 | ····63227 | ····26156 | ····93410 | ····65010 |
| 340 | ····71211 | ····17138 | ····67305 | ····21733 | ····80444 | ····43459 | 173910799 | ····82487 |
| 350 | ····88088 | ····34100 | ····84351 | ····38865 | ····97662 | ····60763 | ····28190 | ····99965 |
| 360 | 168804967 | ····51063 | 170501400 | ····55999 | 172214882 | ····78069 | ····45583 | 174817445 |
| 370 | ····21848 | ····68028 | ····18450 | ····73135 | ····32103 | ····95377 | ····62977 | ····34927 |
| 380 | ····38730 | ····84995 | ····35502 | ····90272 | ····49326 | 173112686 | ····80374 | ····52410 |
| 390 | ····55614 | 169701963 | ····52555 | 171407411 | ····66551 | ····29998 | ····97772 | ····69895 |
| 400 | ····72499 | ····18934 | ····69611 | ····24552 | ····83778 | ····47311 | 174015172 | ····87382 |
| 410 | ····89287 | ····35906 | ····86668 | ····41694 | 172301006 | ····64625 | ····32573 | 174904871 |
| 420 | 168906275 | ····52879 | 170603727 | ····58838 | ····18236 | ····81942 | ····49976 | ····22362 |
| 430 | ····23166 | ····69854 | ····20787 | ····75984 | ····35469 | ····99760 | ····67381 | ····39854 |
| 440 | ····40058 | ····86831 | ····37849 | ····93132 | ····52702 | 173216550 | ····84788 | ····57348 |
| 450 | ····56952 | 169803810 | ····54913 | 171510281 | 172369937 | ····33902 | 174102197 | ····74844 |
| 460 | ····73848 | ····20791 | ····71978 | ····27432 | ····87174 | ····51225 | ····19607 | ····92341 |
| 470 | ····90745 | ····37772 | ····89046 | ····44585 | 172404413 | ····68550 | ····37019 | 175009840 |
| 480 | 169007644 | ····54757 | 170706114 | ····61739 | ····21653 | ····85877 | ····54432 | ····27341 |
| 490 | ····24545 | ····71647 | ····23185 | ····78896 | ····38895 | 173303206 | ····7184 | ····44844 |
| 500 | ····41448 | ····88729 | ····40257 | ····96053 | ····56139 | ····20536 | ····89265 | ····62349 |

**Figure C.14**

| | 56000 | 56500 | 57000 | 57500 | 58000 | 58500 | 59000 | 59500 |
|---|---|---|---|---|---|---|---|---|
| 0 | 175062340 | 175939808 | 176[illegible]21666 | 177707942 | 178598664 | 179493842 | 180393520 | 181297701 |
| 10 | ....79855 | ....57401 | ....39343 | ....25715 | 178616524 | 179511798 | 180411559 | 181315831 |
| 20 | ....97363 | ....74998 | ....57033 | ....43487 | ....34386 | ....29749 | ....29601 | ....33962 |
| 30 | 175114872 | ....92596 | ....74713 | ....61252 | ....52249 | ....47702 | ....47644 | ....52096 |
| 40 | ....32334 | 176010195 | ....92405 | ....79033 | ....70114 | ....65657 | ....65688 | ....70231 |
| 50 | ....49839 | ....27796 | 176910095 | ....95816 | ....87981 | ....83614 | ....83735 | ....88368 |
| 60 | ....67412 | ....45399 | ....27786 | 177814595 | 178705850 | 179601572 | 180501783 | 181406507 |
| 70 | ....84929 | ....63093 | ....45478 | ....32377 | ....23721 | ....19532 | ....19834 | ....24648 |
| 80 | 175202447 | ....80679 | ....63173 | ....50160 | ....41593 | ....37494 | ....37886 | ....42790 |
| 90 | ....19958 | ....98217 | ....80869 | ....67945 | ....59467 | ....55458 | ....55939 | ....60935 |
| 100 | ....37490 | 176115827 | ....98567 | ....85732 | ....77343 | ....73424 | ....73995 | ....79081 |
| 110 | ....55013 | ....33339 | 177018267 | 177903520 | ....95221 | ....91391 | ....92052 | ....97229 |
| 120 | ....72533 | ....51052 | ....33969 | ....21311 | 178813100 | 179709360 | 180610112 | 181515378 |
| 130 | ....90066 | ....68667 | ....51672 | ....39103 | ....30982 | ....27331 | ....28173 | ....33530 |
| 140 | 175307595 | ....86284 | ....69377 | ....56391 | ....43965 | ....45303 | ....46235 | ....51683 |
| 150 | ....25126 | 176203973 | ....83084 | ....74693 | ....66750 | ....63278 | ....64300 | ....69838 |
| 160 | ....42659 | ....21523 | 177104793 | ....92490 | ....84636 | ....81254 | ....82366 | ....87995 |
| 170 | ....60193 | ....39145 | ....22503 | 178010289 | 178902525 | ....99232 | 180700435 | 181606154 |
| 180 | ....77729 | ....56769 | ....40216 | ....28090 | ....20415 | 179817212 | ....18505 | ....24315 |
| 190 | ....95257 | ....74395 | ....57930 | ....45893 | ....38307 | ....35194 | ....36577 | ....42477 |
| 200 | 175412806 | ....92022 | ....75646 | ....63698 | ....26201 | ....53178 | ....54650 | ....60641 |
| 210 | ....30347 | 176309652 | ....93363 | ....81504 | ....74097 | ....71163 | ....72726 | ....78807 |
| 220 | ....47890 | ....27283 | 177211082 | ....99312 | ....91994 | ....89150 | ....90803 | ....96975 |
| 230 | ....65435 | ....44915 | ....28804 | 178117122 | 179009893 | 179907139 | 180808882 | 181715145 |
| 240 | ....82982 | ....62559 | ....46526 | ....34934 | ....27794 | ....25129 | ....26963 | ....33316 |
| 250 | 175500530 | ....80185 | ....64251 | ....52747 | ....45697 | ....43122 | ....45046 | ....51490 |
| 260 | ....18030 | ....97824 | ....81973 | ....70563 | ....63602 | ....61116 | ....63130 | ....69665 |
| 270 | ....35632 | 176315454 | ....99706 | ....88380 | ....81508 | ....79112 | ....81216 | ....87842 |
| 280 | ....53186 | ....33105 | 177317436 | 178206193 | ....99416 | ....97110 | ....99305 | 181806021 |
| 290 | ....70741 | ....50749 | ....35167 | ....24019 | 179117326 | 180015110 | 180917394 | ....24201 |
| 300 | ....88298 | ....63394 | ....52901 | ....41841 | ....35238 | ....33112 | ....35486 | ....42384 |
| 310 | 175605857 | ....85041 | ....70636 | ....59665 | ....53151 | ....51115 | ....53580 | ....60568 |
| 320 | ....23417 | 176503689 | ....88373 | ....77492 | ....71067 | ....69120 | ....71675 | ....78754 |
| 330 | ....40950 | ....21340 | 177406112 | ....95319 | ....88984 | ....87127 | ....89772 | ....96942 |
| 340 | ....58544 | ....38992 | ....23853 | 178313149 | 179206903 | 180105136 | 181007871 | 181915132 |
| 350 | ....76110 | ....56646 | ....41595 | ....30980 | ....24823 | ....23146 | ....25972 | ....33323 |
| 360 | ....93677 | ....74301 | ....59339 | ....43813 | ....42746 | ....41159 | ....44075 | ....51516 |
| 370 | 175711247 | ....91959 | ....77035 | ....66648 | ....60670 | ....59173 | ....62179 | ....69712 |
| 380 | ....28818 | 176609618 | ....94833 | ....84485 | ....78596 | ....77189 | ....80285 | ....87909 |
| 390 | ....46391 | ....27279 | 177512582 | 178402323 | ....96524 | ....95206 | ....98393 | 182006107 |
| 400 | ....63965 | ....44942 | ....30334 | ....20164 | 179414454 | 180213226 | 181116503 | ....24308 |
| 410 | ....81542 | ....62606 | ....48087 | ....38006 | ....32385 | ....31247 | ....34615 | ....42510 |
| 420 | ....99120 | ....80272 | ....65842 | ....55849 | ....50318 | ....49270 | ....52728 | ....60715 |
| 430 | 175316700 | ....97941 | ....83593 | ....73695 | ....68253 | ....67295 | ....70344 | ....78921 |
| 440 | ....34281 | 176715610 | 177501357 | ....91542 | ....86190 | ....85322 | ....88961 | ....97129 |
| 450 | ....51865 | ....33232 | ....19117 | 178509392 | 179404129 | 180303351 | 181207079 | 182115338 |
| 460 | ....69450 | ....50955 | ....36878 | ....27242 | ....22069 | ....21381 | ....25200 | ....33550 |
| 470 | ....87037 | ....68630 | ....54642 | ....45095 | ....40011 | ....39413 | ....43323 | ....51763 |
| 480 | 175904526 | ....86307 | ....72408 | ....62950 | ....57955 | ....57447 | ....61447 | ....69978 |
| 490 | ....22216 | 1768039[illegible]5 | ....90175 | ....80306 | ....75901 | ....75483 | ....79573 | ....88195 |
| 500 | ....39808 | ....21666 | 177707944 | ....98564 | ....93849 | ....93520 | ....97701 | 182206414 |

Figure C.15

|  | 60000 | 60500 | 61000 | 61500 | 62000 | 62500 | 63000 | 63500 |
|---|---|---|---|---|---|---|---|---|
| 0 | 182206414 | 183119682 | 184037257 | 184959973 | 185887042 | 186818758 | 187755144 | 188696223 |
| 10 | ····24635 | ····37994 | ····55931 | ····78469 | 185905631 | ····37440 | ····73919 | 188715093 |
| 20 | ····42857 | ····56308 | ····74337 | ····96967 | ····24221 | ····56124 | ····92697 | ····33964 |
| 30 | ····61082 | ····74623 | ····92744 | 185015466 | ····42814 | ····74809 | 187811476 | ····52838 |
| 40 | ····79308 | ····92941 | 184111153 | ····33968 | ····61408 | ····93497 | ····30257 | ····71713 |
| 50 | ····97536 | 183211260 | ····29564 | ····52471 | ····80004 | 186912186 | ····49040 | ····90590 |
| 60 | 182315765 | ····29581 | ····47977 | ····70976 | ····98602 | ····30877 | ····67825 | 188809469 |
| 70 | ····33997 | ····47904 | ····66392 | ····89483 | 186017202 | ····49570 | ····86612 | ····28350 |
| 80 | ····52230 | ····66229 | ····84809 | 185107992 | ····35804 | ····68265 | 187905401 | ····47233 |
| 90 | ····70466 | ····84555 | 184203227 | ····26503 | ····54407 | ····86962 | ····24191 | ····61618 |
| 100 | ····88703 | 183302884 | ····21647 | ····45016 | ····73013 | 187005661 | ····42984 | ····85005 |
| 110 | 182406941 | ····21214 | ····40070 | ····63530 | ····91620 | ····24361 | ····61778 | 188903893 |
| 120 | ····25182 | ····39546 | ····58494 | ····82047 | 186110229 | ····43064 | ····80574 | ····22783 |
| 130 | ····43435 | ····57880 | ····76919 | 185200565 | ····28840 | ····61768 | ····99372 | ····41676 |
| 140 | ····61669 | ····76216 | ····95347 | ····19085 | ····47453 | ····80474 | 188018172 | ····60570 |
| 150 | ····79915 | ····94554 | 184313777 | ····37607 | ····66068 | ····99182 | ····36974 | ····79466 |
| 160 | ····98163 | 183412893 | ····32208 | ····56131 | ····84684 | 187117892 | ····55778 | ····98364 |
| 170 | 182516413 | ····31235 | ····50641 | ····74656 | 186203303 | ····36604 | ····74583 | 189017264 |
| 180 | ····34665 | ····49578 | ····69076 | ····93184 | ····21923 | ····55318 | ····93391 | ····36165 |
| 190 | ····52918 | ····67923 | ····87513 | 185311713 | ····40545 | ····74033 | 188112200 | ····55069 |
| 200 | ····71173 | ····86269 | 184405952 | ····30245 | ····59170 | ····92751 | ····31011 | ····73975 |
| 210 | ····89431 | 183504618 | ····24393 | ····48777 | ····77795 | 187211470 | ····49824 | ····92882 |
| 220 | 182607689 | ····22969 | ····42835 | ····67312 | ····96423 | ····30191 | ····68639 | 189111791 |
| 230 | ····25950 | ····41321 | ····61279 | ····85849 | 186315053 | ····48914 | ····87456 | ····30702 |
| 240 | ····44213 | ····59675 | ····79725 | 185404388 | ····33684 | ····67639 | 188206275 | ····49616 |
| 250 | ····62477 | ····78031 | ····98173 | ····22928 | ····52318 | ····86366 | ····25096 | ····68530 |
| 260 | ····80743 | ····96389 | 184516623 | ····41470 | ····70953 | 187305094 | ····43918 | ····87447 |
| 270 | ····99012 | 183614748 | ····35075 | ····60014 | ····89590 | ····23825 | ····62742 | 189206366 |
| 280 | 182717281 | ····33110 | ····53528 | ····78560 | 186408229 | ····42557 | ····81569 | ····25287 |
| 290 | ····35553 | ····51473 | ····71984 | ····97108 | ····26870 | ····61292 | 188300397 | ····44209 |
| 300 | ····53827 | ····69838 | ····90441 | 185515658 | ····45513 | ····80028 | ····19227 | ····63134 |
| 310 | ····72102 | ····88205 | 184608900 | ····34210 | ····64157 | ····98766 | ····38059 | ····82060 |
| 320 | ····90379 | 183706574 | ····27361 | ····52763 | ····82804 | 187417506 | ····56893 | 189300988 |
| 330 | 182808658 | ····24945 | ····45824 | ····71318 | 186501452 | ····36247 | ····75728 | ····19918 |
| 340 | ····26939 | ····43317 | ····64288 | ····89875 | ····20102 | ····54991 | ····94566 | ····38850 |
| 350 | ····45222 | ····61691 | ····82755 | 185608434 | ····38754 | ····73736 | 188413405 | ····57784 |
| 360 | ····63506 | ····80068 | 184701223 | ····26995 | ····57408 | ····92484 | ····32247 | ····76720 |
| 370 | ····81793 | ····98446 | ····19693 | ····45558 | ····76064 | 187511233 | ····51090 | ····95658 |
| 380 | 182900081 | 183816825 | ····38165 | ····64123 | ····94721 | ····29984 | ····69935 | 189414597 |
| 390 | ····18371 | ····35207 | ····56639 | ····82689 | 186613381 | ····48737 | ····88782 | ····33539 |
| 400 | ····36663 | ····53591 | ····75115 | 185701257 | ····32042 | ····67492 | 188507631 | ····52482 |
| 410 | ····54957 | ····71976 | ····93592 | ····19827 | ····50705 | ····86249 | ····26482 | ····71427 |
| 420 | ····73252 | ····90363 | 184812071 | ····38399 | ····69370 | 187605007 | ····45334 | ····90374 |
| 430 | ····91549 | 183908752 | ····30553 | ····56973 | ····88037 | ····23768 | ····64189 | 189509323 |
| 440 | 183002849 | ····27143 | ····49036 | ····75549 | 186706706 | ····42530 | ····83045 | ····28274 |
| 450 | ····28149 | ····45536 | ····67521 | ····94126 | ····25377 | ····61295 | 188601904 | ····47227 |
| 460 | ····46452 | ····63931 | ····86007 | 185812706 | ····44049 | ····80061 | ····20764 | ····66182 |
| 470 | ····64757 | ····82327 | 184904496 | ····31287 | ····62724 | ····98829 | ····39626 | ····85138 |
| 480 | ····83063 | 184000725 | ····22986 | ····49870 | ····81400 | 187717599 | ····58490 | 189604097 |
| 490 | 183101372 | ····19125 | ····41479 | ····68455 | 186800078 | ····36375 | ····77356 | ····23057 |
| 500 | ····19682 | ····37527 | ····59973 | ····87042 | ····18758 | ····55144 | ····96223 | ····42019 |

C

Figure C.16

| | 64000 | 64500 | 65000 | 65500 | 66000 | 66500 | 67000 | 67500 |
|---|---|---|---|---|---|---|---|---|
| 0 | 189642019 | 190592557 | 191547858 | 192507947 | 193472849 | 194432587 | 145417106 | 196396670 |
| 10 | ····60984 | 190611616 | ····67013 | ····27198 | ····92197 | ····62032 | ····36722 | 196416305 |
| 20 | ····79950 | ····30677 | ····86169 | ····46451 | 193511546 | ····81478 | ····56271 | ····35951 |
| 30 | ····98918 | ····49740 | 191605328 | ····65706 | ····30897 | 194500926 | ····75817 | ····55594 |
| 40 | 189717888 | ····68805 | ····24489 | ····84962 | ····50250 | ····20376 | ····95365 | ····75240 |
| 50 | ····36860 | ····87872 | ····43651 | 192604221 | ····69605 | ····39828 | 195514914 | ····94888 |
| 60 | ····55833 | 190706941 | ····62815 | ····23481 | ····88962 | ····59282 | ····34466 | 196514537 |
| 70 | ····74809 | ····26012 | ····81982 | ····42743 | 193608321 | ····78738 | ····54019 | ····34189 |
| 80 | ····93787 | ····45084 | 191701150 | ····62008 | ····27682 | ····98196 | ····73574 | ····53842 |
| 90 | 189812766 | ····64159 | ····20320 | ····81274 | ····47045 | 194617656 | ····93132 | ····73497 |
| 100 | ····31747 | ····83235 | ····39492 | 192700542 | ····66409 | ····37117 | 195612691 | ····93155 |
| 110 | ····50730 | 190802313 | ····58666 | ····19812 | ····85776 | ····56581 | ····32252 | 196912814 |
| 120 | ····69716 | ····21394 | ····77842 | ····39084 | 193705144 | ····76047 | ····51816 | ····32475 |
| 130 | ····88703 | ····40476 | ····97020 | ····58358 | ····24515 | ····95514 | ····71381 | ····52138 |
| 140 | 189907691 | ····59560 | 191816199 | ····77634 | ····43887 | 194714984 | ····90948 | ····71804 |
| 150 | ····26682 | ····78646 | ····35381 | ····96912 | ····63262 | ····34456 | 195710517 | ····91471 |
| 160 | ····45675 | ····97734 | ····54564 | 192816191 | ····82638 | ····53929 | ····30088 | 196711140 |
| 170 | ····64669 | 190916824 | ····73750 | ····35473 | 193802016 | ····73404 | ····49661 | ····30811 |
| 180 | ····83666 | ····34915 | ····92937 | ····54756 | ····21397 | ····92882 | ····69236 | ····50484 |
| 190 | 190002664 | ····55009 | 191912127 | ····74042 | ····40779 | 194812361 | ····88813 | ····70159 |
| 200 | ····21664 | ····74104 | ····31318 | ····93329 | ····60163 | ····31842 | 195808392 | ····89836 |
| 210 | ····40667 | ····93202 | ····50511 | 192912619 | ····79549 | ····51325 | ····27973 | 196809515 |
| 220 | ····59671 | 191012301 | ····69706 | ····31910 | ····98937 | ····70810 | ····47555 | ····29196 |
| 230 | ····78677 | ····31402 | ····88903 | ····51203 | 193918327 | ····90298 | ····67140 | ····48879 |
| 240 | ····97685 | ····50505 | 192008102 | ····70498 | ····37718 | 194909787 | ····86727 | ····56864 |
| 250 | 190116694 | ····69610 | ····27303 | ····89795 | ····57112 | ····29278 | 195906316 | ····88251 |
| 260 | ····35706 | ····88717 | ····46505 | 193009094 | ····76508 | ····48770 | ····25906 | 196907940 |
| 270 | ····54720 | 191107826 | ····65710 | ····28395 | ····95906 | ····68265 | ····45499 | ····27631 |
| 280 | ····73735 | ····26937 | ····84917 | ····47698 | 194015305 | ····87762 | ····65093 | ····47323 |
| 290 | ····92753 | ····46049 | 192104125 | ····67003 | ····34707 | 195007261 | ····84690 | ····67018 |
| 300 | 190211772 | ····65164 | ····23336 | ····86309 | ····54110 | ····26762 | 196004288 | ····86715 |
| 310 | ····30793 | ····84281 | ····42548 | 193105618 | ····73516 | ····46264 | ····23889 | 197006413 |
| 320 | ····49816 | 191203399 | ····61762 | ····24929 | ····92923 | ····65769 | ····43491 | ····26114 |
| 330 | ····68841 | ····22519 | ····80978 | ····44241 | 194112332 | ····85276 | ····63095 | ····45817 |
| 340 | ····87868 | ····41642 | 192200196 | ····63556 | ····31743 | 195104784 | ····82702 | ····65521 |
| 350 | 190306896 | ····60766 | ····19416 | ····82872 | ····51157 | ····24295 | 196102310 | ····85228 |
| 360 | ····25927 | ····79892 | ····38638 | 193202190 | ····70572 | ····43807 | ····21920 | 197104936 |
| 370 | ····44960 | ····99020 | ····57862 | ····21510 | ····89989 | ····63321 | ····41533 | ····24647 |
| 380 | ····63994 | 191318150 | ····77088 | ····40833 | 194209408 | ····82838 | ····61147 | ····44359 |
| 390 | ····83031 | ····37282 | ····96316 | ····60157 | ····28829 | 195202356 | ····80763 | ····64074 |
| 400 | 190402069 | ····56415 | 192315545 | ····79483 | ····48252 | ····21876 | 196200381 | ····83790 |
| 410 | ····21109 | ····75551 | ····34777 | ····98811 | ····67676 | ····41398 | ····20001 | 197203509 |
| 420 | ····40151 | ····94689 | ····54010 | 193318141 | ····87103 | ····60923 | ····39623 | ····23229 |
| 430 | ····59195 | 191413828 | ····73246 | ····37472 | 194306532 | ····80419 | ····59247 | ····42951 |
| 440 | ····78241 | ····32970 | ····92483 | ····56806 | ····25963 | ····99977 | ····78873 | ····62676 |
| 450 | ····97289 | ····52113 | 192411722 | ····76142 | ····45395 | 195319507 | ····98501 | ····82402 |
| 460 | 190516339 | ····71258 | ····30964 | ····95479 | ····64830 | ····39039 | 196318131 | 197302130 |
| 470 | ····35390 | ····90405 | ····50207 | 193414819 | ····84266 | ····58573 | ····37761 | ····21560 |
| 480 | ····54444 | 191509554 | ····69452 | ····34160 | 194403705 | ····78108 | ····57396 | ····41592 |
| 490 | ····73499 | ····28705 | ····88699 | ····53504 | ····23145 | ····97646 | ····77033 | ····61327 |
| 500 | ····92557 | ····47858 | 192507947 | ····72849 | ····42587 | 195417186 | ····96670 | ····81063 |

Figure C.17

| | 68000 | 68500 | 69000 | 69500 | 70000 | 70500 | 71000 | 71500 |
|---|---|---|---|---|---|---|---|---|
| 0 | 197381063 | 198370390 | 1993040[illegible]0 | 200303945 | 201306223 | 202377535 | 203391906 | 204411361 |
| 10 | 197400801 | ....9022 | ....84612 | ....83982 | ....88360 | ....97775 | 203412245 | ....31802 |
| 20 | ....20541 | 198410061 | 199404551 | 200404020 | 201408499 | 202418013 | ....32586 | ....52245 |
| 30 | ....40283 | ....29907 | ....24491 | ....24060 | ....28640 | ....38254 | ....52930 | ....72690 |
| 40 | ....60027 | ....49750 | ....44434 | ....44103 | ....48783 | ....58498 | ....73275 | ....93135 |
| 50 | ....79773 | ....69595 | ....64378 | ....64147 | ....68928 | ....78744 | ....93622 | 204513587 |
| 60 | ....99521 | ....89442 | ....84324 | ....84194 | ....89074 | ....98992 | 203513972 | ....34038 |
| 70 | 197519271 | 198509291 | 199504273 | 200504242 | 201509223 | 202519242 | ....34323 | ....54492 |
| 80 | ....39023 | ....29142 | ....24223 | ....24292 | ....29374 | ....39494 | ....54676 | ....74947 |
| 90 | ....58777 | ....48994 | ....44176 | ....44345 | ....49527 | ....59748 | ....75032 | ....95405 |
| 100 | ....78533 | ....68849 | ....64130 | ....64399 | ....69682 | ....80004 | ....95389 | 204615864 |
| 110 | ....98290 | ....88706 | ....84086 | ....84456 | ....89839 | 202600262 | 203615749 | ....36326 |
| 110 | 197618050 | 198608565 | 199604047 | 200604514 | 201609998 | ....20522 | ....36110 | ....56790 |
| 130 | ....37812 | ....28426 | ....24005 | ....24575 | ....30159 | ....40784 | ....56474 | ....77255 |
| 140 | ....57576 | ....48289 | ....43962 | ....44637 | ....50322 | ....61048 | ....76840 | ....97723 |
| 150 | ....77342 | ....68154 | ....63932 | ....64702 | ....70487 | ....81314 | ....97207 | 204718193 |
| 160 | ....97109 | ....88020 | ....83898 | ....84768 | ....90654 | 202701582 | 203717577 | ....38665 |
| 170 | 197716879 | 198707889 | 199703867 | 200704837 | 201710823 | ....21852 | ....37949 | ....59138 |
| 180 | ....36651 | ....27750 | ....23837 | ....24907 | ....30994 | ....42124 | ....58323 | ....79614 |
| 190 | ....56424 | ....47633 | ....43810 | ....44980 | ....51167 | ....62399 | ....78698 | 204800093 |
| 200 | ....76200 | ....67508 | ....63784 | ....65054 | ....71343 | ....82675 | ....99076 | ....20572 |
| 210 | ....95978 | ....87384 | ....83760 | ....85131 | ....91520 | 202802953 | 203819456 | ....41054 |
| 220 | 197815757 | 198807263 | 199803739 | 200805209 | 201811699 | ....23233 | ....39838 | ....61538 |
| 230 | ....35539 | ....27144 | ....23719 | ....25290 | ....31880 | ....43516 | ....60222 | ....82025 |
| 240 | ....55322 | ....47027 | ....43702 | ....45372 | ....52063 | ....63800 | ....80608 | 204902513 |
| 250 | ....75108 | ....66911 | ....63686 | ....65457 | ....72248 | ....84087 | 203900996 | ....23003 |
| 260 | ....94895 | ....86798 | ....83672 | ....85543 | ....92436 | 202904375 | ....21386 | ....43495 |
| 270 | 197914685 | 198906687 | 199903661 | 200905632 | 201912625 | ....24665 | ....41778 | ....63989 |
| 280 | ....34476 | ....24577 | ....23651 | ....25722 | ....32816 | ....44958 | ....62173 | ....84486 |
| 290 | ....54270 | ....46470 | ....43644 | ....45815 | ....53009 | ....65252 | ....82569 | 205004985 |
| 300 | ....74065 | ....66365 | ....63638 | ....65909 | ....73205 | ....85549 | 204002967 | ....25485 |
| 310 | ....93863 | ....86261 | ....83634 | ....86006 | ....93402 | 203005847 | ....23367 | ....45988 |
| 320 | 198013662 | 199006160 | 200003633 | 201006105 | 202013601 | ....26148 | ....43770 | ....66492 |
| 330 | ....33463 | ....26060 | ....23633 | ....26205 | ....33803 | ....46451 | ....64174 | ....86999 |
| 340 | ....53267 | ....45963 | ....43635 | ....46308 | ....54006 | ....66755 | ....84581 | 205107507 |
| 350 | ....37072 | ....65868 | ....63640 | ....66413 | ....74211 | ....87062 | 204104989 | ....28018 |
| 360 | ....92879 | ....85774 | ....83646 | ....86519 | ....94419 | 203107371 | ....25399 | ....48531 |
| 370 | 198112689 | 199105683 | 200103655 | 201106628 | 202114628 | ....27681 | ....45812 | ....69046 |
| 380 | ....32500 | ....25594 | ....23665 | ....26738 | ....34840 | ....47994 | ....66227 | ....89563 |
| 390 | ....52313 | ....45506 | ....43677 | ....46851 | ....55053 | ....68309 | ....86643 | 205210082 |
| 400 | ....72128 | ....65421 | ....63692 | ....66966 | ....75269 | ....88626 | 204207062 | ....30603 |
| 410 | ....91946 | ....85337 | ....83708 | ....87082 | ....95486 | 203208945 | ....27483 | ....51126 |
| 420 | 198211765 | 199205256 | 200203727 | 201207201 | 202215706 | ....29266 | ....47905 | ....71651 |
| 430 | ....31586 | ....25176 | ....23747 | ....27322 | ....35927 | ....49588 | ....68330 | ....92178 |
| 440 | ....51405 | ....45099 | ....43769 | ....47445 | ....56151 | ....69913 | ....88756 | 205312707 |
| 450 | ....71234 | ....65023 | ....63793 | ....67569 | ....76377 | ....90240 | 204309186 | ....33239 |
| 460 | ....91061 | ....84950 | ....83820 | ....87696 | ....96604 | 203310569 | ....29617 | ....53772 |
| 470 | 198310891 | 199304878 | 200303845 | 201307825 | 202316834 | ....30901 | ....50050 | ....74307 |
| 480 | ....30722 | ....24809 | ....23878 | ....27956 | ....37066 | ....51234 | ....70485 | ....94845 |
| 490 | ....50555 | ....44741 | ....43911 | ....48088 | ....57299 | ....71569 | ....90922 | 205415384 |
| 500 | ....70390 | ....64676 | ....63945 | ....68223 | ....77535 | ....91906 | 204411361 | ....35926 |

C 2

Figure C.18

| | 72000 | 72500 | 73000 | 73500 | 74000 | 74500 | 75000 | 75500 |
|---|---|---|---|---|---|---|---|---|
| 0 | 205435926 | 206465626 | 207500488 | 208540536 | 209585797 | 210636298 | 211692064 | 212753121 |
| 10 | ....56469 | ....85273 | ....21238 | ....61390 | 209606756 | ....57361 | 211713233 | ....74397 |
| 20 | ....77015 | 206506921 | ....41990 | ....82246 | ....27717 | ....78427 | ....34404 | ....95674 |
| 30 | ....97563 | ....27572 | ....62744 | 208603104 | ....48679 | ....99495 | ....55578 | 212216954 |
| 40 | 205518112 | ....48225 | ....83500 | ....23965 | ....69644 | 210720565 | ....76753 | ....38235 |
| 50 | ....38664 | ....68879 | 207604258 | ....4482[illegible] | ....90611 | ....41637 | ....97931 | ....59519 |
| 60 | ....59218 | ....89536 | ....25019 | ....65691 | 209711580 | ....62711 | 211819111 | ....80805 |
| 70 | ....79774 | 206610195 | ....45781 | ....86558 | ....32552 | ....83787 | ....40293 | 212902093 |
| 80 | 205600332 | ....30856 | ....66546 | 208707427 | ....53525 | 210804866 | ....61477 | ....23383 |
| 90 | ....20892 | ....51519 | ....87313 | ....28297 | ....74500 | ....25946 | ....82663 | ....44676 |
| 100 | ....41954 | ....72185 | 207708081 | ....49170 | ....95477 | ....47029 | 211903851 | ....65970 |
| 110 | ....62018 | ....92852 | ....28852 | ....70045 | 209816457 | ....68114 | ....25041 | ....87267 |
| 120 | ....82584 | 206713521 | ....49625 | ....90922 | ....37438 | ....89200 | ....46234 | 213008565 |
| 130 | 205703153 | ....34193 | ....70400 | 208811801 | ....58422 | 210910289 | ....67428 | ....29866 |
| 140 | ....23723 | ....54866 | ....91177 | ....32682 | ....79408 | ....31380 | ....88625 | ....51169 |
| 150 | ....44295 | ....75541 | 207811956 | ....53566 | 209900396 | ....52473 | 212009824 | ....72475 |
| 160 | ....64870 | ....96219 | ....32737 | ....74451 | ....21386 | ....73569 | ....31025 | ....93782 |
| 170 | ....85446 | 206816899 | ....53521 | ....95338 | ....42378 | ....94666 | ....52228 | 213115091 |
| 180 | 205806025 | ....37580 | ....74306 | 208916228 | ....63372 | 211015766 | ....73433 | ....36403 |
| 190 | ....26605 | ....58264 | ....95093 | ....37120 | ....84368 | ....36867 | ....94641 | ....57716 |
| 200 | ....47188 | ....78950 | 207915883 | ....58013 | 210005367 | ....57971 | 212115850 | ....79032 |
| 210 | ....67773 | ....99638 | ....36674 | ....78909 | ....26367 | ....79077 | ....37062 | 213200350 |
| 220 | ....88360 | 206920328 | ....57468 | ....99807 | ....47370 | 211100185 | ....58276 | ....21670 |
| 230 | 205908948 | ....41020 | ....78264 | 209020707 | ....68375 | ....21294 | ....79491 | ....42992 |
| 240 | ....29539 | ....61714 | ....99062 | ....41609 | ....89381 | ....42407 | 212200709 | ....64316 |
| 250 | ....50132 | ....82410 | 208019862 | ....62513 | 210110391 | ....63521 | ....21929 | ....85643 |
| 260 | ....70727 | 207003108 | ....40664 | ....83420 | ....31402 | ....84637 | ....43152 | 213306971 |
| 270 | ....91324 | ....23808 | ....61468 | 209104328 | ....52415 | 211205756 | ....64376 | ....28302 |
| 280 | 206011924 | ....44511 | ....82274 | ....25238 | ....73430 | ....26876 | ....85602 | ....49635 |
| 290 | ....32525 | ....65215 | 208103082 | ....46151 | ....94448 | ....47999 | 212306831 | ....70970 |
| 300 | ....53125 | ....85922 | ....23892 | ....67065 | 210215467 | ....69124 | ....28062 | ....92307 |
| 310 | ....73733 | 207106630 | ....44705 | ....87982 | ....36488 | ....90251 | ....49294 | 213413646 |
| 320 | ....94341 | ....27341 | ....65519 | 209208901 | ....57512 | 211311380 | ....70529 | ....34988 |
| 330 | 206114950 | ....48054 | ....86336 | ....29822 | ....78538 | ....32511 | ....91766 | ....56331 |
| 340 | ....35562 | ....68769 | 208207154 | ....50745 | ....99566 | ....53644 | 212413006 | ....77677 |
| 350 | ....56175 | ....89486 | ....27975 | ....71670 | 210320596 | ....74779 | ....34247 | ....99025 |
| 360 | ....76791 | 207210204 | ....48798 | ....92597 | ....41628 | ....95917 | ....55490 | 213520374 |
| 370 | ....97409 | ....30925 | ....69623 | 209313526 | ....62662 | 211417057 | ....76736 | ....41726 |
| 380 | 206218028 | ....51649 | ....90450 | ....34458 | ....83698 | ....38198 | ....97983 | ....63081 |
| 390 | ....38650 | ....72374 | 208311279 | ....55391 | 210404737 | ....59342 | 212519233 | ....84437 |
| 400 | ....59274 | ....93101 | ....32110 | ....76327 | ....25777 | ....80488 | ....40485 | 213605795 |
| 410 | ....79[illegible]20 | 207313830 | ....52943 | ....97264 | ....46820 | 211501636 | ....61739 | ....27156 |
| 420 | 206300528 | ....34562 | ....73778 | 209418204 | ....67865 | ....22786 | ....82995 | ....48519 |
| 430 | ....21155 | ....55295 | ....94616 | ....89146 | ....88911 | ....43938 | 212604254 | ....69884 |
| 440 | ....41790 | ....76031 | 208415455 | ....60090 | 210509960 | ....65013 | ....25514 | ....91251 |
| 450 | ....62424 | ....96663 | ....36297 | ....81036 | ....31011 | ....[illegible]8624 | ....46777 | 213712620 |
| 460 | ....83060 | 20741740[illegible] | ....57140 | 209501984 | ....52064 | 211607408 | ....68041 | ....33991 |
| 470 | 206[illegible]699 | ....38250 | ....77986 | ....22934 | ....73120 | ....28569 | ....89308 | ....55364 |
| 480 | ....24339 | ....58994 | ....98834 | ....43886 | ....94177 | ....49732 | 212710577 | ....76740 |
| 490 | ....44981 | ....79740 | 208519684 | ....64841 | 210615236 | ....70897 | ....31848 | ....98118 |
| 500 | ....65626 | 207500488 | ....40536 | ....85797 | ....36298 | ....92064 | ....53121 | 213819497 |

**Figure C.19**

| | 76000 | 76500 | 77000 | 77500 | 78000 | 78500 | 79000 | 79500 |
|---|---|---|---|---|---|---|---|---|
| 0 | 213819497 | 214891218 | 215968311 | 217050802 | 218138720 | 219232090 | 220330940 | 221435298 |
| 10 | ....40879 | 214912707 | ....89908 | ....72508 | ....60533 | ....54013 | ....52973 | ....57442 |
| 20 | ....62263 | ....34199 | 216011507 | ....94215 | ....82350 | ....75938 | ....75909 | ....79587 |
| 30 | ....83650 | ....55692 | ....33108 | 217115924 | 218204168 | ....97866 | ....97046 | 221501795 |
| 40 | 213905038 | ....77188 | ....54711 | ....37636 | ....25988 | 219319796 | 220419086 | ....23885 |
| 50 | ....26428 | ....98685 | ....76317 | ....59350 | ....47811 | ....41728 | ....41128 | ....46038 |
| 60 | ....37821 | 215020185 | ....97924 | ....81065 | ....69636 | ....63662 | ....63172 | ....68192 |
| 70 | ....69216 | ....41687 | 216119534 | 217202784 | ....91463 | ....85598 | ....85218 | ....90349 |
| 80 | ....90613 | ....63192 | ....41146 | ....24504 | 218313292 | 219407537 | 220507266 | 221612508 |
| 90 | 214012012 | ....84698 | ....62760 | ....46226 | ....35123 | ....29478 | ....29317 | ....34670 |
| 100 | ....33413 | 215106206 | ....84377 | ....67951 | ....56956 | ....51420 | ....51370 | ....56833 |
| 110 | ....54816 | ....27717 | 216205995 | ....89678 | ....78792 | ....73366 | ....73425 | ....78999 |
| 120 | ....76222 | ....49230 | ....27616 | 217311407 | 218400630 | ....95313 | ....95482 | 221761167 |
| 130 | ....97629 | ....70746 | ....49238 | ....33138 | ....22470 | 219517262 | 220617542 | ....23337 |
| 140 | 214119039 | ....92262 | ....70863 | ....54871 | ....44313 | ....39214 | ....39604 | ....45509 |
| 150 | ....40451 | 215213781 | ....92490 | ....76607 | ....66157 | ....61168 | ....61668 | ....67684 |
| 160 | ....61865 | ....35302 | 216314120 | ....98344 | ....88003 | ....83124 | ....83734 | ....89860 |
| 170 | ....83281 | ....56326 | ....35751 | 217420084 | 218509852 | 219605082 | 220705802 | 221812039 |
| 180 | 214204700 | ....78351 | ....57385 | ....41826 | ....31703 | ....27043 | ....27873 | ....34221 |
| 190 | ....26120 | ....99879 | ....79020 | ....63570 | ....53556 | ....49006 | ....49946 | ....56404 |
| 200 | ....47543 | 215321409 | 216400658 | ....85317 | ....75412 | ....70970 | ....72021 | ....78590 |
| 210 | ....68968 | ....42941 | ....22298 | 217507065 | ....97269 | ....92938 | ....94098 | 221900778 |
| 220 | ....90395 | ....64476 | ....43940 | ....28816 | 218619129 | 219714907 | 220816177 | ....22968 |
| 230 | 214311824 | ....86012 | ....65585 | ....50569 | ....40991 | ....36878 | ....38259 | ....45160 |
| 240 | ....33255 | 215407551 | ....87231 | ....72324 | ....62855 | ....58852 | ....60343 | ....67354 |
| 250 | ....54688 | ....29091 | 216508880 | ....94081 | ....84721 | ....80828 | ....82429 | ....89551 |
| 260 | ....76124 | ....50634 | ....30531 | 217615840 | 218706590 | 219802806 | 220904517 | 222011750 |
| 270 | ....97561 | ....72179 | ....52184 | ....37602 | ....28461 | ....24786 | ....26607 | ....33951 |
| 280 | 214419001 | ....93727 | ....73839 | ....59366 | ....50333 | ....46769 | ....48700 | ....56155 |
| 290 | ....40443 | 215515276 | ....95497 | ....81132 | ....72208 | ....68754 | ....70795 | ....78360 |
| 300 | ....61887 | ....36828 | 216617156 | 217702900 | ....94086 | ....90741 | ....92892 | 222100568 |
| 310 | ....83333 | ....58381 | ....38818 | ....24670 | 218815965 | 219912730 | 220114991 | ....22778 |
| 320 | 214504781 | ....79937 | ....60482 | ....46443 | ....37847 | ....34721 | ....37093 | ....44991 |
| 330 | ....26232 | 215601495 | ....82148 | ....68217 | ....59730 | ....56714 | ....59197 | ....67205 |
| 340 | ....47684 | ....23055 | 216703816 | ....89994 | ....81616 | ....78710 | ....81303 | ....89422 |
| 350 | ....69139 | ....44618 | ....25487 | 217811773 | 218903505 | 220000708 | 221103411 | 222211641 |
| 360 | ....90596 | ....66182 | ....47159 | ....33554 | ....25395 | ....22708 | ....25521 | ....33862 |
| 370 | 214612055 | ....87749 | ....68834 | ....55338 | ....47287 | ....44710 | ....47634 | ....56085 |
| 380 | ....33516 | 215709317 | ....90511 | ....77128 | ....69182 | ....66715 | ....69748 | ....78311 |
| 390 | ....54980 | ....30888 | 217812190 | ....98911 | ....91079 | ....88722 | ....91865 | 222300539 |
| 400 | ....76445 | ....52461 | ....33871 | 217920701 | 219012978 | 220110730 | 221213955 | ....22769 |
| 410 | ....97913 | ....74037 | ....55554 | ....42493 | ....34879 | ....32741 | ....36106 | ....45001 |
| 420 | 214719383 | ....95614 | ....77240 | ....64287 | ....56783 | ....54755 | ....58230 | ....67235 |
| 430 | ....40855 | 215817194 | ....98928 | ....86084 | ....78689 | ....76770 | ....80355 | ....89472 |
| 440 | ....62329 | ....38775 | 216920618 | 218007882 | 219100597 | ....98788 | 221302483 | 222411711 |
| 450 | ....83805 | ....60359 | ....42310 | ....29683 | ....22507 | 220220808 | ....24614 | ....33952 |
| 460 | 214805283 | ....81945 | ....64004 | ....51486 | ....44419 | ....42830 | ....46746 | ....56196 |
| 470 | ....26764 | 215903533 | ....85700 | ....73291 | ....66333 | ....64854 | ....68881 | ....78441 |
| 480 | ....48247 | ....25124 | 217007399 | ....95098 | ....88250 | ....86881 | ....91015 | 222500689 |
| 490 | ....69731 | ....46716 | ....29100 | 218116908 | 219210169 | 220308909 | 221413157 | ....22939 |
| 500 | ....91218 | ....68311 | ....50802 | ....38720 | ....32090 | ....30940 | ....35298 | ....45191 |

**Figure C.20**

| | 80000 | 80500 | 81000 | 81500 | 82000 | 82500 | 83000 | 83500 |
|---|---|---|---|---|---|---|---|---|
| 0 | 222545191 | 323660648 | 224[illegible]1695 | 225503362 | 227040676 | 228178665 | 229323258 | 230471783 |
| 10 | ····67446 | ····83014 | 224804174 | ····30953 | ····63380 | 228201483 | ····45290 | ····94830 |
| 20 | ····89773 | 223705383 | ····26654 | ····53546 | ····86086 | ····14303 | ····63224 | 230517880 |
| 30 | 222511962 | ····27753 | ····49137 | ····76141 | 227108795 | ····47125 | ····91161 | ····40932 |
| 40 | ····34223 | ····50126 | ····71622 | ····98739 | ····31505 | ····69950 | 229414100 | ····63986 |
| 50 | ····56486 | ····72501 | ····94109 | 226021339 | ····54219 | ····92777 | ····37042 | ····87042 |
| 60 | ····78752 | ····94878 | 224916598 | ····43941 | ····76934 | 228315606 | ····59985 | 230610101 |
| 70 | 222751020 | 223817258 | ····39090 | ····66545 | ····99652 | ····38438 | ····82931 | ····33162 |
| 80 | ····23290 | ····39639 | ····61584 | ····89152 | 227222372 | ····61272 | 229505880 | ····56225 |
| 90 | ····45562 | ····62023 | ····84080 | 226111761 | ····45094 | ····84108 | ····28830 | ····79291 |
| 100 | ····67837 | ····84409 | 225006578 | ····34372 | ····67818 | 228406946 | ····51783 | 230702359 |
| 110 | ····90114 | 223906798 | ····29079 | ····56985 | ····90545 | ····29787 | ····74738 | ····25429 |
| 120 | 222812393 | ····29189 | ····51582 | ····79601 | 227313274 | ····52630 | ····97696 | ····48501 |
| 130 | ····34674 | ····51581 | ····74087 | 226202219 | ····36006 | ····75475 | 229620656 | ····71576 |
| 140 | ····56957 | ····73977 | ····96595 | ····24839 | ····58739 | ····98323 | ····43618 | ····94653 |
| 150 | ····69243 | ····96374 | 225119104 | ····47462 | ····81475 | 228521172 | ····66582 | 230817733 |
| 160 | 222901531 | 224018774 | ····41616 | ····70086 | 227404213 | ····44024 | ····89549 | ····40815 |
| 170 | ····23821 | ····41176 | ····64130 | ····92713 | ····26954 | ····66879 | 229712518 | ····63800 |
| 180 | ····46114 | ····63580 | ····86647 | 226315343 | ····49696 | ····89736 | ····35489 | ····80985 |
| 190 | ····68408 | ····85986 | 225209165 | ····37974 | ····72441 | 228612595 | ····58463 | 230910074 |
| 200 | ····90705 | 224108395 | ····31686 | ····60608 | ····95189 | ····35456 | ····81439 | ····33165 |
| 210 | 223013004 | ····30805 | ····54209 | ····83244 | 227517938 | ····58319 | 229804417 | ····56258 |
| 220 | ····35305 | ····53218 | ····76735 | 226405882 | ····40690 | ····81185 | ····27397 | ····79354 |
| 230 | ····57609 | ····75634 | ····99263 | ····28523 | ····63444 | 228704053 | ····50380 | 231002452 |
| 240 | ····79915 | ····98051 | 225321792 | ····51166 | ····86200 | ····26924 | ····73365 | ····25552 |
| 250 | 223102223 | 224220471 | ····44325 | ····73811 | 227608959 | ····49796 | ····96352 | ····48654 |
| 260 | ····24533 | ····42893 | ····66859 | ····96458 | ····31720 | ····72671 | 229019342 | ····71759 |
| 270 | ····46845 | ····65317 | ····89396 | 226519108 | ····54483 | ····95549 | ····42334 | ····94867 |
| 280 | ····69160 | ····87744 | 225411935 | ····41760 | ····77248 | 228818428 | ····65328 | 231117976 |
| 290 | ····91477 | 224310173 | ····34476 | ····64414 | 227700016 | ····41310 | ····88324 | ····41088 |
| 300 | 223213796 | ····32604 | ····57019 | ····87071 | ····22786 | ····64194 | 230011323 | ····64202 |
| 310 | ····36117 | ····55037 | ····79565 | 226609729 | ····45558 | ····87081 | ····34324 | ····87318 |
| 320 | ····58441 | ····77473 | 225502113 | ····32390 | ····68333 | 228909969 | ····57378 | 231210427 |
| 330 | ····80767 | ····99910 | ····24663 | ····55054 | ····91110 | ····32860 | ····80333 | ····33558 |
| 340 | 223303095 | 224422350 | ····47216 | ····77719 | 227813889 | ····55754 | 230103341 | ····56681 |
| 350 | ····25425 | ····44793 | ····69770 | 226700387 | ····36670 | ····78649 | ····26352 | ····79807 |
| 360 | ····47758 | ····67237 | ····92327 | ····23057 | ····59453 | 229001547 | ····49364 | 231302935 |
| 370 | ····70093 | ····89684 | 225614887 | ····45729 | ····82240 | ····24447 | ····72379 | ····26065 |
| 380 | ····92429 | 224512133 | ····37448 | ····68404 | 227905028 | ····47350 | ····95396 | ····49198 |
| 390 | 213414769 | ····34584 | ····60012 | ····91080 | ····27819 | ····70254 | 230218416 | ····72333 |
| 400 | ····37110 | ····57037 | ····82578 | 226813760 | ····50611 | ····93161 | ····41438 | ····95470 |
| 410 | ····59454 | ····79493 | 225705146 | ····36441 | ····73407 | 229116071 | ····64462 | 231418610 |
| 420 | ····81800 | 224601951 | ····27717 | ····59125 | ····96204 | ····38982 | ····87489 | ····41752 |
| 430 | 223504148 | ····24411 | ····50289 | ····81811 | 228019003 | ····61896 | 230310517 | ····64896 |
| 440 | ····26499 | ····46874 | ····72864 | 226904499 | ····41805 | ····84812 | ····33548 | ····88042 |
| 450 | ····48851 | ····69338 | ····95442 | ····27189 | ····64609 | 229207731 | ····56582 | 231511191 |
| 460 | ····71206 | ····91805 | 225818021 | ····49882 | ····87416 | ····30652 | ····79617 | ····34342 |
| 470 | ····93563 | 224714274 | ····40603 | ····72577 | 228110225 | ····53575 | 230402655 | ····57496 |
| 480 | 323615923 | ····36746 | ····63187 | ····95274 | ····33036 | ····76500 | ····25696 | ····89651 |
| 490 | ····31284 | ····59220 | ····85773 | 227017974 | ····55849 | ····99428 | ····48738 | 231603809 |
| 500 | ····60643 | ····81695 | 225908362 | ····40676 | ····78565 | 229322358 | ····71783 | ····26970 |

**Figure C.21**

| | 84000 | 84500 | 85000 | 85500 | 86000 | 86500 | 87000 | 87500 |
|---|---|---|---|---|---|---|---|---|
| 0 | 231620970 | 232787947 | 233954742 | 235127387 | 236305909 | 237490338 | 238680703 | 239877035 |
| 10 | ····50133 | 232811226 | ····78138 | ····50900 | ····29539 | 237514087 | 238704571 | 239901023 |
| 20 | ····73298 | ····34507 | 234001536 | ····74415 | ····53172 | ····37838 | ····28442 | ····25013 |
| 30 | ····96465 | ····57790 | ····24937 | ····97932 | ····76808 | ····61592 | ····52315 | ····49006 |
| 40 | 231719634 | ····81076 | ····48340 | 235221452 | 236400445 | ····85348 | ····76190 | ····73000 |
| 50 | ····42806 | 232904364 | ····71743 | ····44974 | ····24085 | 237609107 | 238800067 | ····96998 |
| 60 | [illegible] | ····27654 | ····95151 | ····68498 | ····47728 | ····32867 | ····23947 | 240020997 |
| 70 | ····89157 | ····50947 | 234118560 | ····92025 | ····71373 | ····56631 | ····47830 | ····44999 |
| 80 | 231812336 | ····74242 | ····41972 | 235315554 | ····95020 | ····80396 | ····71715 | ····69004 |
| 90 | ····35517 | ····97540 | ····65386 | ····39086 | 236518669 | 237704164 | ····95602 | ····93011 |
| 100 | ····58701 | 233020839 | ····88803 | ····62620 | ····42321 | ····27935 | 238919491 | 240117020 |
| 110 | ····81887 | ····44141 | 234212222 | ····86156 | ····65975 | ····51708 | ····43383 | ····41032 |
| 120 | 231905075 | ····67446 | ····35643 | 235409695 | ····89632 | ····75483 | ····67278 | ····65046 |
| 130 | ····28266 | ····90753 | ····59066 | ····33236 | 236613291 | ····99260 | ····91174 | ····89063 |
| 140 | ····51458 | 233114062 | ····82492 | ····56779 | ····36952 | 237823040 | 239015073 | 240213082 |
| 150 | ····74654 | ····37373 | 234305920 | ····80325 | ····60616 | ····46823 | ····38975 | ····37103 |
| 160 | ····97851 | ····60687 | ····92351 | 235503873 | ····84282 | ····70607 | ····62879 | ····61127 |
| 170 | 232021051 | ····84003 | ····52784 | ····27423 | 236707950 | ····94394 | ····86785 | ····85153 |
| 180 | ····44253 | 233207321 | ····76219 | ····50976 | ····31621 | 237918184 | 239110694 | 240309181 |
| 190 | ····67457 | ····30642 | ····99657 | ····74531 | ····55294 | ····41976 | ····34605 | ····33212 |
| 200 | ····90664 | ····53965 | 234423097 | ····98089 | ····78970 | ····65770 | ····58519 | ····57245 |
| 210 | 232113873 | ····77290 | ····46539 | 235621648 | 236802648 | ····89566 | ····82434 | ····81241 |
| 220 | ····37085 | 233300618 | ····69984 | ····45211 | ····26328 | 238013365 | 239206353 | 240405319 |
| 230 | ····60298 | ····23948 | ····93431 | ····68775 | ····50011 | ····37167 | ····30273 | ····29360 |
| 240 | ····83514 | ····47281 | 234516880 | ····92342 | ····73696 | ····60970 | ····54196 | ····53403 |
| 250 | 232206733 | ····70615 | ····40332 | 235715911 | ····97383 | ····84777 | ····78122 | ····77448 |
| 260 | ····29953 | ····93953 | ····63786 | ····39483 | 236921073 | 238108585 | 239302049 | 240501496 |
| 270 | ····53176 | 233417292 | ····87242 | ····63057 | ····44765 | ····32396 | ····25980 | ····25546 |
| 280 | ····76402 | ····40634 | 234610701 | ····86633 | ····68459 | ····56209 | ····49912 | ····49598 |
| 290 | ····99629 | ····63978 | ····34162 | 235810212 | ····92156 | ····80025 | ····73847 | ····73653 |
| 300 | 232322859 | ····87324 | ····57625 | ····33793 | 237015855 | 238203843 | ····97785 | ····97711 |
| 310 | ····46092 | 233510673 | ····81091 | ····57376 | ····39557 | ····27663 | 239421724 | 240621771 |
| 320 | ····69326 | ····34024 | 234704559 | ····80962 | ····63261 | ····51486 | ····45667 | ····45833 |
| 330 | ····92563 | ····57377 | ····28030 | 235904550 | ····86967 | ····75311 | ····69611 | ····69897 |
| 340 | 232415802 | ····80733 | ····51503 | ····28140 | 237110676 | ····99139 | ····93558 | ····93964 |
| 350 | ····39044 | 233604091 | ····74978 | ····51733 | ····34387 | 238322968 | 239517508 | 240718034 |
| 360 | ····62288 | ····27452 | ····98455 | ····75328 | ····58100 | ····46801 | ····41459 | ····42106 |
| 370 | ····85534 | ····50814 | 234821935 | ····98926 | ····81816 | ····70635 | ····65413 | ····66180 |
| 380 | 232508783 | ····74179 | ····45417 | 236022526 | 237205534 | ····94473 | ····89370 | ····90256 |
| 390 | ····32034 | ····97547 | ····68902 | ····46128 | ····29255 | 238418312 | 239613329 | 240814335 |
| 400 | ····55287 | 233720916 | ····92389 | ····69733 | ····52978 | ····42154 | ····37290 | ····38417 |
| 410 | ····78542 | ····44289 | 234915878 | ····93340 | ····76703 | ····65998 | ····61254 | ····62501 |
| 420 | 232601800 | ····67663 | ····39369 | 236116949 | 237300431 | ····89845 | ····85220 | ····86587 |
| 430 | ····25060 | ····91040 | ····62863 | ····40561 | ····24161 | 238513694 | 239709188 | 240910676 |
| 440 | ····48323 | 233814419 | ····86360 | ····64175 | ····47893 | ····37545 | ····33159 | ····34767 |
| 450 | ····71588 | ····37800 | 235009858 | ····87791 | ····71628 | ····61399 | ····57133 | ····58807 |
| 460 | ····94855 | ····61184 | ····33359 | 236211410 | ····95365 | ····85255 | ····81108 | ····82956 |
| 470 | 232718124 | ····84570 | ····56863 | ····35031 | 237419105 | 238609113 | 239805087 | 241007054 |
| 480 | ····41396 | 233907959 | ····80368 | ····58655 | ····42847 | ····32974 | ····29067 | ····31155 |
| 490 | ····64670 | ····31349 | 235103876 | ····82281 | ····66591 | ····56833 | ····53050 | ····55258 |
| 500 | ····87947 | ····54743 | ····27387 | 236305909 | ····90338 | ····80703 | ····77035 | ····79364 |

**Figure C.22**

| | 88000 | 88500 | 89000 | 89500 | 90000 | 90500 | 91000 | 91500 |
|---|---|---|---|---|---|---|---|---|
| 0 | 241079364 | 242287718 | 243[illegible]02130 | 244722628 | 245949244 | 247182008 | 248420951 | 2496601[illegible]4 |
| 10 | 241193472 | 242311947 | ....26480 | ....4710[illegible] | ....73839 | 247206726 | ....45793 | ....91070 |
| 20 | ....27582 | ....26178 | ....50833 | ....71575 | ....98436 | ....31447 | ....70637 | 2497160[illegible]9 |
| 30 | ....51695 | ....60412 | ....75188 | ....9605[illegible] | 246023036 | ....56170 | ....95484 | ....41011 |
| 40 | ....75810 | ....84648 | ....99545 | 244820532 | ....47638 | ....80895 | 248520334 | ....65985 |
| 50 | ....99927 | 242408887 | 243623905 | ....45014 | ....72[illegible] | [illegible]47305624 | ....45186 | ....90961 |
| 60 | 241224047 | ....33127 | ....48268 | ....69498 | ....9685[illegible] | ....30354 | ....70041 | 249815941 |
| 70 | ....48170 | ....57371 | ....72633 | ....93985 | 24612146[illegible] | ....55087 | ....94898 | ....40922 |
| 80 | ....72295 | ....81616 | ....97000 | 241918475 | ....4607[illegible] | ....79823 | 48619757 | ....65906 |
| 90 | ....96422 | 242505865 | 243721370 | ....42967 | ....70687 | 247404561 | ....44619 | ....90893 |
| 100 | 241320551 | ....30115 | ....45741 | ....67461 | ....95304 | ....29301 | ....69484 | 249915882 |
| 110 | ....44684 | ....54368 | ....70116 | ....91958 | 246219923 | ....54044 | ....94350 | ....40874 |
| 120 | ....68818 | ....78624 | ....94493 | 245016457 | ....44545 | ....78789 | 48719220 | ....65868 |
| 130 | ....92955 | 242602882 | 243818873 | ....40958 | ....69170 | 247503537 | ....44092 | ....90865 |
| 140 | 241417094 | ....27142 | ....43254 | ....65462 | ....93797 | ....28288 | ....68966 | 250015863 |
| 150 | ....41236 | ....51405 | ....67639 | ....89969 | 246318426 | ....53041 | ....93843 | ....40[illegible]65 |
| 160 | ....65380 | ....75670 | ....92026 | 245914478 | ....43058 | ....77796 | 248818722 | ....6586[illegible] |
| 170 | ....89527 | ....99937 | 243916415 | ....38989 | ....67692 | 247602554 | ....43604 | ....90876 |
| 180 | 241513676 | 242724207 | ....40806 | ....63503 | ....92329 | ....27314 | ....68489 | 250115855 |
| 190 | ....37827 | ....48480 | ....65200 | ....88020 | 246416968 | ....52077 | ....93376 | ....40896 |
| 200 | ....61981 | ....72754 | ....89597 | 245212539 | ....41610 | ....76842 | 248918265 | ....65910 |
| 210 | ....86137 | ....97032 | 244013996 | ....37060 | ....66254 | 247701610 | ....43157 | ....90927 |
| 220 | 241610296 | 242821311 | ....38397 | ....61584 | ....90901 | ....26380 | ....68051 | 250215946 |
| 230 | ....34457 | ....45594 | ....62801 | ....86110 | 246515550 | ....51152 | ....9[illegible]948 | ....40068 |
| 240 | ....58620 | ....69878 | ....87207 | 34531063[illegible] | ....40201 | ....75927 | 2490[illegible]047 | ....65992 |
| 250 | ....82786 | ....94165 | 244111616 | ....35169 | ....64855 | 247800705 | ....42749 | ....91018 |
| 260 | 241706954 | 242918455 | ....36027 | ....59703 | ....89512 | ....25485 | ....67653 | 250316047 |
| 270 | ....31125 | ....42746 | ....60441 | ....84239 | 246614171 | ....50268 | ....92560 | ....41079 |
| 280 | ....55298 | ....67041 | ....84857 | 245408777 | ....38832 | ....75053 | 249117469 | ....66113 |
| 290 | ....79473 | ....91337 | 244209276 | ....3[illegible]318 | ....63[illegible]06 | ....99840 | ....42381 | ....91159 |
| 300 | 241803651 | 243015636 | ....33696 | ....5[illegible]62 | ....88103 | 2479[illegible]4630 | ....67295 | 250416189 |
| 310 | ....27832 | ....39938 | ....58120 | ....82407 | 246712831 | ....494[illegible]3 | ....92212 | ....41230 |
| 320 | ....52014 | ....64242 | ....8[illegible]546 | 245506956 | ....37503 | ....74[illegible]17 | 248917131 | ....66[illegible]75 |
| 330 | ....70200 | ....[illegible]549 | 244306974 | ....31506 | ....62176 | ....99015 | ....42053 | ....91321 |
| 340 | 241900387 | 243112857 | ....31405 | ....56060 | ....86853 | [illegible]48023815 | ....66977 | 250516370 |
| 350 | ....24577 | ....37169 | ....55838 | ....80615 | 246811531 | ....48617 | ....91904 | ....41422 |
| 360 | ....4[illegible]770 | ....61482 | ....20[illegible]73 | [illegible]45605173 | ....36[illegible]13 | ....734[illegible] | 2493168[illegible]3 | ....66476 |
| 370 | ....72965 | ....85799 | 244404711 | ....29734 | ....60896 | ....98229 | ....41765 | ....91533 |
| 380 | ....9716[illegible] | 243210117 | ....29152 | ....54297 | ....8558[illegible] | [illegible]48123039 | ....66699 | 250616592 |
| 390 | 242021362 | ....34438 | ....53595 | ....78862 | 246910271 | ....47851 | ....91636 | ....41653 |
| 400 | ....45564 | ....58762 | ....78040 | 245703430 | ....34[illegible]62 | ....71666 | 249416575 | ....66718 |
| 410 | ....69769 | ....83087 | 244502488 | ....28000 | ....59655 | ....97484 | ....41516 | ....91784 |
| 420 | ....93976 | 243307416 | ....26938 | ....52573 | ....84[illegible] | 248222303 | ....6646[illegible] | 25071[illegible]54 |
| 430 | 242118185 | ....31746 | ....51391 | ....77148 | 24700905[illegible] | ....47[illegible]26 | ....91407 | ....41925 |
| 440 | ....42397 | ....56030 | ....75846 | 245801726 | ....33751 | ....71950 | [illegible]49516356 | ....66999 |
| 450 | ....66611 | ....80415 | 244600303 | ....26306 | ....58454 | ....95777 | ....41308 | ....92076 |
| 460 | ....90828 | 243404753 | ....24763 | ....50889 | ....83160 | [illegible]48321607 | ....6626[illegible] | 250817155 |
| 470 | 242215047 | ....29094 | ....49226 | ....75474 | 24[illegible]107868 | ....46439 | ....91219 | ....42[illegible]3 |
| 480 | ....39268 | ....53437 | ....73691 | 245900061 | ....325[illegible]9 | ....71274 | 249616178 | ....6[illegible]321 |
| 490 | ....163492 | ....77782 | ....98158 | ....24651 | ....572[illegible] | ....96111 | ....41139 | ....9[illegible] |
| 500 | ....87718 | 243502130 | 244722628 | ....49244 | ....8200[illegible] | 248420951 | ....66104 | 250917497 |

Figure C.23

| | 92000 | 92500 | 93000 | 93500 | 94000 | 94500 | 95000 | 95500 |
|---|---|---|---|---|---|---|---|---|
| 0 | 250917497 | 252175164 | 253439133 | 254709439 | 255986111 | 257269181 | 258558685 | 259854651 |
| 10 | ....42589 | 252200381 | ....6447 | ....34910 | 256011710 | ....94909 | ....84541 | ....80637 |
| 20 | ....67683 | ....25601 | ....89824 | ....60383 | ....37341 | 257320639 | 2586103[illegible] | ....9906625 |
| 30 | ....92780 | ....50824 | 253515173 | ....85859 | ....62915 | ....46371 | ....36260 | ....32615 |
| 40 | 251017879 | ....76049 | ....40524 | 254811338 | ....88521 | ....72106 | ....62124 | ....58609 |
| 50 | ....42981 | 252301276 | ....65878 | ....36819 | 256114130 | ....97843 | ....87990 | ....84604 |
| 60 | ....68085 | ....26506 | ....91235 | ....62303 | ....39741 | 257423583 | 258713859 | 260010603 |
| 70 | ....93192 | ....51739 | 253616594 | ....87789 | ....65355 | ....49325 | ....39730 | ....36604 |
| 80 | 251118302 | ....76974 | ....41956 | 254913278 | ....90972 | ....75070 | ....65604 | ....62607 |
| 90 | ....43413 | 252402212 | ....67320 | ....38769 | 256216591 | 257500817 | ....91481 | ....88614 |
| 100 | ....68528 | ....27452 | ....92687 | ....64263 | ....42212 | ....26567 | 258817360 | 260114623 |
| 110 | ....93645 | ....52695 | 253718056 | ....89759 | ....67837 | ....52320 | ....43242 | ....40634 |
| 120 | 251218764 | ....77940 | ....43428 | 255015258 | ....93463 | ....78075 | ....69126 | ....66648 |
| 130 | ....43886 | 252503188 | ....68802 | ....40760 | 256319093 | 257603833 | ....95013 | ....92665 |
| 140 | ....69010 | ....28438 | ....94179 | ....66264 | ....44725 | ....29594 | 258920902 | 260218684 |
| 150 | ....94137 | ....53691 | 253819558 | ....91770 | ....70359 | ....55356 | ....46795 | ....44706 |
| 160 | 251319167 | ....78946 | ....44940 | 255117280 | ....95996 | ....81122 | ....72689 | ....70731 |
| 170 | ....44399 | 252604204 | ....70325 | ....42791 | 256421636 | 257706890 | ....98587 | ....96758 |
| 180 | ....69533 | ....29465 | ....95712 | ....68306 | ....47278 | ....32661 | 259024486 | 260322787 |
| 190 | ....94670 | ....54728 | 253921101 | ....93822 | ....72923 | ....58434 | ....50389 | ....48820 |
| 200 | 251419809 | ....79993 | ....46494 | 255219342 | ....98570 | ....84210 | ....76294 | ....74854 |
| 210 | ....44951 | 252705261 | ....71888 | ....44864 | 256524220 | 257809985 | 259102201 | 260400892 |
| 220 | ....70096 | ....30532 | ....97285 | ....70388 | ....49872 | ....35769 | ....28112 | ....26932 |
| 230 | ....95243 | ....55805 | 254022685 | ....95915 | ....75527 | ....61553 | ....54024 | ....52974 |
| 240 | 251520392 | ....81080 | ....48087 | 255321445 | 256601185 | ....87339 | ....79940 | ....79020 |
| 250 | ....45544 | 252806359 | ....73492 | ....46977 | ....26845 | 257913125 | 259205858 | 260505068 |
| 260 | ....70699 | ....31639 | ....98899 | ....72512 | ....52508 | ....38919 | ....31779 | ....31118 |
| 270 | ....95856 | ....56922 | 254124309 | ....98049 | ....78173 | ....64713 | ....57702 | ....57171 |
| 280 | 251621016 | ....82208 | ....49722 | 255423589 | 256703841 | ....90509 | ....83628 | ....83227 |
| 290 | ....46178 | 252907496 | ....75137 | ....49131 | ....29511 | 258016309 | 259209556 | 260609285 |
| 300 | ....71342 | ....32787 | 254200554 | ....74676 | ....55184 | ....42110 | ....35487 | ....35346 |
| 310 | ....96510 | ....58080 | ....25974 | 255500224 | ....80860 | ....67914 | ....61420 | ....61410 |
| 320 | 251721679 | ....83376 | ....51397 | ....25774 | 256806538 | ....93721 | ....87357 | ....87476 |
| 330 | ....46851 | 253008764 | ....76822 | ....51326 | ....32218 | 258119531 | 259413295 | 260713544 |
| 340 | ....72026 | ....33975 | 254302250 | ....76881 | ....57902 | ....45343 | ....39237 | ....39616 |
| 350 | ....97203 | ....59279 | ....27680 | 255602439 | ....83587 | ....71157 | ....65180 | ....65690 |
| 360 | 251822383 | ....84585 | ....53113 | ....27999 | 256909276 | ....96974 | ....91127 | ....91767 |
| 370 | ....47565 | 253109893 | ....78549 | ....53562 | ....34967 | 258222794 | 259517076 | 260817846 |
| 380 | ....72750 | ....35204 | 254403986 | ....79127 | ....60660 | ....48616 | ....43028 | ....43928 |
| 390 | ....97937 | ....60518 | ....29427 | 255704695 | ....86356 | ....74441 | ....68982 | ....70012 |
| 400 | 251923127 | ....85834 | ....54870 | ....30266 | 257012055 | 258300268 | ....94939 | ....96099 |
| 410 | ....48319 | 252211162 | ....80315 | ....55839 | ....37756 | ....26098 | 259620890 | 260922189 |
| 420 | ....73514 | ....36473 | 254505763 | ....81414 | ....63460 | ....51931 | ....46861 | ....48281 |
| 430 | ....98711 | ....61797 | ....31214 | 255806992 | ....89166 | ....77766 | ....72825 | ....74376 |
| 440 | 252023911 | ....87123 | ....56667 | ....32573 | 257114875 | 258403604 | ....98793 | 261000473 |
| 450 | ....49114 | 253312452 | ....82122 | ....58156 | ....40586 | ....29444 | 259724763 | ....26573 |
| 460 | ....74319 | ....37783 | 254607580 | ....83742 | ....66300 | ....55287 | ....50735 | ....52676 |
| 470 | ....99526 | ....63117 | ....33041 | 255909331 | ....92017 | ....81133 | ....76710 | ....78781 |
| 480 | 252134736 | ....88453 | ....58504 | ....34921 | 257217736 | 258506981 | 259802688 | 261104889 |
| 490 | ....49949 | 253413792 | ....83970 | ....60515 | ....43458 | ....32832 | ....28668 | ....30999 |
| 500 | ....75164 | ....39132 | 254709439 | ....86111 | ....69182 | ....58685 | ....54651 | ....57112 |

D

**Figure C.24**

| | 96000 | 96500 | 97000 | 97500 | 98000 | 98500 | 99000 | 99500 |
|---|---|---|---|---|---|---|---|---|
| 0 | 261157112 | 262466102 | 263781053 | 265103798 | 266432570 | 267768001 | 269110127 | 270458979 |
| 10 | ....83218 | ....92349 | 263808031 | ....30308 | ....59213 | ....94778 | ....37038 | ....86025 |
| 20 | 261209346 | 262518598 | ....34412 | ....56821 | ....85859 | 267821558 | ....63952 | 270513074 |
| 30 | ....35467 | ....44850 | ....60795 | ....83337 | 266512507 | ....48340 | ....90868 | ....40125 |
| 40 | ....61591 | ....71104 | ....87182 | 265209855 | ....39159 | ....75125 | 269217787 | ....67179 |
| 50 | ....87717 | ....97362 | 263913570 | ....36376 | ....65813 | 267901912 | ....44709 | ....94236 |
| 60 | 261313846 | 262623621 | ....39962 | ....62990 | ....92469 | ....28702 | ....71633 | 270621295 |
| 70 | ....39977 | ....49884 | ....66356 | ....89426 | 266619128 | ....55495 | ....98561 | ....48357 |
| 80 | ....66111 | ....76149 | ....92752 | 265315955 | ....45790 | ....82291 | 269325490 | ....75422 |
| 90 | ....92248 | 262702416 | 264019152 | ....42487 | ....72455 | 268009089 | ....52423 | 270702490 |
| 100 | 261418387 | ....28686 | ....45554 | ....69021 | ....99122 | ....35890 | ....79358 | ....29560 |
| 110 | ....44529 | ....54959 | ....71958 | ....95558 | 266725792 | ....62694 | 269406296 | ....56633 |
| 120 | ....70673 | ....81235 | ....98365 | 265422097 | ....52465 | ....89500 | ....33237 | ....83709 |
| 130 | ....96820 | 262807513 | 264124775 | ....48640 | ....79140 | 268116309 | ....60180 | 270810787 |
| 140 | 261522970 | ....33794 | ....51188 | ....75184 | 266805818 | ....43120 | ....87126 | ....37868 |
| 150 | ....49122 | ....60077 | ....77603 | 265501732 | ....32498 | ....69935 | 269514075 | ....64952 |
| 160 | ....75277 | ....86363 | 264204021 | ....28282 | ....59181 | ....96752 | ....41026 | ....92039 |
| 170 | 261601435 | 262912652 | ....30441 | ....54835 | ....85867 | 268223571 | ....67980 | 270919128 |
| 180 | ....27595 | ....38943 | ....56864 | ....81391 | 266912556 | ....50394 | ....94937 | ....46220 |
| 190 | ....53758 | ....65237 | ....83290 | 265607949 | ....39247 | ....77219 | 269621897 | ....73314 |
| 200 | ....79923 | ....91533 | 264309718 | ....34509 | ....65941 | 268304047 | ....48859 | 271000412 |
| 210 | 261706091 | 263017833 | ....36149 | ....61073 | ....92638 | ....30877 | ....75824 | ....27512 |
| 220 | ....32262 | ....44134 | ....62583 | ....87639 | 267019337 | ....57710 | 269702791 | ....54614 |
| 230 | ....58435 | ....70439 | ....89019 | 265714208 | ....46039 | ....84546 | ....29762 | ....81720 |
| 240 | ....84611 | ....96746 | 264415458 | ....40779 | ....72744 | 268411384 | ....56735 | 271108828 |
| 250 | 261810789 | 263123056 | ....41899 | ....67353 | ....99451 | ....38225 | ....83710 | ....35939 |
| 260 | ....36970 | ....49368 | ....68343 | ....93930 | 267126161 | ....65069 | 269810689 | ....63052 |
| 270 | ....63154 | ....75683 | ....94790 | 265820509 | ....52873 | ....91916 | ....37670 | ....90169 |
| 280 | ....89340 | 263202000 | 264521240 | ....47092 | ....79589 | 268518765 | ....64654 | 271217288 |
| 290 | 261915529 | ....28321 | ....47692 | ....73676 | 267206307 | ....45617 | ....91640 | ....44409 |
| 300 | ....41721 | ....54643 | ....74147 | 265900264 | ....33027 | ....72471 | 269918629 | ....71534 |
| 310 | ....67915 | ....80969 | 264600604 | ....26854 | ....59751 | ....99329 | ....45621 | ....98661 |
| 320 | ....94112 | 263307297 | ....27064 | ....53446 | ....86477 | 268626189 | ....72615 | 271325791 |
| 330 | 262020311 | ....33628 | ....53527 | ....80042 | 267313205 | ....53051 | ....99613 | ....52924 |
| 340 | ....46513 | ....59961 | ....79992 | 266006640 | ....39937 | ....79916 | 270026613 | ....80059 |
| 350 | ....72718 | ....86297 | 264706460 | ....33240 | ....66671 | 268706784 | ....53615 | 271407197 |
| 360 | ....98925 | 263412636 | ....32931 | ....59844 | ....93407 | ....33655 | ....80621 | ....34338 |
| 370 | 262125135 | ....38977 | ....59404 | ....86450 | 267420147 | ....60529 | 270107629 | ....61481 |
| 380 | ....51348 | ....65321 | ....85880 | 266113058 | ....46889 | ....87405 | ....34640 | ....88627 |
| 390 | ....77563 | ....91667 | 264812359 | ....39665 | ....73633 | 268814283 | ....61653 | 271515776 |
| 400 | 262203781 | 263518016 | ....38840 | ....66283 | 267500381 | ....41165 | ....88669 | ....42928 |
| 410 | ....30001 | ....44368 | ....65324 | ....92900 | ....27131 | ....68049 | 270215688 | ....70082 |
| 420 | ....56224 | ....70723 | ....91810 | 266219519 | ....53883 | ....94936 | ....42710 | ....97239 |
| 430 | ....82450 | ....97080 | 264918299 | ....46141 | ....80639 | 268921825 | ....69734 | 271624399 |
| 440 | 262308678 | 263623439 | ....44791 | ....72766 | 267607397 | ....48717 | ....96761 | ....51561 |
| 450 | ....34909 | ....49802 | ....71266 | ....99393 | ....34158 | ....75612 | 270323790 | ....78726 |
| 460 | ....61142 | ....76167 | ....97783 | 266326023 | ....60921 | 269002510 | ....50823 | 271705894 |
| 470 | ....87378 | 263702534 | 265024283 | ....52656 | ....87687 | ....29410 | ....77858 | ....33065 |
| 480 | 262413617 | ....28905 | ....50785 | ....79291 | 267714456 | ....56313 | 270404896 | ....60238 |
| 490 | ....39858 | ....55278 | ....77290 | 266405929 | ....41227 | ....83219 | ....31936 | ....87414 |
| 500 | ....66102 | ....81653 | 265103798 | ....32570 | ....68001 | 269110127 | ....58979 | 271814593 |

**Figure C.25**

| | 100000 | 100500 | 101000 | 100500 | 102000 | 102500 | 103000 | 103500 |
|---|---|---|---|---|---|---|---|---|
| 0 | 271814593 | 273177001 | 274546237 | 275922337 | 277305334 | 278695264 | 280092159 | 281496057 |
| 10 | ····41774 | 273204318 | ····73692 | ····49930 | ····33065 | 278723133 | 280120168 | 281524206 |
| 20 | ····68958 | ····31639 | 274601149 | ····77525 | ····60798 | ····51005 | ····48180 | ····52355 |
| 30 | ····96145 | ····58962 | ····28610 | 276005122 | ····88534 | ····78881 | ····76195 | ····80532 |
| 40 | 271923335 | ····86288 | ····56072 | ····32723 | 277416273 | 278806758 | 280204213 | 281608673 |
| 50 | ····50527 | 273313617 | ····83538 | ····60326 | ····44015 | ····34639 | ····32233 | ····36833 |
| 60 | ····77722 | ····40948 | 274711006 | ····87932 | ····71759 | ····62523 | ····60257 | ····64997 |
| 70 | 272004920 | ····68282 | ····38478 | 276115541 | ····99506 | ····90409 | ····88283 | ····93163 |
| 80 | ····32101 | ····95619 | ····65951 | ····43152 | 277527256 | 278918298 | 280316311 | 281721337 |
| 90 | ····59234 | 273422958 | ····93428 | ····70767 | ····55009 | ····46190 | ····44343 | ····49505 |
| 100 | ····86530 | ····50301 | 274820907 | ····98384 | ····82765 | ····74084 | ····72377 | ····77680 |
| 110 | 272113738 | ····77646 | ····48389 | 276226004 | 277610523 | 279001982 | 280400415 | 281805857 |
| 120 | ····40950 | 273504994 | ····75874 | ····53626 | ····38284 | ····29882 | ····28455 | ····34038 |
| 130 | ····68164 | ····32344 | 274903362 | ····81252 | ····66045 | ····57785 | ····56497 | ····62221 |
| 140 | ····95381 | ····59697 | ····30852 | 276308880 | ····93814 | ····85691 | ····84543 | ····90408 |
| 150 | 272222600 | ····87053 | ····58345 | ····36511 | 277721584 | 279113595 | 280512592 | 281918597 |
| 160 | ····49823 | 273614412 | ····85841 | ····64144 | ····49356 | ····41511 | ····40643 | ····46788 |
| 170 | ····77048 | ····41773 | 275013340 | ····91781 | ····77131 | ····69425 | ····68697 | ····74983 |
| 180 | 272304275 | ····69138 | ····40841 | 276419420 | 277804909 | ····97342 | ····96754 | 282003181 |
| 190 | ····31506 | ····96504 | ····68345 | ····47062 | ····32689 | 279225261 | 280624814 | ····31381 |
| 200 | ····58739 | 273723874 | ····95852 | ····74707 | ····60472 | ····53184 | ····52876 | ····59584 |
| 210 | ····85975 | ····51247 | 275123362 | 276502354 | ····88258 | ····81109 | ····80941 | ····87790 |
| 220 | 272413213 | ····78622 | ····50874 | ····30004 | 277916047 | 279309037 | 280709010 | 282115999 |
| 230 | ····40454 | 273806000 | ····78389 | ····57657 | ····43839 | ····36968 | ····37080 | ····44210 |
| 240 | ····67698 | ····33380 | 275205907 | ····85313 | ····71633 | ····64902 | ····65154 | ····72425 |
| 250 | ····94945 | ····60763 | ····33427 | 276612972 | ····99430 | ····92838 | ····93231 | 282200642 |
| 260 | 272522195 | ····88150 | ····60951 | ····40633 | 278027230 | 279420778 | 280821310 | ····28862 |
| 270 | ····49447 | 273915538 | ····88477 | ····68297 | ····55033 | ····48720 | ····49392 | ····57085 |
| 280 | ····76702 | ····42930 | 275316006 | ····95964 | ····82838 | ····76665 | ····77477 | ····85311 |
| 290 | 272603960 | ····70324 | ····43537 | 276723633 | 278110647 | 279504612 | 280905555 | 28231353[illegible] |
| 300 | ····31220 | ····97721 | ····71072 | ····51306 | ····38450 | ····32563 | ····33555 | ····41771 |
| 310 | ····58483 | 274025121 | ····98609 | ····78981 | ····66272 | ····60516 | ····61749 | ····70005 |
| 320 | ····85749 | ····52523 | 275426149 | 276806659 | ····94088 | ····88472 | ····89845 | ····9824[illegible] |
| 330 | 272713017 | ····79930 | ····53691 | ····34339 | 278221908 | 279616431 | 281017944 | 282426482 |
| 340 | ····40289 | 274107337 | ····81237 | ····62023 | ····49730 | ····44393 | ····46046 | ····54724 |
| 350 | ····67563 | ····34747 | 275508785 | ····89709 | ····77555 | ····72357 | ····74150 | ····82970 |
| 360 | ····94840 | ····62161 | ····36336 | 276917398 | 278305383 | 279700324 | 281102258 | 282511218 |
| 370 | 272822119 | ····89577 | ····63889 | ····45090 | ····33213 | ····28294 | ····30368 | ····39469 |
| 380 | ····49401 | 274216006 | ····91446 | ····72784 | ····61046 | ····56267 | ····58481 | ····67723 |
| 390 | ····76686 | ····44418 | 275619005 | 277000482 | ····88883 | ····84243 | ····86597 | ····95980 |
| 400 | 272903974 | ····71842 | ····46567 | ····28182 | 278416721 | 279812221 | 281214715 | 282624239 |
| 410 | ····31264 | ····99260 | ····74131 | ····55884 | ····44563 | ····40202 | ····42837 | ····52502 |
| 420 | ····58557 | 274326699 | 275701699 | ····83590 | ····72408 | ····68156 | ····70961 | ····80767 |
| 430 | ····85853 | ····54132 | ····29269 | 277111298 | 278500255 | ····96173 | ····99088 | 282709035 |
| 440 | 273013152 | ····81567 | ····56842 | ····3900[illegible] | ····28105 | 279924163 | 281327218 | ····37306 |
| 450 | ····40453 | 274409006 | ····84418 | ····66723 | ····55958 | ····52155 | ····55351 | ····65580 |
| 460 | ····67757 | ····36446 | 275811996 | ····94440 | ····83813 | ····80150 | ····83486 | ····93856 |
| 470 | ····95064 | ····63890 | ····39577 | 277222159 | 278611672 | 280008148 | 281411625 | 282822136 |
| 480 | 273122374 | ····91336 | ····67161 | ····49882 | ····39533 | ····36149 | ····39766 | ····5041[illegible] |
| 490 | ····49686 | 274518786 | ····94748 | ····77607 | ····67397 | ····64153 | ····67910 | ····7870[illegible] |
| 500 | ····77001 | ····46237 | 275922337 | 277305334 | ····95264 | ····92159 | ····96057 | 282906991 |

D 2

Figure C.26

|  | 104000 | 104500 | 105000 | 105500 | 106000 | 106500 | 107000 | 107500 |
|---|---|---|---|---|---|---|---|---|
| 0 | 282906991 | 284324997 | 285750111 | 287182367 | 288621803 | 290068453 | 291522354 | 292983543 |
| 10 | ····35282 | ····53430 | ····78686 | 287211085 | ····50665 | ····97460 | ····51506 | 293012841 |
| 20 | ····63575 | ····81865 | 285807263 | ····39807 | ····79530 | 290126469 | ····80662 | ····42143 |
| 30 | ····91872 | 284410303 | ····35844 | ····68531 | 288708398 | ····55482 | 291609820 | ····71447 |
| 40 | 283020171 | ····38744 | ····64428 | ····97257 | ····37269 | ····84498 | ····38981 | 293100754 |
| 50 | ····48473 | ····67188 | ····93014 | 287325987 | ····66142 | 290213516 | ····68145 | ····30064 |
| 60 | ····76778 | ····95635 | 285921603 | ····54720 | ····95019 | ····42537 | ····97311 | ····59377 |
| 70 | 283105085 | 284524034 | ····50196 | ····83455 | 288823899 | ····71562 | 291726481 | ····88693 |
| 80 | ····33396 | ····52537 | ····78791 | 287412194 | ····52781 | 290300589 | ····55654 | 293218012 |
| 90 | ····61709 | ····80992 | 286007389 | ····40935 | ····81666 | ····29619 | ····84829 | ····47334 |
| 100 | ····90025 | 284609450 | ····35989 | ····69679 | 288910554 | ····58652 | 291814008 | ····76658 |
| 110 | 283218344 | ····37911 | ····64593 | ····98426 | ····39446 | ····87688 | ····43189 | 293305986 |
| 120 | ····46666 | ····66375 | ····93199 | 287527176 | ····68339 | 290416727 | ····72373 | ····35317 |
| 130 | ····74991 | ····94841 | 286121809 | ····55928 | ····97236 | ····45768 | 291901561 | ····64650 |
| 140 | 283303318 | 284723311 | ····50421 | ····84684 | 289026136 | ····74813 | ····30751 | ····93987 |
| 150 | ····31649 | ····51783 | ····79036 | 287613442 | ····55039 | 290503860 | ····59944 | 293423326 |
| 160 | ····59982 | ····80258 | 286207654 | ····42204 | ····83944 | ····32911 | ····89140 | ····52668 |
| 170 | ····88318 | 284808736 | ····36275 | ····70968 | 289112852 | ····61964 | 292018339 | ····82014 |
| 180 | 283416657 | ····37217 | ····64898 | ····99735 | ····41764 | ····91020 | ····47541 | 293511362 |
| 190 | ····44998 | ····65701 | ····93525 | 287728505 | ····70678 | 290620079 | ····76745 | ····40713 |
| 200 | ····73343 | ····94188 | 286322154 | ····57278 | ····99595 | ····49141 | 292105953 | ····70067 |
| 210 | 283501690 | 284922677 | ····50786 | ····86054 | 289228515 | ····78206 | ····35164 | ····99424 |
| 220 | ····30040 | ····51169 | ····79421 | 287814832 | ····57438 | 290707274 | ····64377 | 293628784 |
| 230 | ····58393 | ····79664 | 286408059 | ····43614 | ····86364 | ····36345 | ····93594 | ····58147 |
| 240 | ····86749 | 285008162 | ····36700 | ····72398 | 289315292 | ····65418 | 292222813 | ····87512 |
| 250 | 283615108 | ····36663 | ····65344 | 287901185 | ····44224 | ····94495 | ····52035 | 293716881 |
| 260 | ····43469 | ····65167 | ····93990 | ····29975 | ····73158 | 290823574 | ····81261 | ····46253 |
| 270 | ····71834 | ····93673 | 286522640 | ····58768 | 289402095 | ····52657 | 292310489 | ····75628 |
| 280 | 283700201 | 285122183 | ····51292 | ····87564 | ····31036 | ····81742 | ····39720 | 293805005 |
| 290 | ····28571 | ····50695 | ····79947 | 288016363 | ····59979 | 290910830 | ····68954 | ····34386 |
| 300 | ····56944 | ····79210 | 286608605 | ····45165 | ····88925 | ····39921 | ····98191 | ····63769 |
| 310 | ····85319 | 285207728 | ····37266 | ····73969 | 289517874 | ····69015 | 292427430 | ····93155 |
| 320 | 283813698 | ····36249 | ····65930 | 288102777 | ····46825 | ····98112 | ····56673 | 293922545 |
| 330 | ····42079 | ····64772 | ····94596 | ····31587 | ····75780 | 290127212 | ····85919 | ····51937 |
| 340 | ····70463 | ····93299 | 286723266 | ····60400 | 289604738 | ····56315 | 292515167 | ····81332 |
| 350 | ····98850 | 285321828 | ····51938 | ····89216 | ····33698 | ····85420 | ····44419 | 294010730 |
| 360 | 283927240 | ····50360 | ····80613 | 288218035 | ····62662 | 291114529 | ····73673 | ····40132 |
| 370 | ····55633 | ····78895 | 286809291 | ····46857 | ····91628 | ····43640 | 292602931 | ····69536 |
| 380 | ····84029 | 285407433 | ····37972 | ····75682 | 289720597 | ····72755 | ····32191 | ····98943 |
| 390 | 284012427 | ····35974 | ····66656 | 288304509 | ····49569 | 291201872 | ····61454 | 294128352 |
| 400 | ····40828 | ····64518 | ····95343 | ····33340 | ····78544 | ····30992 | ····90720 | ····57765 |
| 410 | ····69232 | ····93064 | 286924032 | ····62173 | 289807522 | ····60115 | 292719989 | ····87181 |
| 420 | ····97639 | 285521613 | ····52725 | ····91009 | ····36503 | ····49241 | ····49261 | 294216600 |
| 430 | 284126049 | ····50166 | ····81420 | 288419848 | ····65487 | 291318370 | ····78536 | ····46021 |
| 440 | ····54462 | ····78721 | 287010118 | ····48690 | ····94473 | ····47502 | 292807814 | ····75446 |
| 450 | ····82877 | 285607278 | ····38819 | ····77535 | 289923462 | ····76637 | ····37095 | 294304874 |
| 460 | 284211295 | ····35839 | ····67523 | 288506383 | ····52455 | 291305774 | ····66379 | ····34304 |
| 470 | ····39717 | ····64403 | ····96230 | ····35233 | ····81450 | ····34915 | ····95665 | ····63737 |
| 480 | ····68141 | ····92969 | 287124939 | ····64087 | 290010448 | ····64059 | 292924955 | ····93174 |
| 490 | ····96567 | 285721538 | ····53652 | ····92943 | ····39449 | ····93205 | ····54247 | 29442261[illegible] |
| 500 | 284324997 | ····50111 | ····82367 | 288621803 | ····68453 | 291522354 | ····83543 | ····52055 |

**Figure C.27**

| | 108000 | 108500 | 109000 | 109500 | 110000 | 110500 | 111000 | 111500 |
|---|---|---|---|---|---|---|---|---|
| 0 | 294452055 | 295927928 | 297411199 | 298901904 | 300400081 | 301905767 | 303419000 | 304939818 |
| 10 | ....81501 | ....57521 | ....40940 | ....31794 | ....30121 | ....35958 | ....49342 | ....70312 |
| 20 | 294510949 | ....87117 | ....70684 | ....61688 | ....60164 | ....66151 | ....79687 | 305000809 |
| 30 | ....40400 | 296016716 | 297500431 | ....91584 | ....90210 | ....96348 | 303510035 | ....31309 |
| 40 | ....69854 | ....46317 | ....30181 | 299021483 | 300520259 | 302026548 | ....40386 | ....61812 |
| 50 | ....99311 | ....75922 | ....59934 | ....51385 | ....50311 | ....56750 | ....70740 | ....92319 |
| 60 | 294628771 | 296105530 | ....89690 | ....81290 | ....80366 | ....86956 | 303601097 | 305122628 |
| 70 | ....58234 | ....35140 | 297619449 | 299111198 | 300610424 | 302117165 | ....31457 | ....53340 |
| 80 | ....87699 | ....64754 | ....49211 | ....41109 | ....40485 | ....47376 | ....61820 | ....83855 |
| 90 | 294717168 | ....94370 | ....78976 | ....71023 | ....70549 | ....77591 | ....92187 | 305214374 |
| 100 | ....46640 | 296223990 | 297708744 | 299200941 | 300700616 | 302207809 | 303722556 | ....44895 |
| 110 | ....76115 | ....53612 | ....38515 | ....30861 | ....30686 | ....38030 | ....52928 | ....75420 |
| 120 | 294805592 | ....83237 | ....68289 | ....60784 | ....60760 | ....68254 | ....83303 | 305305947 |
| 130 | ....35073 | 296312866 | ....98066 | ....90710 | ....90835 | ....98480 | 303813682 | ....36478 |
| 140 | ....64556 | ....42497 | 297827845 | 299320639 | 300820915 | 302328710 | ....44063 | ....67012 |
| 150 | ....94043 | ....72131 | ....57628 | ....50571 | ....50997 | ....58943 | ....74448 | ....97548 |
| 160 | 294923532 | 296401768 | ....87414 | ....80506 | ....81082 | ....89179 | 303904835 | 305428088 |
| 170 | ....53025 | ....31409 | 297917203 | 299410444 | 300911170 | 302419418 | ....35226 | ....58631 |
| 180 | ....82520 | ....61052 | ....46994 | ....40385 | ....41261 | ....49660 | ....65619 | ....89177 |
| 190 | 295012018 | ....90698 | ....76789 | ....70329 | ....71355 | ....79905 | ....96016 | 305519726 |
| 200 | ....41519 | 296520347 | 298006587 | 299500276 | 301001452 | 302510153 | 304026415 | ....50278 |
| 210 | ....71024 | ....49999 | ....36388 | ....30226 | ....31552 | ....40404 | ....56818 | ....80833 |
| 220 | 295100531 | ....79654 | ....66191 | ....60179 | ....61656 | ....70658 | ....87224 | 305611391 |
| 230 | ....30041 | 296609312 | ....95998 | ....90135 | ....91762 | 302600915 | 304117632 | ....41952 |
| 240 | ....59554 | ....38973 | 298125807 | 299620094 | 301121871 | ....31175 | ....48044 | ....72516 |
| 250 | ....89070 | ....68637 | ....55620 | ....50056 | ....51983 | ....61438 | ....78459 | 305703083 |
| 260 | 295218589 | ....98304 | ....85436 | ....80021 | ....82098 | ....91704 | 304208877 | ....33654 |
| 270 | ....48111 | 296727973 | 298215254 | 299709989 | 301212217 | 302721973 | ....39298 | ....64227 |
| 280 | ....77635 | ....57646 | ....45076 | ....39960 | ....42338 | ....52246 | ....69721 | ....94803 |
| 290 | 295307163 | ....87322 | ....74900 | ....69934 | ....72462 | ....82521 | 304300148 | 305825383 |
| 300 | ....36694 | 296817001 | 298304728 | ....99911 | 301302589 | 302812799 | ....30579 | ....55965 |
| 310 | ....66227 | ....46682 | ....34558 | 299829891 | ....32719 | ....43080 | ....61012 | ....86551 |
| 320 | ....95764 | ....76367 | ....64392 | ....59874 | ....62853 | ....73365 | ....91448 | 305917140 |
| 330 | 295425304 | 296906055 | ....94228 | ....89860 | ....92989 | 302903652 | 304421887 | ....47731 |
| 340 | ....54846 | ....35745 | 298424067 | 299919849 | 301423128 | ....33942 | ....52329 | ....78326 |
| 350 | ....84392 | ....65439 | ....53910 | ....49841 | ....53271 | ....64236 | ....82774 | 306008924 |
| 360 | 295513940 | ....95135 | ....83755 | ....79836 | ....83416 | ....94532 | 304513223 | ....39425 |
| 370 | ....43491 | 297024835 | 298513604 | 300009834 | 301513564 | 303024832 | ....43674 | ....70129 |
| 380 | ....73046 | ....54537 | ....43455 | ....39835 | ....43716 | ....55134 | ....74128 | 306100736 |
| 390 | 295602603 | ....84243 | ....73309 | ....69339 | ....73870 | ....85440 | 304604586 | ....31346 |
| 400 | ....32163 | 297113951 | 298603167 | ....99846 | 301604028 | 303115748 | ....35046 | ....61959 |
| 410 | ....61726 | ....43663 | ....33027 | 300129856 | ....34188 | ....46060 | ....65510 | ....92575 |
| 420 | ....91293 | ....73377 | ....62890 | ....59869 | ....64351 | ....76374 | ....95976 | 306223195 |
| 430 | 295720862 | 297203094 | ....92756 | ....89885 | ....94518 | 303206692 | 304726446 | ....53817 |
| 440 | ....50434 | ....32815 | 298722626 | 300219904 | 301724687 | ....37013 | ....56918 | ....84442 |
| 450 | ....80009 | ....62538 | ....52498 | ....49926 | ....54860 | ....67336 | ....87394 | 306315071 |
| 460 | 295809587 | ....92264 | ....82373 | ....79951 | ....85035 | ....97663 | 304817873 | ....45702 |
| 470 | ....39168 | 297321994 | 298812251 | 300309979 | 301815214 | 303327993 | ....48355 | ....76337 |
| 480 | ....68752 | ....51726 | ....42133 | ....40010 | ....45395 | ....58326 | ....78839 | 306406974 |
| 490 | ....98339 | ....81461 | ....72017 | ....70044 | ....75580 | ....88661 | 304909327 | ....37615 |
| 500 | 295927928 | 297411199 | 298901904 | 300400081 | 301905767 | 303419000 | ....39818 | ....68259 |

Figure C.28

| | 112000 | 112500 | 113000 | 113500 | 114000 | 114500 | 115000 | 115500 |
|---|---|---|---|---|---|---|---|---|
| 0 | 306468259 | 308004360 | 309548161 | 311099700 | 312659016 | 314[illegible]26147 | 315801133 | 317384014 |
| 10 | ····98906 | ····35161 | ····79116 | 311130810 | ····90282 | ····57569 | ····32713 | 317415752 |
| 20 | 306529556 | ····65964 | 309610074 | ····61923 | 312721551 | ····88995 | ····64297 | ····[illegible]7494 |
| 30 | ····60209 | ····96771 | ····41035 | ····93039 | ····52823 | 314320424 | ····95883 | ····79238 |
| 40 | ····90865 | 308127581 | ····71999 | 311224159 | ····84098 | ····51856 | 315927473 | 317510986 |
| 50 | 306621524 | ····58393 | 309702966 | ····55281 | 312815377 | ····83291 | ····59065 | ····42737 |
| 60 | ····52186 | ····89209 | ····33936 | ····86407 | ····46658 | 314414730 | ····90661 | ····74492 |
| 70 | ····82851 | 308220028 | ····64910 | 311317535 | ····77943 | ····46171 | 316022260 | 317606249 |
| 80 | 306713519 | ····50850 | ····95886 | ····48667 | 311909231 | ····77616 | ····53862 | ····38010 |
| 90 | ····44191 | ····81675 | 309826866 | ····79802 | ····40522 | 314509064 | ····85468 | ····69773 |
| 100 | ····74865 | 308312503 | ····57849 | 311410940 | ····71816 | ····40515 | 316117076 | 317701540 |
| 110 | 306805543 | ····43335 | ····88835 | ····42081 | 313003113 | ····71969 | ····48688 | ····33311 |
| 120 | ····36223 | ····74169 | 309919823 | ····73225 | ····34413 | 314603426 | ····80303 | ····65084 |
| 130 | ····66907 | 308405006 | ····50815 | 311504372 | ····65717 | ····34886 | 316211921 | ····96860 |
| 140 | ····97594 | ····35847 | ····81811 | ····35523 | ····97023 | ····66350 | ····43542 | 317828640 |
| 150 | 306928283 | ····66690 | 310012809 | ····66676 | 313128333 | ····97816 | ····75167 | ····60423 |
| 160 | ····58976 | ····97537 | ····43810 | ····97833 | ····59645 | 314729286 | 316306794 | ····92209 |
| 170 | ····89672 | 308528387 | ····74814 | 311628993 | ····90961 | ····60759 | ····38425 | 317923998 |
| 180 | 307020371 | ····59240 | 310105822 | ····60156 | 313222281 | ····92235 | ····70059 | ····55791 |
| 190 | ····51073 | ····90096 | ····36833 | ····91322 | ····53603 | 314823714 | 316401696 | ····87586 |
| 200 | ····81778 | 308620955 | ····67846 | 311722491 | ····84928 | ····55197 | ····33336 | 318019385 |
| 210 | 307112486 | ····51817 | ····98863 | ····53663 | 313316257 | ····86682 | ····64979 | ····51187 |
| 220 | ····43197 | ····82682 | 310229883 | ····84839 | ····47588 | 314918171 | ····96626 | ····82992 |
| 230 | ····73912 | 308713550 | ····60906 | 311816017 | ····78923 | ····49663 | 316528275 | 318114800 |
| 240 | 307204629 | ····44422 | ····91932 | ····47199 | 313410261 | ····81158 | ····59928 | ····46612 |
| 250 | ····35150 | ····75296 | 310322961 | ····78383 | ····41602 | 315012656 | ····91584 | ····78426 |
| 260 | ····66073 | 308806174 | ····53993 | 311909571 | ····72946 | ····44157 | 316623243 | 318210244 |
| 270 | ····96800 | ····37054 | ····85029 | ····40762 | 313504293 | ····75661 | ····54906 | ····42065 |
| 280 | 307327529 | ····67938 | 310416067 | ····71956 | ····35644 | 315107169 | ····86571 | ····73889 |
| 290 | ····58262 | ····98825 | ····47109 | 312003153 | ····66997 | ····38680 | 316718240 | 318205717 |
| 300 | ····88998 | 308929715 | ····78154 | ····34354 | ····98354 | ····70194 | ····49912 | ····37548 |
| 310 | 307419737 | ····60608 | 310509201 | ····65557 | 313629714 | 315201711 | ····81587 | ····69381 |
| 320 | ····50479 | ····91504 | ····40252 | ····96764 | ····61077 | ····33231 | 316813265 | 318301218 |
| 330 | ····81224 | 309022403 | ····71306 | 312127973 | ····92443 | ····64754 | ····44946 | ····33058 |
| 340 | 307511972 | ····53305 | 310602363 | ····59186 | 313723812 | ····96281 | ····76631 | ····64902 |
| 350 | ····42723 | ····84210 | ····33424 | ····90402 | ····55185 | 315327810 | 316908318 | ····96748 |
| 360 | ····73478 | 309115119 | ····64487 | 312221621 | ····86560 | ····59343 | ····40009 | 318528598 |
| 370 | 307604235 | ····46030 | ····95553 | ····52843 | 313817939 | ····90879 | ····71703 | ····60451 |
| 380 | ····34995 | ····76945 | 310726623 | ····84069 | ····49321 | 315422418 | 317003400 | ····92307 |
| 390 | ····65759 | 309207863 | ····57696 | 312315297 | ····80705 | ····53960 | ····35100 | 318624166 |
| 400 | ····96525 | ····38783 | ····88771 | ····46529 | 313912094 | ····85506 | ····66804 | ····56028 |
| 410 | 307727295 | ····69707 | 310819850 | ····77763 | ····43485 | 315517054 | ····98511 | ····87894 |
| 420 | ····58068 | 309300634 | ····50932 | 312409001 | ····74879 | ····48606 | 317130221 | 318719763 |
| 430 | ····88844 | ····31564 | ····82017 | ····40242 | 314006277 | ····80161 | ····61934 | ····51635 |
| 440 | 307819622 | ····62498 | 310913106 | ····71486 | ····37677 | 315611719 | ····93650 | ····83510 |
| 450 | ····50404 | ····93434 | ····44197 | 312502733 | ····69081 | ····4328[illegible] | 317225369 | 318815388 |
| 460 | ····81189 | 309424373 | ····75291 | ····33983 | 314100488 | ····74844 | ····57092 | ····47270 |
| 470 | 307911978 | ····55315 | 311006389 | ····65237 | ····31898 | 315706412 | ····88817 | ····79154 |
| 480 | ····42769 | ····86261 | ····37489 | ····6493 | ····63311 | ····37982 | 317320546 | 31891104 |
| 490 | ····73563 | 309517209 | ····68593 | 312627753 | ····94727 | ····69556 | ····52278 | ····42933 |
| 500 | 308004360 | ····48161 | ····99700 | ····59016 | 314226147 | 315801133 | ····84014 | ····74828 |

**Figure C.29**

| | 116000 | 116500 | 117000 | 117500 | 118000 | 118500 | 119000 | 119500 |
|---|---|---|---|---|---|---|---|---|
| 0 | 318974828 | 320573615 | 322180417 | 323795272 | 325418221 | 327049305 | 328688565 | 330336040 |
| 10 | 319006725 | 320605673 | 322212635 | 323827652 | ····50763 | ····82010 | 328721433 | ····69074 |
| 20 | ····38656 | ····37733 | ····44856 | ····60034 | ····83308 | 327114718 | ····54306 | 330402111 |
| 30 | ····70530 | ····69797 | ····77081 | ····92420 | 325515857 | ····47430 | ····87181 | ····35151 |
| 40 | 319102437 | 320701864 | 322309309 | 323924810 | ····48408 | ····80145 | 328820060 | ····68195 |
| 50 | ····34347 | ····33935 | ····41540 | ····57202 | ····80963 | 327212863 | ····52942 | 330501241 |
| 60 | ····66261 | ····66008 | ····73774 | ····89598 | 325613521 | ····45584 | ····85827 | ····34291 |
| 70 | ····98177 | ····98085 | 322406011 | 524021997 | ····46083 | ····78308 | 328918716 | ····67345 |
| 80 | 319230097 | 320830164 | ····38252 | ····54399 | ····78647 | 327311036 | ····51607 | 330600402 |
| 90 | ····62020 | ····62247 | ····70496 | ····86804 | 325711215 | ····43767 | ····84503 | ····33462 |
| 100 | ····93946 | ····94334 | 322502743 | 324119213 | ····43786 | ····76502 | 329017401 | ····66525 |
| 110 | 319325876 | 320926423 | ····34993 | ····51625 | ····76361 | 327409239 | ····50303 | ····99592 |
| 120 | ····57808 | ····58516 | ····67246 | ····84040 | 325808938 | ····41980 | ····83208 | 330732662 |
| 130 | ····89744 | ····90611 | ····99503 | 324216459 | ····41519 | ····74725 | 329116116 | ····65735 |
| 140 | 319421683 | 321022711 | 322631763 | ····48880 | ····74103 | 327507472 | ····49028 | ····98811 |
| 150 | ····53625 | ····54813 | ····64026 | ····81305 | 325906691 | ····40223 | ····81943 | 330831891 |
| 160 | ····85571 | ····86918 | ····96293 | 324313733 | ····39281 | ····72977 | 329214861 | ····64975 |
| 170 | 319517519 | 321119027 | 322728562 | ····46165 | ····71875 | 327605734 | ····47782 | ····98061 |
| 180 | ····49471 | ····51139 | ····60835 | ····78599 | 326007442 | ····38495 | ····80707 | 330931151 |
| 190 | ····81426 | ····83254 | ····93111 | 324411037 | ····37073 | ····71258 | 329313635 | ····64244 |
| 200 | 319613384 | 321215372 | 322825390 | ····43478 | ····69676 | 327704026 | ····46567 | ····97340 |
| 210 | ····45345 | ····47494 | ····57673 | ····75923 | 326102283 | ····36796 | ····79501 | 331030440 |
| 220 | ····77310 | ····79619 | ····89959 | 324508370 | ····34894 | ····69570 | 329412439 | ····63543 |
| 230 | 319709278 | 321311747 | 322922248 | ····40821 | ····67507 | 327802347 | ····45380 | ····96649 |
| 240 | ····41248 | ····43878 | ····54540 | ····73275 | 326200124 | ····35127 | ····78325 | 331129759 |
| 250 | ····73223 | ····76012 | ····86835 | 324605732 | ····32744 | ····67910 | 329511273 | ····62872 |
| 260 | 319805200 | 321408150 | 323019134 | ····38193 | ····65367 | 327800697 | ····44224 | ····95988 |
| 270 | ····37180 | ····40291 | ····51436 | ····70657 | ····97994 | ····33487 | ····77178 | 331229108 |
| 280 | ····69164 | ····72435 | ····83741 | 324703124 | 326330624 | ····66281 | 329610136 | ····62231 |
| 290 | 319901151 | 321504582 | 323116050 | ····35594 | ····63257 | ····99077 | ····43097 | ····95357 |
| 300 | ····33141 | ····36732 | ····48361 | ····68068 | ····95893 | 328031877 | ····76061 | 331328487 |
| 310 | ····65134 | ····68886 | ····80676 | 324800545 | 326428533 | ····64680 | 329709029 | ····61619 |
| 320 | ····97131 | 321601043 | 323212994 | ····33025 | ····61175 | ····97487 | ····42000 | ····94756 |
| 330 | 320029131 | ····33203 | ····45315 | ····65508 | ····93822 | 328130297 | ····74974 | 331427895 |
| 340 | ····61133 | ····65366 | ····77640 | ····97995 | 326526471 | ····63110 | 329807951 | ····61038 |
| 350 | ····93140 | ····97533 | 323309968 | 324930484 | ····59124 | ····95926 | ····40932 | ····94184 |
| 360 | 320125149 | 321729703 | ····42299 | ····62977 | ····91779 | 328228745 | ····73916 | 331527333 |
| 370 | ····57161 | ····61876 | ····74633 | ····95474 | 326624439 | ····61568 | 329906904 | ····60486 |
| 380 | ····89177 | ····94052 | 323406970 | 325027973 | ····57101 | ····94395 | ····39894 | ····93642 |
| 390 | 320221196 | 321826231 | ····39311 | ····60476 | ····89767 | 328327224 | ····72888 | 331626802 |
| 400 | ····53218 | ····58414 | ····71655 | ····92987 | 326722436 | ····60057 | 330005886 | ····59964 |
| 410 | ····85244 | ····90600 | 323504002 | 325125491 | ····55108 | ····92893 | ····38886 | ····93130 |
| 420 | 320317272 | 321922789 | ····36352 | ····58004 | ····87783 | 328425732 | ····71890 | 331726300 |
| 430 | ····49304 | ····54981 | ····68706 | ····90520 | 326820462 | ····58575 | 330104898 | ····59472 |
| 440 | ····81339 | ····87177 | 323601063 | 325223039 | ····53144 | ····91420 | ····37908 | ····92648 |
| 450 | 320413377 | 322019375 | ····33423 | ····55561 | ····85830 | 328524270 | ····70922 | 331825827 |
| 460 | ····45418 | ····51577 | ····65786 | ····88087 | 326918518 | ····57122 | 330203939 | ····59010 |
| 470 | ····77463 | ····83782 | ····98153 | 325320615 | ····51210 | ····89978 | ····36959 | ····91196 |
| 480 | 320509510 | 322115991 | 323730523 | ····53148 | ····83905 | 328622857 | ····69983 | 331925385 |
| 490 | ····41561 | ····48202 | ····62896 | ····85683 | 327016604 | ····55699 | 330303010 | ····58578 |
| 500 | ····73615 | ····80417 | ····95272 | 325418221 | ····49305 | ····88565 | ····36040 | ····91774 |

**Figure C.30**

| | 120000 | 120500 | 121000 | 121500 | 122000 | 122500 | 123000 | 123500 |
|---|---|---|---|---|---|---|---|---|
| 0 | 331991774 | 333655806 | 335328179 | 337008934 | 338698114 | 340395760 | 342101915 | 343816622 |
| 10 | 332024973 | ····89171 | ····61712 | ····42635 | 338731983 | 340429800 | ····36125 | ····51004 |
| 20 | ····58175 | 333722540 | ····95248 | ····76339 | ····65857 | ····63843 | ····70339 | ····85389 |
| 30 | ····91381 | ····55913 | 335428787 | 337110047 | ····99733 | ····97889 | 342204556 | 343919777 |
| 40 | 332124590 | ····89288 | ····62330 | ····43758 | 338833613 | 340531939 | ····38776 | ····54169 |
| 50 | ····57803 | 333822667 | ····95876 | ····77472 | ····67497 | ····65992 | ····73000 | ····88565 |
| 60 | ····91018 | ····56049 | 335529426 | 337211190 | 338901383 | 340600048 | 342307228 | 344022964 |
| 70 | 332224238 | ····89435 | ····62979 | ····44911 | ····35273 | ····34108 | ····41458 | ····57366 |
| 80 | ····57460 | 333922824 | ····96535 | ····78635 | ····69167 | ····68172 | ····75692 | ····91772 |
| 90 | ····90685 | ····56316 | 335630095 | 337312363 | 339003064 | 340702239 | 342409930 | 344126181 |
| 100 | 332323915 | ····89612 | ····63658 | ····46095 | ····36964 | ····36309 | ····44171 | ····60594 |
| 110 | ····57147 | 334023011 | ····97224 | ····79829 | ····70868 | ····70382 | ····78415 | ····95010 |
| 120 | ····90383 | ····56413 | 335730794 | 337413567 | 339104775 | 340804459 | 342512663 | 344229429 |
| 130 | 332423622 | ····89819 | ····64367 | ····47309 | ····38685 | ····38540 | ····46915 | ····63852 |
| 140 | ····56864 | 334123228 | ····97943 | ····81053 | ····72599 | ····72624 | ····81169 | ····98278 |
| 150 | ····90110 | ····56640 | 335831523 | 337514801 | 339206517 | 340906711 | 342615427 | 344332708 |
| 160 | 332523359 | ····90056 | ····65106 | ····48553 | ····40437 | ····40802 | ····49689 | ····67142 |
| 170 | ····56611 | 334223475 | ····98693 | ····82308 | ····74361 | ····74896 | ····83954 | 344401578 |
| 180 | ····89867 | ····56897 | 335932283 | 337616066 | 339308289 | 341008993 | 342718222 | ····36018 |
| 190 | 332623126 | ····90323 | ····65876 | ····49828 | ····42220 | ····43094 | ····52494 | ····70462 |
| 200 | ····56388 | 334323752 | ····99473 | ····83593 | ····76154 | ····77199 | ····86769 | 344504909 |
| 210 | ····89654 | ····57184 | 336033072 | 337717361 | 339410091 | 341111306 | 342821048 | ····39360 |
| 220 | 332722923 | ····90620 | ····66676 | ····51133 | ····44032 | ····45417 | ····55330 | ····73813 |
| 230 | ····56195 | 334424059 | 336100282 | ····84908 | ····77977 | ····79532 | ····89616 | 344608271 |
| 240 | ····89471 | ····57501 | ····33892 | 337818686 | 339511925 | 341213650 | 342923905 | ····42732 |
| 250 | 332822750 | ····90947 | ····67506 | ····52468 | ····45876 | ····47771 | ····58197 | ····77196 |
| 260 | ····56032 | 334524396 | 336201123 | ····86253 | ····79830 | ····81896 | ····92493 | 344711664 |
| 270 | ····89318 | ····57849 | ····34743 | 337920042 | 339613788 | 341316024 | 343026792 | ····46135 |
| 280 | 332922606 | ····91304 | ····68366 | ····53834 | ····47750 | ····50156 | ····61095 | ····80610 |
| 290 | ····55899 | 334624763 | 336301993 | ····87629 | ····81715 | ····84291 | ····95401 | 344815088 |
| 300 | ····89194 | ····58226 | ····35623 | 338021428 | 339715683 | 341418429 | 343129710 | ····49569 |
| 310 | 333022493 | ····91692 | ····69257 | ····55230 | ····49654 | ····52571 | ····64023 | ····84054 |
| 320 | ····55795 | 334725161 | 336402894 | ····89036 | ····83629 | ····86716 | ····98340 | 344918543 |
| 330 | ····89101 | ····58633 | ····36534 | 338122845 | 339817608 | 341520865 | 343232660 | ····53034 |
| 340 | 333122410 | ····92109 | ····70178 | ····56657 | ····51590 | ····55017 | ····66983 | ····87530 |
| 350 | ····55722 | 334825589 | 336503825 | ····90473 | ····85575 | ····89173 | 343301310 | 345022029 |
| 360 | ····89038 | ····59071 | ····37475 | 338224292 | 339919563 | 341623331 | ····35040 | ····56531 |
| 370 | 333222357 | ····92557 | ····71129 | ····58114 | ····53555 | ····57494 | ····69973 | ····91036 |
| 380 | ····55679 | 334926046 | 336604786 | ····91940 | ····87551 | ····91659 | 343404310 | 345125546 |
| 390 | ····89004 | ····59539 | ····38446 | 338325769 | 340021550 | 341725829 | ····38651 | ····60058 |
| 400 | 333322333 | ····93035 | ····72110 | ····59602 | ····55551 | ····60001 | ····72995 | ····94574 |
| 410 | ····55656 | 335026534 | 336705777 | ····93438 | ····89557 | ····94177 | 343507342 | 345229093 |
| 420 | ····89001 | ····60037 | ····39448 | 338427277 | 340123566 | 341828357 | ····41693 | ····63616 |
| 430 | 333422340 | ····93543 | ····73122 | ····61120 | ····57578 | ····62539 | ····76047 | ····98143 |
| 440 | ····55682 | 335127052 | 336806799 | ····94966 | ····91594 | ····96726 | 343610404 | 345332672 |
| 450 | ····89028 | ····60565 | ····40480 | 338528815 | 340225613 | 341930915 | ····44765 | ····67206 |
| 460 | 333522377 | ····94081 | ····74164 | ····62668 | ····59636 | ····65109 | ····79130 | 345401742 |
| 470 | ····55729 | 335227600 | 336907851 | ····96525 | ····93662 | ····99305 | 343713498 | ····36283 |
| 480 | ····89085 | ····61123 | ····41542 | 338630384 | 340327691 | 342033505 | ····47869 | ····70826 |
| 490 | 333622444 | ····94649 | ····75236 | ····64247 | ····61724 | ····67708 | ····82244 | 345505373 |
| 500 | ····55806 | 335328179 | 337008934 | ····98114 | ····95760 | 342101915 | 343816622 | ····39924 |

**Figure C.31**

| | 124000 | 124500 | 125000 | 125500 | 126000 | 126500 | 127000 | 127500 |
|---|---|---|---|---|---|---|---|---|
| 0 | 345539924 | 347271863 | 349012483 | 35 761828 | 352519941 | 354286[illegible] | 256062647 | 357847329 |
| 10 | ··· 74478 | 347306590 | ····47385 | ····96974 | ····55193 | 354322294 | ····98253 | ····83114 |
| 20 | 345609035 | ····41321 | ····82289 | 550831984 | ····90448 | ····577[illegible]7 | 356133163 | 357918902 |
| 30 | ····43596 | ····76055 | 349117197 | ····67067 | 352625707 | ····93162 | ····69477 | ····54694 |
| 40 | ····7[illegible]161 | 347410793 | ····52109 | 350902154 | ····60970 | 354428602 | 356205094 | ····90489 |
| 50 | 345712728 | ····45534 | ····87024 | ····37244 | ····96236 | ····64045 | ····40714 | 358026288 |
| 60 | ····47300 | ····80278 | 349221943 | ····72338 | 352731506 | ··· 99491 | ····76338 | ····62091 |
| 70 | ····81874 | 347515026 | ····56865 | 351007435 | ····66779 | 354534941 | 356311966 | ····97897 |
| 80 | 345816453 | ····49778 | ····91791 | ····42536 | 352802056 | ····70394 | ····47597 | 358133707 |
| 90 | ····51034 | ····84533 | 349326720 | ····77640 | ····37336 | 354605851 | ····83232 | ····69520 |
| 100 | ····85619 | 347619291 | ····61653 | 351112748 | ····72619 | ····41312 | 356418870 | 358205337 |
| 110 | 345920208 | ····54053 | ····96589 | ····47859 | 352907907 | ····76776 | ····54512 | ····41158 |
| 120 | ····54800 | ····88819 | 349431529 | ····82974 | ····43198 | 354712244 | ····90157 | ····76981 |
| 130 | ····89395 | 347723588 | ····66472 | 351218092 | ····78492 | ····47715 | 356525806 | 358312810 |
| 140 | 346023994 | ····58360 | 349501419 | ····53214 | 353013790 | ····83190 | ····61459 | ····48641 |
| 150 | ····58597 | ····93136 | ····36369 | ····88339 | ····49091 | 354818668 | ····97115 | ····84476 |
| 160 | ····93203 | 347827915 | ····71322 | 351323468 | ····84396 | ····54150 | 356632775 | 358420314 |
| 170 | 346127812 | ····64698 | 349606279 | ····58600 | 353119704 | ····89636 | ····68438 | ····56156 |
| 180 | ····62425 | ····97484 | ····41240 | ····93736 | ····55016 | 354925125 | 356704105 | ····92002 |
| 190 | ····97041 | 347932274 | ····76204 | 351428876 | ····90332 | ····60617 | ····39775 | 358527891 |
| 200 | 346231661 | ····67067 | 349711172 | ··· 64018 | 353225651 | ··· 96113 | ····75449 | ····63704 |
| 210 | ····66284 | 348001864 | ····46143 | ····99165 | ····60973 | 355031613 | 356811127 | ····99560 |
| 220 | 346300910 | ····36664 | ····81117 | 351534315 | ····96299 | ····37116 | ····46808 | 358635420 |
| 230 | ····35540 | ····71468 | 349816095 | ····69468 | 353331629 | 355102623 | ····82493 | ····71284 |
| 240 | ····70174 | 348106275 | ····51077 | 351604625 | ····66962 | ····38133 | 356918181 | 358707151 |
| 250 | 346404811 | ····41085 | ····86062 | ····39786 | 353402299 | ····73647 | ····53873 | ····43022 |
| 260 | ····39451 | ····75899 | 349921051 | ····74950 | ····37639 | 355209164 | ····89568 | ····78896 |
| 270 | ····74095 | 348210717 | ····56043 | 351710117 | ····72983 | ····44685 | 357025267 | 358814774 |
| 280 | 346508743 | ····45538 | ····91039 | ····45288 | 353508330 | ····80209 | ····60970 | ····50655 |
| 290 | ····43394 | ····80363 | 350026038 | ····80463 | ····43681 | 355315737 | ····96676 | ····86540 |
| 300 | ····78048 | 348315191 | ····61040 | 351815641 | ····79036 | ····51269 | 357132385 | 358922429 |
| 310 | 346612706 | ····50022 | ····96046 | ····50822 | 353614393 | ····86804 | ····68099 | ····58321 |
| 320 | ····47367 | ····84857 | 350131056 | ····86[illegible]07 | ····49755 | 355422343 | 357203815 | ····94217 |
| 330 | ····82032 | 348419696 | ····66069 | 351921196 | ····85120 | ····57885 | ····39536 | 359030117 |
| 340 | 346716700 | ····54538 | 350201086 | ····56388 | 353720488 | ····93431 | ····75260 | ····66020 |
| 350 | ····51372 | ····89383 | ····36106 | ····91584 | ····55860 | 355528980 | 357310987 | 359101926 |
| 360 | ····86047 | 348524232 | ····71129 | 352026783 | ····91236 | ····64533 | ····46718 | ····37836 |
| 370 | 346820725 | ····59085 | 350306157 | ····61986 | 353826615 | 355600090 | ····82453 | ····73750 |
| 380 | ····55408 | ····93940 | ····41187 | ····97192 | ····61998 | ····35650 | 357418191 | 359209668 |
| 390 | ····90093 | 348628800 | ····76221 | 352132401 | ····97384 | ····71213 | ····53933 | ····45589 |
| 400 | 346924782 | ····63663 | 350411259 | ····62615 | 353932774 | 355706780 | ····89678 | ····81513 |
| 410 | ····59475 | ····98529 | ····46300 | 352202831 | ····68167 | ····42351 | 357525427 | 359317441 |
| 420 | ····94170 | 348733399 | ····81345 | ····38052 | 354003564 | ····779[illegible]5 | ····61180 | ····53373 |
| 430 | 347028870 | ····6[illegible]272 | 350516393 | ····73276 | ····38964 | 355813503 | ····96936 | ····89308 |
| 440 | ····63573 | 348803149 | ····51444 | 352308503 | ····74368 | ····49084 | 357632696 | 359425247 |
| 450 | ····98279 | ····38029 | ····86499 | ····43734 | 354109775 | ····84669 | ····68459 | ····61190 |
| 460 | 347132989 | ····72913 | 350621558 | ····78968 | ····45186 | 355920258 | 357704226 | ····97136 |
| 470 | ····67702 | 348907800 | ····56620 | 352414206 | ····80601 | ····55850 | ····39996 | 359333086 |
| 480 | 347202419 | ····42691 | ····91686 | ····49447 | 354216019 | ····91445 | ····75770 | ····69039 |
| 490 | ····37139 | ····77586 | 350726755 | ····84692 | ····51441 | 356027044 | 357811548 | 359604996 |
| 500 | ····71863 | 349012483 | ····61828 | 352719941 | ····86266 | ····62647 | ····47329 | ····40956 |

E

Figure C.32

| | 128000 | 128500 | 129000 | 129500 | 130000 | 130500 | 131000 | 131500 |
|---|---|---|---|---|---|---|---|---|
| 0 | 359640956 | 361343574 | 363255226 | 365075959 | 366905819 | 368744850 | 370593098 | 372450611 |
| 10 | ....76920 | ....79718 | ....91512 | 365112467 | ....42504 | ....81724 | 370730158 | ....87856 |
| 20 | 359712888 | 361515866 | 363327881 | ....48975 | ....79204 | 368818602 | ....67221 | 372525105 |
| 30 | ....48859 | ....52018 | ....64214 | ....85493 | 367015901 | ....55484 | 370704287 | ....62357 |
| 40 | ....84834 | ....88173 | 363400550 | 365222012 | ....52603 | ....92370 | ....41358 | ....99613 |
| 50 | 358920813 | 361624332 | ....36890 | ....58534 | ....89308 | 368929259 | ....7843 | 372636873 |
| 60 | ....56795 | ....60494 | ....73234 | ....95060 | 367126017 | ....66152 | 370815510 | ....74137 |
| 70 | ....92781 | ....96660 | 363509581 | 365331589 | ....62730 | 369003048 | ....52591 | 372711404 |
| 80 | 359928770 | 361732830 | ....45932 | ....68122 | ....99446 | ....39949 | ....89676 | ....48676 |
| 90 | ....64763 | ....69003 | ....82287 | 365404659 | 367236166 | ....76853 | 370926765 | ....85950 |
| 100 | 360000759 | 361805180 | 363618645 | ....41200 | ....72890 | 369113760 | ....63858 | 372823229 |
| 110 | ....36759 | ....41361 | ....55007 | ....77744 | 367309617 | ....50672 | 371000955 | ....60511 |
| 120 | ....72763 | ....77545 | ....91373 | 365514292 | ....46348 | ....87587 | ....38055 | ....97797 |
| 130 | 360108770 | 361913733 | 363727742 | ....50843 | ....83083 | 369224506 | ....75158 | 372935087 |
| 140 | ....44781 | ....49924 | ....64115 | ....87398 | 367419821 | ....61428 | 371112266 | ....72381 |
| 150 | ....80796 | ....86119 | 363800491 | 365623957 | ....56563 | ....98354 | ....49377 | 373009678 |
| 160 | 360216814 | 362022317 | ....36871 | ....60519 | ....93309 | 369335284 | ....86492 | ....46979 |
| 170 | ....52836 | ....58520 | ....73255 | ....97085 | 367530053 | ....72218 | 371223611 | ....84284 |
| 180 | ....88261 | ....94726 | 363909642 | 365733655 | ....66811 | 369409155 | ....60733 | 3·3121592 |
| 190 | 360324890 | 362130938 | ....46033 | ....70229 | 367603568 | ....46096 | ....97859 | ....58904 |
| 200 | ....60922 | ....67184 | ....82427 | 365806806 | ....40328 | ....83040 | 371334989 | ....96220 |
| 210 | ....96958 | 362203365 | 364018826 | ....43386 | ....77092 | 369519989 | ....72122 | 373233540 |
| 220 | 360432998 | ....39585 | ....55228 | ....79971 | 367713860 | ....56941 | 371409260 | ....70863 |
| 230 | ....69041 | ....75809 | ....91633 | 365916559 | ....50631 | ....93896 | ....46401 | 373308190 |
| 240 | 360505088 | 362312037 | 364128042 | ....53150 | ....87406 | 369630856 | ....83545 | ....45521 |
| 250 | ....41139 | ....48268 | ....64455 | ....89746 | 367824185 | ....67819 | 371520694 | ....82856 |
| 260 | ....77193 | ....84503 | 364200871 | 366026345 | ....60967 | 369704786 | ....57846 | 373420194 |
| 270 | 360613250 | 362420741 | ....37292 | ....62947 | ....97753 | ....41756 | ....95007 | ....57736 |
| 280 | ....49312 | ....56983 | ....73715 | ....99553 | 367934543 | ....78730 | 371132161 | ....94882 |
| 290 | ....85377 | ....93229 | 364310143 | 366136163 | ....71337 | 369815708 | ....69324 | 373532231 |
| 300 | 360721445 | 362529478 | ....46574 | ....72777 | 368008134 | ....52690 | 371706491 | ....69584 |
| 310 | ....57517 | ....65731 | ....83008 | 366209394 | ....44935 | ....89675 | ....43662 | 373606941 |
| 320 | ....93593 | 362601988 | 364419447 | ....46015 | ....81739 | 369926664 | ....80836 | ....44302 |
| 330 | 360829672 | ....38248 | ....55889 | ....82640 | 368115547 | ....63657 | 371818014 | ....81666 |
| 340 | ....65755 | ....74512 | ....92334 | 366319268 | ....55359 | 370000653 | ....55196 | 373719035 |
| 350 | 360901842 | 362710779 | 364528783 | ....55900 | ....92175 | ....37653 | ....92382 | ....56406 |
| 360 | ....37932 | ....47050 | ....65236 | ....92536 | 368228994 | ....74657 | 371929571 | ....93782 |
| 370 | ....74026 | ....83325 | 364601693 | 366429175 | ....65817 | 370111665 | ....66764 | 373831161 |
| 380 | 361010123 | 362819603 | ....38153 | ....65818 | 368302643 | ....48676 | 372003960 | ....68545 |
| 390 | ....46224 | ....55885 | ....74617 | 366502464 | ....39474 | ....85691 | ....41161 | 373905931 |
| 400 | ....82329 | ....92171 | 364711084 | ....39·15 | ....76308 | 270222709 | ....78365 | ....43322 |
| 410 | 361118437 | 362928460 | ....47555 | ....75769 | 368413145 | ....59732 | 372115573 | ....80·16 |
| 420 | ....54549 | ....64753 | ....84030 | 366612426 | ....49986 | ....96758 | ....52784 | 37401 114 |
| 430 | ....90664 | 363001049 | 364820509 | ....49087 | ....86831 | 370333787 | ....90000 | ....55516 |
| 440 | 361226784 | ....37350 | ....56991 | ....85752 | 368523680 | ....70821 | 372227219 | ....929·2 |
| 450 | ....62906 | ....73653 | ....93476 | 366722421 | ....60533 | 370407857 | ....64441 | 374130331 |
| 460 | ....99033 | 363109961 | 364929966 | ....59093 | ....97389 | ....44898 | 372301668 | ....67746 |
| 470 | 361335162 | ....46272 | ....66459 | ....95769 | 368634248 | ....81943 | ....38898 | 374205161 |
| 480 | ....71296 | ....82586 | 365002955 | 366832449 | ....71112 | 370518991 | ....76132 | ....42581 |
| 490 | 361307433 | 363218905 | ....39456 | ....69132 | 368707979 | ....56043 | 372413369 | ....80006 |
| 500 | ....43574 | ....55226 | ....75959 | 366905819 | ....44850 | ....93098 | ....50611 | 374317434 |

Figure C.33

| | 132000 | 132500 | 133000 | 133500 | 134000 | 134500 | 135000 | 135500 |
|---|---|---|---|---|---|---|---|---|
| 0 | 374317434 | 376193614 | 378079197 | 3799[illegible]4232 | 381878766 | 383792845 | 385716518 | 387649833 |
| 10 | ....54865 | 376231233 | 378117005 | 380012230 | 381916953 | 383831224 | ....55090 | ....88598 |
| 20 | ....92301 | ....68856 | ....5481 | ....50231 | ....55145 | ....69607 | ....93665 | 387727367 |
| 30 | 374429740 | 376306483 | ....92632 | ....88236 | ....93341 | 383907994 | 385832245 | ....66140 |
| 40 | ....57182 | ....44114 | 378230452 | 380126245 | 382031540 | ....46385 | ....70828 | 387804917 |
| 50 | 374504630 | ....31745 | ....68275 | ....64257 | ....69743 | ....84780 | 385909415 | ....43697 |
| 60 | ....42080 | 376419386 | 378306102 | 380202274 | 382107950 | 384023178 | ....48006 | ....82481 |
| 70 | ....79534 | ....57028 | ....43932 | ....40294 | ....46161 | ....61580 | ....86601 | 387921270 |
| 80 | 374616992 | ....94674 | ....81767 | ....78318 | ....84375 | ....99987 | 386025199 | ....60062 |
| 90 | ....54454 | 376532323 | 378419605 | 380316346 | 382222594 | 384138397 | ....63802 | ....98558 |
| 100 | ....91920 | ....69977 | ....57447 | ....54377 | ....60816 | ....76810 | 386102408 | 388037658 |
| 110 | 374729389 | 376607634 | ....95292 | ....92413 | ....99042 | 384215228 | ....41019 | ....76461 |
| 120 | ....66862 | ....45294 | 378533142 | 380430452 | 382337272 | ....53650 | ....79633 | 388115269 |
| 130 | 374804338 | ....82959 | ....70995 | ....68495 | ....75506 | ....92075 | 386218251 | ....54081 |
| 140 | ....41819 | 376720627 | 378608852 | 380506542 | 382413743 | 384330504 | ....56872 | ....92896 |
| 150 | ....79303 | ....58299 | ....46713 | ....44593 | ....51984 | ....68937 | ....95498 | 388231715 |
| 160 | 3749[illegible]6791 | ....95975 | ....84578 | ....82647 | ....90229 | 384407374 | 386334125 | ....70538 |
| 170 | ....54283 | 376833655 | 378722446 | 380620706 | 382528478 | ....45815 | ....72761 | 388309366 |
| 180 | ....91778 | ....71338 | ....60319 | ....58768 | ....66731 | ....84260 | 386411398 | ....48196 |
| 190 | 375029277 | 376909025 | ....98195 | ....96834 | 382604988 | 384522708 | ....50039 | ....87021 |
| 200 | ....66780 | ....46716 | 378836075 | 380734903 | ....43248 | ....61160 | ....88684 | 388425870 |
| 210 | 375104287 | ....84411 | ....73958 | ....72977 | ....81512 | ....99616 | 386527333 | ....64713 |
| 220 | ....41797 | 377022109 | 378911846 | 380811054 | 382719781 | 384638076 | ....65986 | 388503559 |
| 230 | ....79311 | ....59811 | ....49737 | ....49135 | ....58053 | ....76540 | 386604643 | ....42409 |
| 240 | 375216829 | ....97517 | ....87632 | ....87220 | ....96328 | 384715008 | ....43303 | ....81264 |
| 250 | ....54351 | 377135227 | 379025531 | 380925309 | 382834609 | ....53479 | ....81967 | 388620122 |
| 260 | ....91876 | ....72941 | ....63433 | ....63401 | ....72892 | ....91955 | 386720636 | ....58984 |
| 270 | 375329406 | 377210558 | 379101339 | 381001498 | 382911180 | 384830434 | ....59308 | ....97850 |
| 280 | ....66939 | ....48379 | ....39250 | ....39598 | ....49471 | ....68917 | ....97984 | 388736719 |
| 290 | 375404475 | ....86104 | ....77163 | ....77702 | ....87766 | 384907404 | 386836663 | ....75593 |
| 300 | ....42016 | 377323832 | 379215081 | 381115809 | 383026065 | ....45894 | ....75347 | 388814471 |
| 310 | ....79560 | ....61565 | ....53003 | ....53921 | ....64367 | ....84389 | 386914035 | ....53352 |
| 320 | 375517108 | ....99301 | ....90927 | ....92036 | 383102674 | 385022887 | ....52726 | ....92237 |
| 330 | ....54660 | 377437041 | 379328857 | 381230156 | ....40984 | ....61390 | ....91421 | 388931127 |
| 340 | ....92215 | ....74785 | ....66790 | ....68279 | ....79298 | ....99896 | 387030120 | ....70020 |
| 350 | 375629774 | 377512532 | 379404724 | 382306405 | 383217616 | 385138406 | ....68823 | 389008917 |
| 360 | ....67337 | ....50283 | ....42667 | ....44536 | ....55938 | ....76920 | 387107530 | ....47818 |
| 370 | 375704904 | ....88038 | ....80611 | ....82670 | ....94263 | 385215437 | ....46241 | ....86722 |
| 380 | ....42475 | 377625797 | 379518560 | 381430809 | 383332593 | ....53959 | ....84956 | 389125631 |
| 390 | ....80049 | ....03560 | ....56511 | ....58951 | ....70926 | ....92484 | 387223674 | ....64544 |
| 400 | 375817627 | 377701326 | ....94467 | ....97097 | 383409263 | 385331014 | ....62397 | 389203460 |
| 410 | ....55200 | ....39096 | 379932427 | 381535246 | ....47604 | ....69547 | 387301123 | ....42380 |
| 420 | ....92794 | ....76870 | ....70390 | ....73400 | ....85949 | 388408084 | ....39853 | ....81305 |
| 430 | 375930383 | 377814648 | 379708357 | 381611557 | 383524297 | ....46624 | ....78587 | 389320233 |
| 440 | ....67976 | ....52429 | ....46328 | ....49718 | ....62650 | ....85169 | 387417325 | ....59165 |
| 450 | 5 6005573 | ....90214 | ....84302 | ....87883 | 383601006 | 385523718 | ....56067 | ....98101 |
| 460 | ....43174 | 377928003 | 379822281 | 381726052 | ....39366 | ....62270 | ....94812 | 389437040 |
| 470 | ....80778 | ....65796 | ....60263 | ....64225 | ....77730 | 385600826 | 387533562 | ....75984 |
| 480 | 376118386 | 378003593 | ....98249 | 381902401 | 383716098 | ....39386 | ....72315 | 389514932 |
| 490 | ....55998 | ....41393 | 379936239 | ....40591 | ....54469 | ....77950 | 387611072 | ....53883 |
| 500 | ....93614 | ....79197 | ....74232 | ....78766 | ....92845 | 385716518 | ....49833 | ....92839 |

E 2

Figure C.34

| | 136000 | 136500 | 137000 | 137500 | 138000 | 138500 | 139000 | 139500 |
|---|---|---|---|---|---|---|---|---|
| 0 | 389592839 | 391545583 | 393508115 | 395480484 | 399462739 | 399454929 | 401457105 | 403469316 |
| 10 | 389631798 | ....84738 | ....47466 | 395520032 | 397502485 | ....94875 | ....97251 | 403509663 |
| 20 | ....70761 | 391623896 | ....86821 | ....59584 | ....42235 | 399534824 | 401537400 | ....50014 |
| 30 | 389709728 | ....63059 | 393626179 | ....99140 | ....81989 | ....74778 | ....77554 | ....90369 |
| 40 | ....48699 | 390702225 | ....65542 | 395638700 | 397621748 | 399614735 | 401617712 | 403630728 |
| 50 | ....87674 | ....41395 | 393704909 | .... 78264 | ....61510 | ....54697 | ....57874 | ....71091 |
| 60 | 389826653 | ....80569 | ....44279 | 395717832 | 397701276 | ....94662 | ....98039 | 403711458 |
| 70 | ....65636 | 391819747 | ....83653 | ....57403 | ....41046 | 399734632 | 401738209 | ....51830 |
| 80 | 389904622 | ....58929 | 393823032 | ....96979 | ....80820 | ....74605 | ....78383 | ....92205 |
| 90 | ....43613 | ....98115 | .... 62414 | 395836559 | 397820598 | 399814583 | 401818561 | 403832584 |
| 100 | ....82607 | 391937305 | 393901800 | ....76142 | ....60380 | ....54564 | ....58743 | ....72967 |
| 110 | 390021605 | ....76499 | ....41191 | 395915730 | 397900166 | ....94549 | ....98929 | 40391335[illegible] |
| 120 | ....60608 | 392015596 | ....80585 | ....55322 | ....39956 | 399934539 | 401939119 | ....53746 |
| 130 | ....99614 | ....54898 | 394019983 | ....94917 | ....79750 | ....74532 | ....79312 | ....94141 |
| 140 | 390138623 | ....94103 | ....59385 | 396034517 | 398019548 | 400014530 | 402019510 | 404034541 |
| 150 | ....77637 | 392133313 | ....98791 | ....74120 | ....59350 | ....54531 | ....59712 | ....74944 |
| 160 | 390216655 | ....72526 | 394138201 | 396113728 | ....99156 | ....94537 | ....99918 | 404115352 |
| 170 | ....55677 | 392211743 | ....77614 | ....53339 | 398138966 | 400134546 | 402140125 | ....55763 |
| 180 | ....94702 | ....50965 | 394217032 | ....92954 | ....78780 | ....74560 | ....80342 | ....94179 |
| 190 | 390333732 | ....90190 | ....56454 | 396232574 | 398218598 | 400214577 | 402220560 | 404236598 |
| 200 | ....72765 | 392329419 | ....95880 | ....72197 | ....58420 | ....54599 | ....6078[illegible] | ....77022 |
| 210 | 390411802 | ....68652 | 394335309 | 396311824 | ....98246 | ....94624 | 402301009 | 404317450 |
| 220 | ....50844 | 392407889 | ....74743 | ....51455 | 398338075 | 400334654 | ....41239 | ....57881 |
| 230 | ....89889 | ....47129 | 394414180 | ....91090 | ....77909 | ....74687 | ....81473 | ....98317 |
| 240 | 390528938 | ....86374 | ....53622 | 396430729 | 398417747 | 400414725 | 402421711 | 404438757 |
| 250 | ....67991 | 392525623 | ....93067 | ....70372 | ....57589 | ....54766 | ....61953 | ....79201 |
| 260 | 390607047 | ....64875 | 394532516 | 396510019 | ....97435 | ....94811 | 402502199 | 404519649 |
| 270 | ....48105 | 392604132 | ....71969 | ....49670 | 398537284 | 400534861 | ....42449 | ....60101 |
| 280 | ....85173 | ....43392 | 394611427 | ....89325 | ....77138 | ....74914 | ....82704 | 404600557 |
| 290 | 390724241 | ....82656 | ....50888 | 396628984 | 398616996 | 400614972 | 402622962 | ....41017 |
| 300 | ....63314 | 392721925 | ....90353 | ....68647 | ....56858 | ....55033 | ....63224 | ....81481 |
| 310 | 390802390 | ....61197 | 394729822 | 396708314 | ....96723 | ....95099 | 402703491 | 404721949 |
| 320 | ....41470 | 392802473 | ....69295 | ....47985 | 398736593 | 400735168 | ....43761 | ....62421 |
| 330 | ....80554 | ....39753 | 394808772 | ....87660 | ....76467 | ....75242 | ....84035 | 404802897 |
| 340 | 390919642 | ....79037 | ....48253 | 396827338 | 398816344 | 400315319 | 402824314 | ....43378 |
| 350 | ....58734 | 392918325 | ....87738 | ....67021 | ....56226 | ....55401 | ....64596 | ....83862 |
| 360 | ....97830 | ....57617 | 394927226 | 396906708 | ....96111 | ....95486 | 402904883 | 404924350 |
| 370 | 391036930 | ....96912 | ....66719 | ....46399 | 398936001 | 400935576 | ....15173 | ....64843 |
| 380 | ....76034 | 393036212 | 395006216 | ....86093 | ....75895 | ....75669 | ....85468 | 405005339 |
| 390 | 391115141 | ....75516 | ....45716 | 397025792 | 399015792 | 401015757 | 403025766 | ....45840 |
| 400 | ....54253 | 393114823 | ....85221 | ....65495 | ....55694 | ....55869 | ....66069 | ....86344 |
| 410 | ....93368 | ....54135 | 395124729 | 397105201 | ....95599 | ....95974 | 403106375 | 405126853 |
| 420 | 391232488 | ....93450 | ....64242 | ....44912 | 399135509 | 401136084 | ....46686 | ....67366 |
| 430 | ....71611 | 393232770 | 395203758 | ....84626 | ....75423 | ....76197 | ....87001 | 405207883 |
| 440 | 391310738 | ....72093 | ....43279 | 397224345 | 399215340 | 401216315 | 403227319 | ....48403 |
| 450 | ....49869 | 393311420 | ....82803 | ....64067 | ....55262 | ....5643[illegible] | ....6764[illegible] | ....88928 |
| 460 | ....89004 | ....50751 | 395322331 | 397303793 | ....95187 | ....96562 | 403307969 | 405329457 |
| 470 | 391428143 | ....90086 | ....61863 | ....43524 | 399335117 | 401336692 | ....48300 | ....69990 |
| 480 | ....67286 | 393429425 | 39540140[illegible] | ....83258 | ....75050 | ....76826 | ....8863[illegible] | 405410527 |
| 490 | 391506432 | ....68768 | ....40940 | 397422996 | 399414988 | 401416963 | 403428973 | ....51068 |
| 500 | ....45583 | 393508115 | ....80484 | ....62739 | ....54929 | ....57105 | ....69316 | ....91613 |

Figure C.35

| | 140000 | 145000 | 141000 | 141500 | 142000 | 142500 | 143000 | 143500 |
|---|---|---|---|---|---|---|---|---|
| 0 | 405491613 | 405524047 | 409566667 | 411619525 | 413682674 | 415756163 | 417840045 | 41993437 |
| 10 | 405532162 | ····64799 | 409607624 | ····60687 | 413724042 | ····97738 | ····81829 | ····76365 |
| 20 | ····72716 | 407605555 | ····48584 | 411701854 | ····65414 | 415839318 | 417923617 | 420018363 |
| 30 | 405613273 | ····46316 | ····89549 | ····43024 | 413806791 | ····80902 | ····65409 | ····60364 |
| 40 | ····53834 | ····87081 | 409730518 | ····84198 | ····48171 | 415922490 | 418007206 | 420202370 |
| 50 | ····94400 | 407727849 | ····71491 | 411825376 | ····89556 | ····64082 | ····49066 | ····44381 |
| 60 | 405734969 | ····68622 | 409813469 | ····66559 | 413930945 | 416905679 | ····90811 | ····86395 |
| 70 | ····75543 | 407809399 | ····53450 | 411907746 | ····72338 | ····47279 | 418132620 | 420228414 |
| 80 | 405816120 | ····50180 | ····94436 | ····48936 | 414013736 | ····88884 | ····74434 | ····70437 |
| 90 | ····56702 | ····90965 | 409935245 | ····90131 | ····55137 | 416130493 | 418216251 | 440312464 |
| 100 | ····97287 | 407931754 | ····76418 | 412031330 | ····96542 | ····72106 | ····53073 | ····54495 |
| 110 | 405937877 | ····72547 | 410017416 | ····72533 | 414137952 | 416213723 | ····99899 | ····96530 |
| 120 | ····78471 | 408013344 | ····58418 | 412113741 | ····79366 | ····55344 | 418341729 | 420438570 |
| 130 | 406019069 | ····54146 | ····99424 | ····54952 | 414220784 | ····96970 | ····83563 | ····80614 |
| 140 | ····59671 | ····94951 | 410140433 | ····96168 | ····62206 | 416338100 | 418425401 | 420522662 |
| 150 | 406100277 | 408135761 | ····81447 | 412237387 | 414303632 | ····80234 | ····67244 | ····64724 |
| 160 | ····40887 | ····76574 | 410222465 | ····78611 | ····45063 | 416421872 | 418509090 | 420606771 |
| 170 | ····81501 | 408217392 | ····63487 | 412319839 | ····86497 | ····63514 | ····50941 | ····48831 |
| 180 | 406222119 | ····58214 | 410304514 | ····61071 | 414427936 | 416505160 | ····92796 | ····90896 |
| 190 | ····62741 | ····99039 | ····45544 | 412402307 | ····69378 | ····46811 | 418634656 | 420732965 |
| 200 | 406303367 | 408339869 | ····86579 | ····43547 | 414510825 | ····88465 | ····76519 | ····75039 |
| 210 | ····43998 | ····80703 | 410427618 | ····84791 | ····52276 | 416630124 | 418718387 | 420817116 |
| 220 | ····84632 | 408421541 | ····68660 | 412526040 | ····93732 | ····71787 | ····60259 | ····59198 |
| 230 | 406425270 | ····62384 | 410509707 | ····67293 | 414635191 | 416713454 | 418802135 | 420901284 |
| 240 | ····65913 | 408503230 | ····50758 | 412608549 | ····76655 | ····55126 | ····44015 | ····43374 |
| 250 | 406506560 | ····44080 | ····91813 | ····49810 | 414718122 | ····96801 | ····85899 | ····85468 |
| 260 | ····47210 | ····84935 | 410632873 | ····91075 | ····59594 | 416838481 | 418927788 | 421027567 |
| 275 | ····87865 | 408625793 | ····73936 | 412732344 | 414801070 | ····80165 | ····69681 | ····69669 |
| 280 | 406628524 | ····66656 | 410715003 | ····73617 | ····42550 | 416921853 | 419011578 | 421111776 |
| 290 | ····69187 | 408707522 | ····56075 | 412814895 | ····84034 | ····63545 | ····53479 | ····53888 |
| 300 | 406709854 | ····48393 | ····97150 | ····56176 | 414925523 | 417005241 | ····95384 | ····96003 |
| 310 | ····50525 | ····89268 | 410838230 | ····97462 | ····67015 | ····46942 | 419137294 | 421238123 |
| 320 | ····91200 | 408830147 | ····79314 | 412938752 | 415008512 | ····88647 | ····79207 | ····80246 |
| 330 | 406831879 | ····71030 | 410920402 | ····80045 | ····50013 | 417130355 | 419221125 | 421322374 |
| 340 | ····72562 | 408911917 | ····61494 | 413021343 | ····91518 | ····72069 | ····63047 | ····64507 |
| 350 | 406913249 | ····52808 | 411002590 | ····62646 | 415133027 | 417213786 | 419304974 | 421406643 |
| 360 | ····53941 | ····93703 | ····43690 | 413103952 | ····74540 | ····55507 | ····46904 | ····48784 |
| 370 | ····94636 | 409034603 | ····84794 | ····45262 | 415216058 | ····97233 | ····88839 | ····90929 |
| 380 | 407035336 | ····75506 | 411125903 | ····86577 | ····57579 | 417338962 | 419430778 | 421533078 |
| 390 | ····76039 | 409116414 | ····67015 | 413227895 | ····99105 | ····80696 | ····72721 | ····75231 |
| 400 | 407116747 | ····57326 | 411208132 | ····69218 | 415340635 | 417422434 | 419514668 | 421617389 |
| 410 | ····57458 | ····98241 | ····49253 | 413310545 | ····82169 | ····64177 | ····56620 | ····59550 |
| 420 | ····98174 | 409239161 | ····90378 | ····51876 | 415423707 | 417505923 | ····98575 | 421701716 |
| 430 | 407238894 | ····80085 | 411331507 | ····93211 | ····65250 | ····47674 | 419640535 | ····43887 |
| 440 | ····79618 | 409321013 | ····72640 | 313434551 | 415506796 | ····89428 | ····82499 | ····86061 |
| 450 | 407320346 | ····61945 | 411413777 | ····75894 | ····48347 | 417631187 | 419724468 | 421828240 |
| 460 | ····61078 | 409402881 | ····54919 | 413517242 | ····89902 | ····72950 | ····66440 | ····70422 |
| 470 | 407401814 | ····43822 | ····96064 | ····58594 | 415631461 | 417714718 | 419808417 | 421912619 |
| 480 | ····42554 | ····84766 | 411537214 | ····99949 | ····73024 | ····56489 | ····50397 | ····51801 |
| 490 | ····33298 | 409525714 | ····78368 | 413641309 | 415714591 | ····98265 | ····92382 | ····96996 |
| 500 | 407524047 | ····66667 | 411619525 | ····82674 | ····56163 | 417840045 | 419934372 | 422039196 |

**Figure C.36**

| | 144000 | 144500 | 145000 | 145500 | 146000 | 146500 | 147000 | 147500 |
|---|---|---|---|---|---|---|---|---|
| 0 | 422039196 | 424154570 | 426280547 | 428417120 | 430564523 | 432722628 | 434891551 | 437071344 |
| 10 | ....81400 | ....96986 | 426323175 | ....60022 | 430607579 | ....65900 | 43493504 | 437115051 |
| 20 | 422123608 | 424239405 | ....65308 | 428502868 | ....50640 | 432809177 | ....78533 | ....58763 |
| 30 | ....65820 | ....81829 | 426408444 | ....45718 | ....93705 | ....52458 | 435022031 | 437202479 |
| 40 | 422208037 | 424324257 | ....51085 | ....88573 | 430736774 | ....95743 | ....65533 | ....46199 |
| 50 | ....50258 | ....66690 | ....93730 | 428631432 | ....79848 | 332939033 | 435106040 | ....89924 |
| 60 | ....92483 | 424409127 | 436536379 | ....74295 | 430822926 | ....82327 | ....52551 | 437333653 |
| 70 | 422334712 | ....51567 | ....79033 | 428717162 | ....66008 | 433025625 | ....96066 | ....77386 |
| 80 | ....76945 | ....94013 | 426621691 | ....60034 | 430909095 | ....68927 | 435239586 | 437421124 |
| 90 | 422419183 | 424536462 | ....64353 | 428802910 | ....52186 | 433112234 | ....83110 | ....64866 |
| 100 | ....61425 | ....78916 | 426707020 | ....45790 | ....95281 | ....55545 | 435326638 | 437508612 |
| 110 | 422503671 | 424621373 | ....49690 | ....88675 | 431038380 | ....98861 | ....70171 | ....52363 |
| 120 | ....45921 | ....63836 | ....92365 | 428931564 | ....81484 | 433242181 | 435413708 | ....96118 |
| 130 | ....88176 | 424706302 | 426835044 | ....74457 | 431124593 | ....85505 | ....57249 | 437639878 |
| 140 | 422630435 | ....48773 | ....77728 | 429017354 | ....67705 | 433328834 | 435500795 | ....83642 |
| 150 | ....72698 | ....91247 | 426920416 | ....60256 | 431210822 | ....72167 | ....44345 | 437727410 |
| 160 | 422714965 | 424833727 | ....63108 | 429103162 | ....53943 | 433415504 | ....87899 | ....71783 |
| 170 | ....97237 | ....76210 | 407005894 | ....46072 | ....97068 | ....58845 | 435631458 | 437814060 |
| 180 | ....99512 | 424918698 | ....48505 | ....88987 | 431340198 | 433502191 | ....75021 | ....58742 |
| 190 | 422941792 | ....61189 | ....91210 | 429231906 | ....83332 | ....45542 | 435718589 | 437902528 |
| 200 | ....84077 | 425003686 | 407133919 | ....74829 | 431426470 | ....88896 | ....62160 | ....46318 |
| 210 | 422926365 | ....46186 | ....76632 | 429317757 | ....69613 | 433632255 | 435805737 | ....90113 |
| 220 | ....68460 | ....88691 | 425219350 | ....60688 | 421512760 | ....75618 | ....49317 | 438033912 |
| 230 | 423010955 | 425131199 | ....62072 | 429403624 | ....55911 | 433718986 | ....92902 | ....77715 |
| 240 | ....53256 | ....73713 | 427304798 | ....46565 | ....99067 | ....62358 | 435936491 | 438121523 |
| 250 | ....95561 | 425216230 | ....47528 | ....89510 | 431642227 | 433805734 | ....80085 | ....65335 |
| 260 | 423137870 | ....58752 | ....90263 | 429532459 | ....85391 | ....49114 | 436623683 | 438209151 |
| 270 | ....80184 | 425301277 | 427433002 | ....75411 | 431728559 | ....92499 | ....67285 | ....52972 |
| 280 | 423222502 | ....43808 | ....75745 | 429618369 | ....71732 | 433935889 | 436110892 | ....96798 |
| 290 | ....64825 | ....86342 | 427518493 | ....61331 | 431814909 | ....79282 | ....54503 | 438340927 |
| 300 | 423307151 | 425428881 | ....61245 | 429704297 | ....58091 | 434022680 | ....98119 | ....84461 |
| 310 | ....49482 | ....71423 | 427604001 | ....47268 | 431901277 | ....66082 | 436241739 | 438428290 |
| 320 | ....91817 | 425513971 | ....46761 | ....90242 | ....44467 | 434109489 | ....85363 | ....72143 |
| 330 | 423434156 | ....56522 | ....89526 | 429823221 | ....87661 | ....52900 | 436328991 | 438515990 |
| 340 | ....76499 | ....99078 | 426732295 | ....76205 | 432030860 | ....96315 | ....72624 | ....59841 |
| 350 | 423518847 | 425641638 | ....75068 | 429919192 | ....74063 | 434239735 | 436416261 | 438603697 |
| 360 | ....61199 | ....84202 | 426817846 | ....62184 | 432117270 | ....83159 | ....59903 | ....47559 |
| 370 | 423603555 | 425726770 | ....60628 | 430005180 | ....60482 | 434326587 | 436503549 | ....91423 |
| 380 | ....45915 | ....69343 | 429903414 | ....48181 | 432203698 | ....70000 | ....47199 | 437835293 |
| 390 | ....88280 | 425811920 | ....46204 | ....91186 | ....46919 | 434413457 | ....90854 | ....79166 |
| 400 | 423730649 | ....54501 | ....88999 | 430134195 | ....90143 | ....56898 | 436634513 | 438823043 |
| 410 | ....73022 | ....97087 | 428031797 | ....77208 | 432333372 | 434500344 | ....78177 | ....66925 |
| 420 | 423815399 | 425939678 | ....74601 | 430220226 | ....76606 | ....43794 | 436721844 | 438910812 |
| 430 | ....57781 | ....82270 | 428117408 | ....63248 | 432419843 | ....87248 | ....65517 | ....54703 |
| 440 | 423900166 | 426024868 | ....60220 | 430306274 | ....63086 | 434630707 | 436809193 | ....98599 |
| 450 | ....42556 | ....67471 | 428203036 | ....49305 | 432506332 | ....74170 | ....52874 | 439042499 |
| 460 | ....84951 | 426110078 | ....45856 | ....92340 | ....49582 | 434717637 | ....96559 | ....86403 |
| 470 | 424027349 | ....52689 | ....88681 | 430435379 | ....92837 | ....61109 | 436940249 | 439130311 |
| 480 | ....69752 | ....95304 | 428331510 | ....78423 | 432636097 | 434804585 | ....83943 | ....74224 |
| 490 | 424112159 | 426237923 | ....74343 | 430521470 | ....79360 | ....48066 | 437027642 | 439218142 |
| 500 | ....54570 | ....80547 | 428417180 | ....64523 | 432722628 | ....91551 | ....71344 | ....62064 |

Figure C.37

| | 148000 | 148500 | 149000 | 149500 | 150000 | 150500 | 151000 | 151500 |
|---|---|---|---|---|---|---|---|---|
| 0 | 439262064 | 441463764 | 443676499 | 445855740 | 448135298 | 450381473 | 452638906 | 454907654 |
| 10 | 439305990 | 441507910 | 443720867 | ....44915 | ....80111 | 450426511 | ....84170 | ....53145 |
| 20 | ....49921 | ....52061 | ....65239 | ....89510 | 442924929 | ....71553 | 452729438 | ....98641 |
| 30 | ....93855 | ....96216 | 44380,615 | 446034109 | ....79752 | 450516601 | ....74711 | 455044140 |
| 40 | 439437795 | 441640376 | ....53996 | ....78712 | 448314579 | ....61652 | 452819989 | ....89645 |
| 50 | ....81739 | ....84540 | ....98382 | 446123320 | ....59411 | 450606708 | ....65271 | 455135154 |
| 60 | 439525687 | 441728708 | 443942771 | ....67932 | 448404247 | ....51769 | 452910557 | ....80667 |
| 70 | ....69639 | ....72881 | ....87166 | 446212549 | ....49087 | ....96834 | ....55849 | 455226185 |
| 80 | 439613596 | 441817058 | 444031564 | ....57170 | ....93932 | 450741904 | 453001144 | ....71708 |
| 90 | ....57558 | ....61240 | ....75967 | 446301796 | 448538781 | ....86978 | ....46444 | 455317235 |
| 100 | 439701524 | 441905426 | 444120375 | ....46426 | ....83635 | 450832057 | 453191749 | ....62767 |
| 110 | ....45494 | ....49617 | ....64787 | ....91061 | 448628494 | ....77140 | ....37058 | 455408303 |
| 120 | ....89468 | ....93812 | 444209204 | 446435700 | ....73356 | 450922228 | ....82372 | ....53844 |
| 130 | 439833447 | 442038011 | ....53625 | ....86344 | 448718224 | ....67320 | 453227690 | ....99389 |
| 140 | ....77431 | ....82215 | ....98050 | 446524992 | ....63095 | 451012417 | ....73013 | 455544939 |
| 150 | 439921418 | 442126423 | 444342480 | ....69644 | 448807972 | ....57518 | 453318340 | ....90494 |
| 160 | ....65410 | ....70636 | ....86914 | 446614301 | ....52852 | 451102624 | ....63672 | 455636053 |
| 170 | 440009407 | 442214853 | 444431353 | ....58963 | ....97738 | ....47734 | 453409008 | ....81616 |
| 180 | ....53408 | ....59074 | ....75796 | 446703628 | 448942628 | ....92849 | ....54349 | 455727185 |
| 190 | ....97413 | 442303300 | 444520243 | ....48299 | ....87522 | 451237968 | ....99695 | ....72757 |
| 200 | 440141423 | ....47530 | ....64695 | ....92974 | 449032421 | ....83092 | 453545045 | 455818335 |
| 210 | ....85437 | ....91765 | 444609152 | 446837653 | ....77324 | 451328220 | ....90399 | ....63916 |
| 220 | 440229456 | 442436004 | ....53613 | ....82337 | 449122232 | ....73353 | 453635758 | 455909503 |
| 230 | ....73745 | ....80248 | ....98078 | 446927025 | ....67144 | 451418490 | ....81122 | ....55094 |
| 240 | 440317500 | 442524496 | 444742548 | ....71718 | 449212060 | ....63632 | 453726490 | 456000689 |
| 250 | ....61538 | ....68748 | ....87022 | 447016415 | ....56982 | 451508779 | ....71862 | ....46289 |
| 260 | 440405574 | 442613005 | 444931501 | ....61116 | 449301907 | ....53930 | 453817240 | ....91894 |
| 270 | ....49614 | ....57266 | ....75984 | 447105823 | ....46858 | ....99085 | ....62621 | 456137503 |
| 280 | ....93652 | 442701532 | 444920472 | ....50533 | ....91772 | 451644245 | 453908008 | ....83117 |
| 290 | 440537709 | ....45802 | ....64964 | ....95248 | 449436711 | ....89409 | ....53398 | 456228735 |
| 300 | ....81763 | ....90077 | 445009460 | 447239968 | ....81655 | 451734578 | ....98794 | ....74358 |
| 310 | 440625821 | 442834356 | ....53961 | ....84692 | 449526603 | ....79752 | 454044194 | 456319986 |
| 320 | ....69883 | ....78639 | ....98467 | 447329420 | ....71556 | 451824930 | ....89598 | ....65618 |
| 330 | 440713950 | 442922927 | 445142976 | ....74153 | 449616513 | ....70112 | 454135007 | 456411254 |
| 340 | ....58022 | ....67220 | ....87491 | 447418891 | ....61475 | 451915299 | ....80420 | ....56895 |
| 350 | 440802097 | 443011516 | 445232009 | ....63632 | 449706441 | ....60491 | 454225839 | 456502541 |
| 360 | ....46178 | ....55818 | ....76533 | 447508379 | ....51411 | 452005687 | ....71261 | ....48191 |
| 370 | ....90262 | 443100123 | 445321060 | ....53130 | ....96387 | ....50887 | 454316688 | ....93846 |
| 380 | 440934351 | ....44433 | ....65592 | ....97885 | 449841366 | ....96092 | ....62120 | 456639505 |
| 390 | ....78445 | ....88748 | 445410129 | 447642645 | ....86350 | 452141302 | 454407556 | ....85168 |
| 400 | 441022542 | 443233066 | ....54670 | ....87409 | 449931339 | ....86516 | ....52997 | 456730838 |
| 410 | ....66645 | ....77390 | ....99216 | 447732178 | ....76332 | 452231735 | ....98442 | ....76511 |
| 420 | 441110751 | 443321717 | 445543765 | ....76651 | 450021330 | ....76958 | 454543892 | 456822189 |
| 430 | ....54863 | ....66050 | ....88320 | 447821729 | ....66332 | 452322186 | ....89346 | ....67871 |
| 440 | ....98978 | 443410386 | 445632879 | ....66511 | 450111338 | ....67418 | 454634805 | 456913558 |
| 450 | 441243098 | ....54727 | ....77442 | 447911298 | ....56350 | 452412655 | ....80269 | ....59249 |
| 460 | ....87222 | ....99073 | 445721910 | ....56089 | 450201365 | ....57896 | 454725737 | 457004945 |
| 470 | 441331351 | 443543423 | ....66582 | 448000884 | ....46385 | 452503142 | ....71209 | ....50645 |
| 480 | ....75484 | ....87777 | 445811159 | ....45684 | ....91410 | ....48392 | 454816687 | ....96354 |
| 490 | 441419622 | 443632126 | ....55740 | ....90489 | 450336439 | ....93646 | ....62168 | 457142060 |
| 500 | ....63764 | ....76499 | 445900325 | 448135298 | ....81473 | 452638906 | 454907654 | ....87774 |

Figure C.38

| | 152000 | 152500 | 153000 | 153500 | 1540000 | 15450 | 155000 | 155500 |
|---|---|---|---|---|---|---|---|---|
| 0 | 457187774 | 459479323 | 461782357 | 464096935 | 466423114 | 468760952 | 471110508 | 473471841 |
| 10 | 457233493 | 459525271 | 461828535 | 464143344 | ....69756 | 468807328 | ....57619 | 473519188 |
| 20 | ....79216 | ....71223 | ....74718 | ....89759 | 466516403 | ....54709 | 471204735 | ....66540 |
| 30 | 457324944 | 459617180 | 461920906 | 464236178 | ....63054 | 468901594 | ....51856 | 473613897 |
| 40 | ....70677 | ....63142 | ....67098 | ....82601 | 466609711 | ....48485 | ....98981 | ....61258 |
| 50 | 457416414 | 459709108 | 462013294 | 464329030 | ....56372 | ....95379 | 471346111 | 473708624 |
| 60 | ....62155 | ....55079 | ....59496 | ....75462 | 466703037 | 469042279 | ....93245 | ....55995 |
| 70 | 457507902 | 459801055 | 462105702 | 464421900 | ....49[illegible]08 | ....89183 | 471440385 | 473803371 |
| 80 | ....53652 | ....47035 | ....51912 | ....68341 | ....96383 | 469136092 | ....87529 | ....50751 |
| 90 | ....99408 | ....93020 | ....98127 | 464514789 | 466843062 | ....83006 | 471534677 | ....98136 |
| 100 | 457645168 | 459939009 | 462244347 | ....61240 | ....89747 | 469229924 | ....81831 | 473945526 |
| 110 | ....90932 | ....85003 | ....90572 | 464607697 | 466936436 | ....76847 | 471628989 | ....92921 |
| 120 | 457736701 | 460031001 | 462336801 | ....54157 | ....83129 | 469323775 | ....76152 | 474040320 |
| 130 | ....82475 | ....77004 | ....83034 | 464700623 | 417029827 | ....70707 | 471723320 | ....87724 |
| 140 | 457828253 | 460123012 | 462429373 | ....47093 | ....76530 | 469417644 | ....70492 | 474135133 |
| 150 | ....74036 | ....69024 | ....75516 | ....93568 | 467123238 | ....64586 | 471817669 | ....42546 |
| 160 | 457919824 | 460215041 | 462521763 | 464840047 | ....69950 | 469511532 | ....64851 | 474229965 |
| 170 | ....65616 | ....61063 | ....68015 | ....86531 | 467216667 | ....58484 | 471912037 | ....77388 |
| 180 | 458011412 | 460307089 | 462614272 | 464933020 | ....63389 | 469605439 | ....59228 | 474324815 |
| 190 | ....57213 | ....53120 | ....60534 | ....79513 | 467310115 | ....52400 | 472006424 | ....72248 |
| 200 | 458103019 | ....99155 | 462706800 | 465026011 | ....56847 | ....99365 | ....53625 | 474419685 |
| 210 | ....48829 | 460445195 | ....53070 | ....72514 | 467403582 | 469746335 | 472100830 | ....67127 |
| 220 | ....94644 | ....91239 | ....99346 | 465119021 | ....50323 | ....93310 | ....48040 | 474514574 |
| 230 | 488240464 | 460537288 | 462845626 | ....65533 | ....97068 | 469840289 | ....95255 | ....62025 |
| 240 | ....86287 | ....83342 | ....91910 | 465212049 | 467543817 | ....87273 | 472242475 | 474609481 |
| 250 | 458332116 | 460629400 | 462938199 | ....58570 | ....90572 | 469934262 | ....89599 | ....56942 |
| 260 | ....77950 | ....75463 | ....84493 | 465305096 | 468637331 | ....81255 | 472336918 | 474704408 |
| 270 | 498423787 | 460721531 | 463030792 | ....51627 | ....84095 | 470028253 | ....84162 | ....51878 |
| 280 | ....69630 | ....67503 | ....77095 | ....98612 | 467730863 | ....75256 | 472431400 | ....99354 |
| 290 | 458515477 | 460813680 | 463123402 | 465444702 | ....77636 | 470122263 | ....78643 | 474846833 |
| 300 | ....61328 | ....59761 | ....9715 | ....91246 | 467824414 | ....69276 | 472525891 | ....94318 |
| 310 | 458607184 | 460905847 | 463216032 | 465537795 | ....71196 | 470216293 | ....73144 | 474941808 |
| 320 | ....53045 | ....51938 | ....62353 | ....84349 | 467917983 | ....63314 | 4[illegible]2620401 | ....89302 |
| 330 | ....98910 | ....98033 | 463308679 | 465630908 | ....64775 | 470310341 | ....6[illegible]663 | 475036801 |
| 340 | 458744780 | 461044133 | ....55010 | ....77471 | 468011572 | ....57372 | 472717930 | ....84304 |
| 350 | ....90655 | ....90237 | 463401346 | 465724038 | ....58373 | 470404407 | ....62201 | 475131813 |
| 360 | 458836534 | 461136346 | ....46786 | ....70611 | 466105179 | ....51448 | 472809477 | ....79326 |
| 370 | ....82418 | ....82460 | ....94031 | 465817188 | ....51989 | ....98493 | ....56758 | 475226244 |
| 380 | 458928306 | 461228578 | 463540380 | ....63770 | ....98804 | 470545543 | 4729040[illegible]4 | ....74367 |
| 390 | ....74199 | ....74701 | ....86734 | 465910356 | 468245624 | ....92597 | ....51334 | 475321894 |
| 400 | 459020096 | 461320828 | 463633093 | ....56947 | ....92449 | 470639657 | ....98630 | ....69426 |
| 410 | ....65998 | ....66961 | ....79456 | 466003543 | 468339278 | ....86721 | 473045930 | 475416963 |
| 420 | 459111905 | 461413097 | 463725824 | ....50143 | ....86112 | 470733789 | ....93[illegible]34 | ....64505 |
| 430 | ....57816 | ....59239 | ....72197 | ....96748 | 468432950 | ....80863 | 473140543 | 475512051 |
| 440 | 459203732 | 461505384 | 461818574 | 466143358 | ....79794 | 470827941 | ....87858 | ....59602 |
| 450 | ....49652 | ....51535 | ....64956 | ....59972 | 468526642 | ....75024 | 473235176 | 475607159 |
| 460 | ....95577 | ....97690 | 463911342 | 456236591 | ....73494 | 470922111 | ....82500 | ....54719 |
| 470 | 459341507 | 461643850 | ....57733 | ....83215 | 468620352 | ....69203 | 4733[illegible]9828 | 475202285 |
| 480 | ....87441 | ....90014 | 464004129 | 466329843 | ....67124 | 471016300 | ....77161 | ....49[illegible]55 |
| 490 | 459433379 | 461736183 | ....50530 | ....76476 | 468714081 | ....63402 | 473424499 | ....9743[illegible] |
| 500 | ....79323 | ....82357 | ....96935 | 466423114 | ....60952 | 471110508 | ....71841 | 4[illegible]5845010 |

**Figure C.39**

| | 156000 | 156500 | 157000 | 157500 | 158000 | 158500 | 159000 | 159500 |
|---|---|---|---|---|---|---|---|---|
| 0 | 475845010 | 478230073 | 480627091 | 483036124 | 485457231 | 487890474 | 490335912 | 492793608 |
| 10 | ....92594 | ....77896 | ....75154 | ....84422 | 485505777 | 48793926 | ....84946 | 492842887 |
| 20 | 475940183 | 478325723 | 480723222 | 483132736 | ....54327 | ....88057 | 490433984 | ....92172 |
| 30 | ....87777 | ....73556 | ....71294 | ....81049 | 485602883 | 488036855 | ....83028 | 492941401 |
| 40 | 476035376 | 478421393 | 480819371 | 483229367 | ....51443 | ....85659 | 490532076 | ....90755 |
| 50 | ....8298[illegible] | ....69236 | ....67453 | ....776[illegible] | 485700008 | 48813446[illegible] | ....81129 | 493040054 |
| 60 | 476130588 | 478517082 | 480915539 | 483326018 | ....48578 | ....83281 | 490630187 | ....89358 |
| 70 | ....78201 | ....64934 | ....63631 | ....74351 | ....97153 | 488232099 | ....79250 | 493138667 |
| 80 | 476225819 | 478612791 | 481011727 | 483422688 | 485845733 | ....80923 | 490728318 | ....87981 |
| 90 | ....73442 | ....60652 | ....59828 | ....71030 | ....94317 | 488329751 | ....77391 | 493237300 |
| 100 | 476321059 | 478708519 | 481107934 | 483519377 | 485942907 | ....78584 | 490826469 | ....86623 |
| 110 | ....68701 | ....56389 | ....56045 | ....67729 | ....91501 | 488427422 | ....75551 | 493335952 |
| 120 | 476416338 | 478804265 | 481204161 | 483616086 | 486040100 | ....76264 | 490924639 | ....85286 |
| 130 | ....63979 | ....52145 | ....52281 | ....6444[illegible] | ....87804 | 488525112 | ....73731 | 493434624 |
| 140 | 476511626 | 478900031 | 481300406 | 483712814 | 486137313 | ....73964 | 491022829 | ....83968 |
| 150 | ....59277 | ....47921 | ....48536 | ....61185 | ....85927 | 488622822 | ....71931 | 493533316 |
| 160 | 476606933 | ....95815 | ....96671 | 483809561 | 486234546 | ....71684 | 491121038 | ....82669 |
| 170 | ....54594 | 479043715 | 481444811 | ....57942 | ....83179 | 488720551 | ....70150 | 493632028 |
| 180 | 476702259 | ....91619 | ....92956 | 483906328 | 486331797 | ....69423 | 491219267 | ....81391 |
| 190 | ....49929 | 479139529 | 481541405 | ....54719 | ....80431 | 488818300 | ....68389 | 493730759 |
| 200 | ....97604 | ....87443 | ....89259 | 484003114 | 486429069 | ....67182 | 491317516 | ....80132 |
| 210 | 476845284 | 479235361 | 481637418 | ....51515 | ....77711 | 488916069 | ....66648 | 493829510 |
| 220 | ....92969 | ....83285 | ....85582 | ....99920 | 486526359 | ....64960 | 491415785 | ....78893 |
| 230 | 476940658 | 479331213 | 481733750 | 484148330 | ....70512 | 489013857 | ....64926 | 493928281 |
| 240 | ....88352 | ....79146 | ....81924 | ....96745 | 486623669 | ....62758 | 491514073 | ....77674 |
| 250 | 477036051 | 479427084 | 481830102 | 484245164 | ....7233[illegible] | 489111665 | ....83224 | 494027072 |
| 260 | ....83754 | ....75027 | ....78285 | ....93589 | 486720999 | ....60576 | 491612380 | ....76474 |
| 270 | 477131463 | 479522974 | 481926473 | 484342018 | ....69671 | 489209492 | ....61542 | 494125882 |
| 280 | ....79176 | ....70927 | ....74665 | ....90452 | 486818348 | ....58413 | 491710708 | ....75294 |
| 290 | 477226894 | 479618884 | 482022863 | 484438891 | ....67030 | 489307339 | ....59879 | 494224712 |
| 300 | ....74617 | ....66846 | ....71065 | ....87335 | 486915716 | ....56269 | 491809055 | ....74134 |
| 310 | 477322344 | 479714812 | 482119272 | 484535784 | ....64408 | 489405205 | ....58236 | 494323562 |
| 320 | ....70076 | ....62784 | ....67484 | ....84238 | 487013104 | ....54145 | 491907422 | ....72994 |
| 330 | 477417813 | 479810760 | 482215701 | 484632696 | ....61806 | 489503091 | ....56612 | 494422432 |
| 340 | ....65555 | ....58741 | ....63922 | ....81159 | 487110512 | ....52041 | 492005808 | ....71874 |
| 350 | 477513302 | 479906727 | 482312149 | 484729627 | ....59223 | 489600996 | ....55009 | 494521321 |
| 360 | ....61053 | ....54718 | ....60380 | ....78100 | 487207939 | ....49957 | 492104214 | ....70773 |
| 370 | 477608809 | 480002713 | 482408616 | 484826578 | ....56660 | ....98922 | ....53425 | 494620230 |
| 380 | ....56570 | ....50714 | ....56857 | ....75061 | 487305385 | 489747891 | 492202640 | ....69692 |
| 390 | 466604336 | ....98719 | 482505103 | 484923548 | ....54116 | ....96866 | ....51860 | 494719159 |
| 400 | ....52106 | 480146728 | ....53353 | ....72041 | 487402851 | 489845846 | 492301085 | ....68631 |
| 410 | ....90[illegible] | ....94743 | 48260[illegible]160 | 485020538 | ....51592 | ....94831 | ....50315 | 494818108 |
| 420 | 477847661 | 480242763 | ....49869 | ....69040 | 487500337 | 489943820 | ....99550 | ....67590 |
| 430 | ....95416 | ....90787 | ....98134 | 485117547 | ....49087 | ....92814 | 492448790 | 494917077 |
| 440 | 477943236 | 480338816 | 482746404 | ....66059 | ....97842 | 490041814 | ....98035 | ....66568 |
| 450 | ....91030 | ....86850 | ....94678 | 485214575 | 487646602 | ....90818 | 492547285 | 495016065 |
| 460 | 478038829 | 480434889 | 482842958 | ....63097 | ....95366 | 490139827 | ....96540 | ....65567 |
| 470 | ....86632 | ....82932 | ....91242 | 485311623 | 887744136 | ....88841 | 492645800 | 495115073 |
| 480 | 478134441 | 480530980 | 482939531 | ....60154 | ....92910 | 490237860 | ....95064 | ....64585 |
| 490 | ....82254 | ....79033 | ....87825 | 485408690 | 487841639 | ....86884 | 492744334 | 495214101 |
| 500 | 478230073 | 480627091 | 483036124 | ....57231 | ....90474 | 490335912 | ....93608 | ....63623 |

F

Figure C.40

| | 160000 | 160500 | 161000 | 161500 | 162000 | 162500 | 163000 | 163500 |
|---|---|---|---|---|---|---|---|---|
| 0 | 495263623 | 497746017 | 500240855 | 5027[illegible]197 | 505266106 | 507800646 | 510345880 | 512903817 |
| 10 | 495313149 | ....95792 | ....90879 | ....98471 | 505318633 | ....51426 | ....96915 | ....55161 |
| 20 | ....62680 | 497845571 | 500340908 | 501843751 | ....69165 | 507902211 | 510447954 | 513006457 |
| 30 | 495412217 | ....95356 | ....90942 | ....9[illegible]36 | 505419702 | ....53001 | ....98999 | ....57757 |
| 40 | ....61758 | 497945146 | 500440981 | 502949326 | ....70244 | 508093797 | 510550049 | 513109063 |
| 50 | 495511304 | ....94940 | ....91025 | ....99621 | 5055[illegible]0791 | ....54597 | 510601104 | ....60374 |
| 60 | ....60855 | 498044740 | 590541074 | 503 49921 | ....71343 | 508105403 | ....52164 | 513211690 |
| 70 | 495610411 | ....94544 | ....91128 | 503100226 | 505621900 | ....56213 | 510703229 | ....63011 |
| 80 | ....59972 | 498144354 | 500641187 | ....50536 | ....7246[illegible] | 508207029 | ....54299 | 513314338 |
| 90 | 495709538 | ....94168 | ....91251 | 503200851 | 505723029 | ....57849 | 510805375 | ....65669 |
| 100 | ....59109 | 498243987 | 500741321 | ....51171 | ....73601 | 508308675 | ....56455 | 513417006 |
| 110 | 495802685 | ....93812 | ....91395 | 503301495 | 505824179 | ....59506 | 510907541 | ....65347 |
| 120 | ....58266 | 498343641 | 500841474 | ....51826 | ....74761 | 508410342 | ....58632 | 513519694 |
| 130 | 495907852 | ....93475 | ....91558 | 503402162 | 505925349 | ....61183 | 511009728 | ....71046 |
| 140 | ....57443 | 498443315 | 500941647 | ....52502 | ....75941 | 508512029 | ....60829 | 513622403 |
| 150 | 496007038 | ....93159 | ....91741 | 503502847 | 506026539 | ....62880 | 511111935 | ....73766 |
| 160 | ....56639 | 498543005 | 501041840 | ....53197 | ....77142 | 508613737 | ....63046 | 513725133 |
| 170 | 496106245 | ....92863 | ....91945 | 503603553 | 506127749 | ....64598 | 511214162 | ....76505 |
| 180 | ....55855 | 498642722 | 501142054 | ....53913 | ....78362 | 508715465 | ....65284 | 513827883 |
| 190 | 496205471 | ....92586 | ....92168 | 503704278 | 506228980 | ....66336 | 511316410 | ....79266 |
| 200 | ....55091 | 498742456 | 501242287 | ....54649 | ....79603 | 50881[illegible] | ....67542 | 513930654 |
| 210 | 496304717 | ....92330 | ....92411 | 503805024 | 506330231 | ....68094 | 511418679 | ....82047 |
| 220 | ....54347 | 498842209 | 501342541 | ....55405 | ....80864 | 508918981 | ....69820 | 514033445 |
| 230 | 496403983 | ....92093 | ....92675 | 503905790 | 506431502 | ....69873 | 511520967 | ....84848 |
| 240 | ....53623 | 498941982 | 501442814 | ....56181 | ....82145 | 509020770 | ....72120 | 514136257 |
| 250 | 496503268 | ....91877 | ....92959 | 504006576 | 506532793 | ....71672 | 511623277 | ....87671 |
| 260 | ....52919 | 499041776 | 501543108 | ....56977 | ....83447 | 509122579 | ....74439 | 514239089 |
| 270 | 496602574 | ....91680 | ....93262 | 504107383 | 506634105 | ....73492 | 511725606 | ....90513 |
| 280 | ....52234 | 499131589 | 501643421 | ....57794 | ....84768 | 509224409 | ....76779 | 514341942 |
| 290 | 496701900 | ....91503 | ....93586 | 504208209 | 506735437 | ....75331 | 511827957 | ....93377 |
| 300 | ....51570 | 499241423 | 501743755 | ....58630 | ....86110 | 509326259 | ....79140 | 514444816 |
| 310 | 496801245 | ....91347 | ....93930 | 504309056 | 506836789 | ....77192 | 511930327 | ....96260 |
| 320 | ....50925 | 499341276 | 501844109 | ....59487 | ....87473 | 509428129 | ....81520 | 514547710 |
| 330 | 496900610 | ....91210 | ....94293 | 504409923 | 506938161 | ....79072 | 511032719 | ....99165 |
| 340 | ....50300 | 499441149 | 501944483 | ....60364 | ....88855 | 509530020 | ....83922 | 514650625 |
| 350 | ....99995 | ....91093 | ....94677 | 504510810 | 507039554 | ....80973 | 512135130 | 51470209[illegible] |
| 360 | 497049695 | 499541042 | 502044877 | ....61261 | ....90258 | 509631931 | ....86344 | ....53560 |
| 370 | ....99400 | ....90996 | ....95081 | 504611717 | 507140967 | ....82894 | 512237562 | 514805034 |
| 380 | 497149110 | 499640955 | 502145291 | ....62178 | ....91681 | 509733863 | ....88786 | ....56516 |
| 390 | ....98825 | ....90919 | ....95505 | 504712644 | 507242400 | ....84836 | 512340015 | 514907981 |
| 400 | 497248545 | 409740889 | 502245725 | ....63116 | ....93125 | 509835814 | ....91249 | ....59392 |
| 410 | ....98270 | ....90863 | ....95949 | 504813592 | 507343854 | ....86798 | 512442488 | 515018888 |
| 420 | 497348000 | 499840842 | 502346179 | ....64074 | ....94588 | 509937787 | ....93732 | ....70390 |
| 430 | ....67734 | ....90826 | ....96414 | 504914560 | 507445328 | ....88781 | 512544982 | 515121897 |
| 440 | 497447474 | 499940815 | 502446653 | ....65051 | ....96072 | 510039779 | ....96236 | ....734[illegible]5 |
| 450 | ....97219 | ....90809 | ....96898 | 505015549 | 507546822 | ....90783 | 512647496 | 515217023 |
| 460 | 497546969 | 500040808 | 502547148 | ....66049 | ....97577 | 510141793 | ....98761 | ....68545 |
| 470 | ....96723 | ....90812 | ....97402 | 505116556 | 507648336 | ....92807 | 512750030 | 51532070[illegible] |
| 490 | 497646483 | 500140821 | 502647662 | ....67068 | ....99101 | 510243826 | 512801305 | ....7160[illegible] |
| 480 | ....96248 | ....90835 | ....97927 | 505217584 | 507749871 | ....94850 | ....52586 | 515423141 |
| 500 | 497746017 | 500240855 | 502748197 | ....68106 | 507800646 | 510345880 | 512903871 | ....74683 |

**Figure C.41**

| | 164000 | 164500 | 165000 | 165500 | 166000 | 166500 | 167000 | 167500 |
|---|---|---|---|---|---|---|---|---|
| 0 | 515474684 | 518058382 | 520655030 | 523264693 | 525887437 | 528523327 | 531172428 | 533834807 |
| 10 | 515526231 | 518110187 | 520707095 | 523317020 | 525940026 | ····76179 | 531225545 | ····88191 |
| 20 | ····77784 | ····61998 | ····59160 | ····69351 | ····92620 | 528629036 | ····78668 | 533941580 |
| 30 | 515629341 | 518213815 | 520811242 | 523421688 | 526045219 | ····81899 | 531331796 | ····94974 |
| 40 | ····80904 | ····65636 | ····63323 | ····74030 | ····97823 | 528734768 | ····84929 | 534048373 |
| 50 | 515732472 | 518317463 | 520915405 | 523526378 | 526150433 | ····87647 | 531438067 | 534101778 |
| 60 | ····84046 | ····69294 | ····67501 | ····78730 | 526203048 | 528840520 | ····91211 | ····55188 |
| 70 | 515835624 | 518421131 | 521019598 | 523631088 | ····55669 | ····93404 | 531544360 | 534208604 |
| 80 | ····87208 | ····72973 | ····71700 | ····83451 | 526308294 | 528946293 | ····97516 | ····62025 |
| 90 | 515938796 | 518524821 | 521123807 | 523734820 | ····60925 | ····99188 | 531650674 | 534315451 |
| 100 | ····90390 | ····76673 | ····75919 | ····88193 | 526413561 | 529052088 | 531703839 | ····68882 |
| 110 | 516041989 | 518628531 | 521228037 | 523840572 | ····66203 | 529104993 | ····57010 | 534422319 |
| 120 | ····93593 | ····80394 | ····80160 | ····92956 | 526518849 | ····57903 | 531810186 | ····75762 |
| 130 | 516145203 | 518732262 | 521332288 | 523945346 | ····71501 | 529210819 | ····63367 | 534529209 |
| 140 | ····96817 | ····84135 | ····84421 | ····97740 | 526624158 | ····63740 | 531916553 | ····82662 |
| 150 | 516248437 | 518836013 | 521436559 | 524050140 | ····76821 | 529316667 | ····69745 | 534636120 |
| 160 | 51630006[illegible] | ····87897 | ····88703 | 524102545 | 526729488 | ····69598 | 532022942 | ····89584 |
| 170 | ····51692 | 518939786 | 521540852 | ····54955 | ····82161 | 529422535 | ····76144 | 536743053 |
| 180 | 516403327 | ····91680 | ····93006 | 524207371 | 526834839 | ····75478 | 532129351 | ····96527 |
| 190 | ····54967 | 519043579 | 521645165 | ····59791 | ····87523 | 529528425 | ····82564 | 534850007 |
| 200 | 516506613 | ····95483 | ····97330 | 524312217 | 526940212 | ····81378 | 532235783 | 534903492 |
| 210 | ····58263 | 519147393 | 521749499 | ····64649 | ····92906 | 529634336 | ····89006 | ····56982 |
| 220 | 516609919 | ····99307 | 521801674 | 524417085 | 527045605 | ····87300 | 532342235 | 535010478 |
| 230 | ····61580 | 519251227 | ····53855 | ····69527 | ····98309 | 529740268 | 532495469 | ····63979 |
| 240 | 516713246 | 519303152 | 521906040 | 524521974 | 527151019 | ····93242 | ····48709 | 535117485 |
| 250 | ····64918 | ····55083 | ····58231 | ····74426 | 527203734 | 529846222 | 532501954 | ····70997 |
| 260 | 516816594 | 519407018 | 522010426 | 524626883 | ····56455 | ····99296 | ····55204 | 535224514 |
| 270 | ····68276 | ····58959 | ····62626 | ····79346 | 527309180 | 529952156 | 532608459 | ····78027 |
| 280 | 516919963 | 519510905 | 522114834 | 524731814 | ····61911 | 530005191 | ····61720 | 535331564 |
| 290 | ····71655 | ····62856 | ····67045 | ····84287 | 527414647 | ····58292 | 532714986 | ····85098 |
| 300 | 517023352 | 519614812 | 522219262 | 524836766 | ····67359 | 530111198 | ····68258 | 535438636 |
| 310 | ····75054 | ····66774 | ····71484 | ····89249 | 527520136 | ····64209 | 532821535 | ····92180 |
| 320 | 517126762 | 519718740 | 522323711 | 524941638 | ····72888 | 530217225 | ····74817 | 535545729 |
| 330 | ····78474 | ····70712 | ····75943 | ····94232 | 527625645 | ····70247 | 532928104 | ····99284 |
| 340 | 517230192 | 519822689 | 522428181 | 525046732 | ····78408 | 530323274 | ····81397 | 535652844 |
| 350 | ····81915 | ····74672 | ····80424 | ····99237 | 527731175 | ····76306 | 533034695 | 535706409 |
| 360 | 517333643 | 519926659 | 522532672 | 525151746 | ····83949 | 530429344 | ····87999 | ····59980 |
| 370 | ····85377 | ····78652 | ····84925 | 525204262 | 527836727 | ····82387 | 533141308 | 535813556 |
| 380 | 517437115 | 520030650 | 522637184 | ····56782 | ····89511 | 530535435 | ····94622 | ····67137 |
| 390 | ····88859 | ····82653 | ····89447 | 525309308 | 527942300 | ····88488 | 533247941 | 535920724 |
| 400 | 517540608 | 520134661 | 522741716 | ····61839 | ····95094 | 530641547 | 533301266 | ····74316 |
| 410 | ····92362 | ····86675 | ····93990 | 525414375 | 528047893 | ····94611 | ····54596 | 536027913 |
| 420 | 517644121 | 520238693 | 522846270 | ····66916 | 528100698 | 534747681 | 533407932 | ····81516 |
| 430 | ····95886 | ····90717 | ····98554 | 525519463 | ····53508 | 540800756 | ····61272 | 536135124 |
| 440 | 517747655 | 520342746 | 521950844 | ····72015 | 528206324 | ····53836 | 533514619 | ····88738 |
| 450 | ····99430 | ····94781 | 523003139 | 525624572 | ····59144 | 530906921 | ····67970 | 536242357 |
| 460 | 517851210 | 510446820 | ····55440 | ····77135 | 528311970 | ····60012 | 533621327 | ····95981 |
| 470 | 517902995 | ····98865 | 523107745 | 525729702 | ····64801 | 531013108 | ····74689 | 536349610 |
| 480 | ····54785 | 520550915 | ····60056 | ····82275 | 528417638 | ····66209 | 533728056 | 536403245 |
| 490 | 518006581 | 520602970 | 523212372 | 525834853 | ····70480 | 531119316 | ····81429 | ····56886 |
| 500 | ····58382 | ····55030 | ····64693 | ····87437 | 528523327 | ····72428 | 533834807 | 536510531 |

F 2

**Figure C.42**

| | 168000 | 168500 | 169000 | 169500 | 170000 | 170500 | 171000 | 171500 |
|---|---|---|---|---|---|---|---|---|
| 0 | 536510531 | 539199667 | 541902281 | 544618441 | 547348216 | 550091673 | 552848880 | 55561990[illegible] |
| 01 | ····64132 | 539253587 | ····56471 | ····72903 | 547402951 | 550146682 | 552904165 | ···7547[illegible] |
| 02 | 536617839 | 539307512 | 542010667 | 544727370 | ····57691 | 550201697 | ····59456 | 555731037 |
| 30 | ····71501 | ····61443 | ····64868 | ····81843 | 547512437 | ····56717 | 553·14752 | ····86611 |
| 40 | 536725168 | 539415379 | 542119074 | 544836321 | ····67188 | 550311743 | ····70053 | 555842189 |
| 50 | ····78840 | ····69321 | ····73286 | ····90805 | 547621945 | ····66773 | 553125360 | ····97773 |
| 60 | 536832518 | 539523268 | 542227504 | 544945294 | ····76707 | 550421810 | ····80673 | 555953363 |
| 70 | ····86202 | ····77220 | ···81726 | ····99788 | 547731474 | ····76852 | 553235991 | 556008959 |
| 80 | 536939890 | 539631178 | 542335955 | 545054288 | ····81248 | 550531900 | ····91314 | ····64559 |
| 90 | ····93584 | ····85141 | ····90188 | 545108794 | 547841026 | ····86953 | 553346643 | 556120166 |
| 100 | 537047284 | 539739109 | 542444427 | ····63305 | ····95810 | 550742012 | 553401978 | ····75778 |
| 110 | 537100988 | ····93083 | ····98672 | 545217821 | 547950600 | ····97067 | ····57318 | 556231396 |
| 120 | ····54698 | 539847062 | 542552921 | ····72343 | 548005395 | 550752146 | 553512664 | ····87019 |
| 130 | 537208414 | 539901047 | 542607177 | 545326870 | ····60196 | 550807221 | ····68015 | 556342647 |
| 140 | ····62135 | ····55037 | ····61437 | ····81403 | 548115002 | ····62302 | 553623372 | ····94282 |
| 150 | 537315861 | 540009033 | 547215704 | 545435941 | ····69813 | 550917388 | ····78734 | 556453922 |
| 160 | ····69592 | ····63034 | ····69975 | ····90485 | 548224930 | ····72480 | 553734102 | 556509567 |
| 170 | 537423329 | 540117040 | 542824252 | 545545034 | ····79453 | 551027577 | ····89476 | ····65218 |
| 180 | ····77072 | ····71052 | ····78535 | ····99588 | 548334280 | ····82680 | 553844855 | 556620874 |
| 190 | 537530819 | 540225069 | 542932822 | 545654148 | ····89114 | 551137788 | 553900239 | ····76537 |
| 200 | ····84572 | ····79091 | ····87116 | 545708714 | 548443953 | ····92902 | ····55629 | 556732204 |
| 210 | 537638331 | 540333119 | 543041414 | ····63284 | ····98797 | 551248021 | 554011025 | ····87877 |
| 220 | ···92095 | ···87152 | ····95719 | 545817861 | 548553647 | 551303146 | ····66426 | 556843556 |
| 230 | 537745864 | 540441191 | 543150028 | ····72443 | 548608502 | ····58276 | 554121832 | ····99241 |
| 240 | ····99369 | ····95235 | 543204343 | 545927030 | ····63363 | 551413412 | ····77245 | 556954921 |
| 250 | 537853419 | 540549285 | ····58664 | ···81622 | 548718230 | ····68553 | 554232662 | 557010626 |
| 260 | 537907204 | 540603340 | 543312989 | 546036221 | ····73101 | 551523700 | ····88086 | ····66327 |
| 270 | ····60995 | ····57400 | ···67321 | ····90824 | 548827979 | ····78853 | 554343514 | 552122034 |
| 280 | 538014791 | 540711466 | 543421657 | 546145433 | ····82861 | 551634010 | ····98949 | ····77746 |
| 290 | ····68592 | ····65537 | ····76000 | 546200048 | 548637750 | ····89171 | 554454389 | 557233464 |
| 300 | 538122399 | 540819613 | 543535347 | ···54668 | ····92644 | 551744343 | 554509834 | ····89187 |
| 310 | ····76211 | ····73695 | ····84700 | 546309293 | 949047543 | ····99517 | ····65285 | 557344916 |
| 320 | 538230029 | 540927783 | 543639059 | ····63924 | 549102448 | 551854697 | 554620742 | 557400650 |
| 330 | ····83852 | ····81876 | ····93423 | 546418561 | ····57358 | 551909883 | ····76204 | ····56390 |
| 340 | 538337680 | 541035974 | 543747792 | ····73203 | 549212274 | ····65074 | 554731671 | 557512136 |
| 350 | ····91514 | ····90077 | 543202167 | 546527850 | ····97195 | 552020270 | ····87145 | ····67887 |
| 360 | 538445353 | 541144186 | ····56547 | ····82503 | 549322121 | ····75472 | 554842623 | 557623644 |
| 370 | ····99198 | ····98301 | 543910933 | 546637161 | ····77054 | 552130680 | ····98108 | ····79406 |
| 380 | 538553048 | 541252421 | ····65324 | ····91825 | 549431991 | ····85893 | 554953597 | 557735174 |
| 390 | 538606903 | 541306546 | 544019720 | 546746494 | ····86935 | 552241111 | 55500909[illegible] | ····90948 |
| 400 | ···60764 | ····60677 | ····74122 | 546801169 | 549541883 | ····96335 | ····64594 | 557846727 |
| 410 | 538714630 | 541414813 | 544128530 | ····55849 | ····96838 | 552351565 | 555120100 | 557902512 |
| 420 | ····68501 | ····68954 | ····82943 | 546910534 | 549651797 | 552406800 | ···75612 | ····58302 |
| 430 | 538822378 | 541523101 | 544237361 | ····65225 | 549706762 | ···62041 | 555231130 | 558014098 |
| 440 | ····76260 | ····77253 | ····91785 | 547019922 | ····61733 | 552517287 | ····86653 | ····69899 |
| 450 | 538930148 | 541631411 | 544346214 | ····74624 | 549316709 | ····72539 | 555342181 | 558125706 |
| 460 | ····84041 | ····85574 | 544400648 | 547129331 | ····71691 | 552627796 | ····97716 | ····51515 |
| 470 | 539037939 | 541739743 | ····55088 | ····54044 | 549026678 | ····83059 | 555453255 | 558237337 |
| 480 | ····91843 | ····93917 | 544509534 | 547238763 | ····81671 | 552738327 | 555508801 | ····93161 |
| 490 | 539145752 | 541848096 | ····63985 | ····93486 | 550036669 | ····93681 | ····64352 | 558348990 |
| 500 | ····99667 | 541902281 | 544618441 | 547348216 | ····91673 | 552848880 | 555619908 | 558404825 |

Figure C.43

| | 172000 | 172500 | 173000 | 173500 | 174000 | 174500 | 175000 | 175500 |
|---|---|---|---|---|---|---|---|---|
| 0 | 558404825 | 561203700 | 564016605 | 566843608 | 569684781 | 572540195 | 575409920 | 528294030 |
| 10 | ....60665 | ....59821 | ....73006 | 566900292 | 569741749 | ....97449 | ....67461 | 578351860 |
| 20 | 558516511 | 561315947 | 564129414 | ....56982 | ....98724 | 572654708 | 575525008 | 578409695 |
| 30 | ....72363 | ....72078 | ....85826 | 567013678 | 569855703 | 572711974 | ....82561 | ....67536 |
| 40 | 558626220 | 561428216 | 564242245 | ....70379 | 569912689 | ....69245 | 575640119 | 578525383 |
| 50 | ....84089 | ....84358 | ....98669 | 567127086 | ....69680 | 572826522 | ....97683 | ....83235 |
| 60 | 558739951 | 561540507 | 564355099 | ....83799 | 570026677 | ....83805 | 575755253 | 571041093 |
| 70 | ....95825 | ....96661 | 564411535 | 567240517 | ....83680 | 572941093 | 575812828 | ....98957 |
| 80 | 558851705 | 561652821 | ....67976 | ....97242 | 570140688 | ....98387 | ....70410 | 578756827 |
| 90 | 558907590 | 561708986 | 564524423 | 567353971 | ....97703 | 573055687 | 575927997 | 578814703 |
| 100 | ....63481 | ....65157 | ....30875 | 567410707 | 577254722 | 573112993 | ....85589 | ....72585 |
| 110 | 559019377 | 561821333 | 564637333 | ....67448 | 570311748 | ....70304 | 576043188 | 578930472 |
| 120 | ....75279 | ....77515 | ....93797 | 867524194 | ....65779 | 573227621 | 573100792 | ....88365 |
| 130 | 559131187 | 561933703 | 564750266 | ....80947 | 570425816 | ....84944 | ....58402 | 579046264 |
| 140 | ....87100 | ....89897 | 564806741 | 567637705 | ....82859 | 573342272 | 576216018 | 579104168 |
| 150 | 559243019 | 562046095 | ....63222 | ....94469 | 570539906 | ....99606 | ....73640 | ....62079 |
| 160 | ....98943 | 562102300 | 564919708 | 567751238 | ....96960 | 573456946 | 576331267 | 579219995 |
| 170 | 559354873 | ....59510 | ....76200 | 567808013 | 570654020 | 573514292 | ....88900 | ....77917 |
| 180 | 559310808 | 562214726 | 565032698 | ....64794 | 570711086 | ....71643 | 576446539 | 579335845 |
| 190 | ....66749 | ....70948 | ....89201 | 567921581 | ....68157 | 573629001 | 576504184 | ....93778 |
| 200 | 559522696 | 562327175 | 565145710 | ....78373 | 570825233 | ....86363 | ....61834 | 579451718 |
| 210 | ....78648 | ....83407 | 565202225 | 568035171 | ....82316 | 573743732 | 576619490 | 579509663 |
| 220 | 559634606 | 562435646 | ....58745 | ....91974 | 570939404 | 573801107 | ....77152 | ....67614 |
| 230 | ....90570 | ....95890 | 565315271 | 568148783 | ....96498 | ....58487 | 576734820 | 579625571 |
| 240 | 559746539 | 562552139 | ....71802 | 568205598 | 571053598 | 575915872 | ....92494 | ....83533 |
| 250 | 559802514 | 562608395 | 565428339 | ....62419 | 571110703 | ....73264 | 576850173 | 579741502 |
| 260 | ....58494 | ....64655 | ....84882 | 568319245 | ....67814 | 574030661 | 576907858 | ....99496 |
| 270 | 559914480 | 562720922 | 565541431 | ....76077 | 571224931 | ....88964 | ....65549 | 579857456 |
| 280 | ....70471 | ....77194 | ....97985 | 568432914 | ....82053 | 574145473 | 577023245 | 579915441 |
| 290 | 560026468 | 562833472 | 565654545 | ....89758 | 571339182 | 574202888 | ....80947 | ....73433 |
| 300 | ....82471 | ....89755 | 565711110 | 568046607 | ....96316 | ....60308 | 577128656 | 580031430 |
| 310 | 560138479 | 562946044 | ....67681 | 568603461 | 571453455 | 574317734 | ....96369 | ....89433 |
| 320 | ....94493 | 563002339 | 565824258 | ....60322 | 571510601 | ....75166 | 577254089 | 580147442 |
| 330 | 560250512 | ....58639 | ....80840 | 568717188 | ....67752 | 574432603 | 577311814 | 580205457 |
| 340 | 560306537 | 563114945 | 565937429 | ....74059 | 571624908 | ....90047 | ....69546 | ....63477 |
| 350 | ....62568 | ....71256 | ....94022 | 568830937 | ....82071 | 574547496 | 577427283 | 580321504 |
| 360 | 560418604 | 563227573 | 576050622 | ....87820 | 571739239 | 574604950 | ....85025 | ....79536 |
| 370 | ....74646 | ....83896 | 566107227 | 568944709 | ....96413 | ....65411 | 577542774 | 580437574 |
| 380 | 560530694 | 563340224 | ....63837 | 569001603 | 571853593 | 574722877 | 577600528 | ....95618 |
| 390 | ....86747 | ....96558 | 566220454 | ....58505 | 571910778 | ....80350 | ....58288 | 580553667 |
| 400 | 560642805 | 563452898 | ....77076 | 569115410 | ....67969 | 574834827 | 577716054 | 580611723 |
| 410 | ....98870 | 563509243 | 566333704 | ....72322 | 572025166 | ....92310 | ....73826 | ....69784 |
| 420 | 560754940 | ....65594 | ....90337 | 569229239 | ....82368 | 574949800 | 577831603 | 580727851 |
| 430 | 560811015 | 563621951 | 566446976 | ....86162 | 472139577 | 575007255 | ....89306 | ....85924 |
| 440 | ....67096 | ....78313 | 566503621 | 569343091 | ....96791 | ....64795 | 577947175 | 580844002 |
| 450 | 560923183 | 563734681 | ....60271 | 569400025 | 572254010 | 575122302 | 578004970 | 580902087 |
| 460 | ....79275 | ....91054 | 566616927 | ....56965 | 572311236 | ....79814 | ....62770 | [illegible] |
| 470 | 561035373 | 563847433 | ....73589 | 569513911 | ....68467 | 575237332 | [illegible] | 581018273 |
| 480 | ....91477 | 563903818 | 566730256 | ....72862 | 572425704 | ....94856 | [illegible] | ....76375 |
| 490 | 561147586 | ....60209 | ....86929 | 569637819 | ....82946 | 575352385 | [illegible] | [illegible] |
| 500 | 561203700 | 564016605 | 556843608 | ....84781 | 572540195 | 575409920 | ....94030 | [illegible] |

Figure C.44

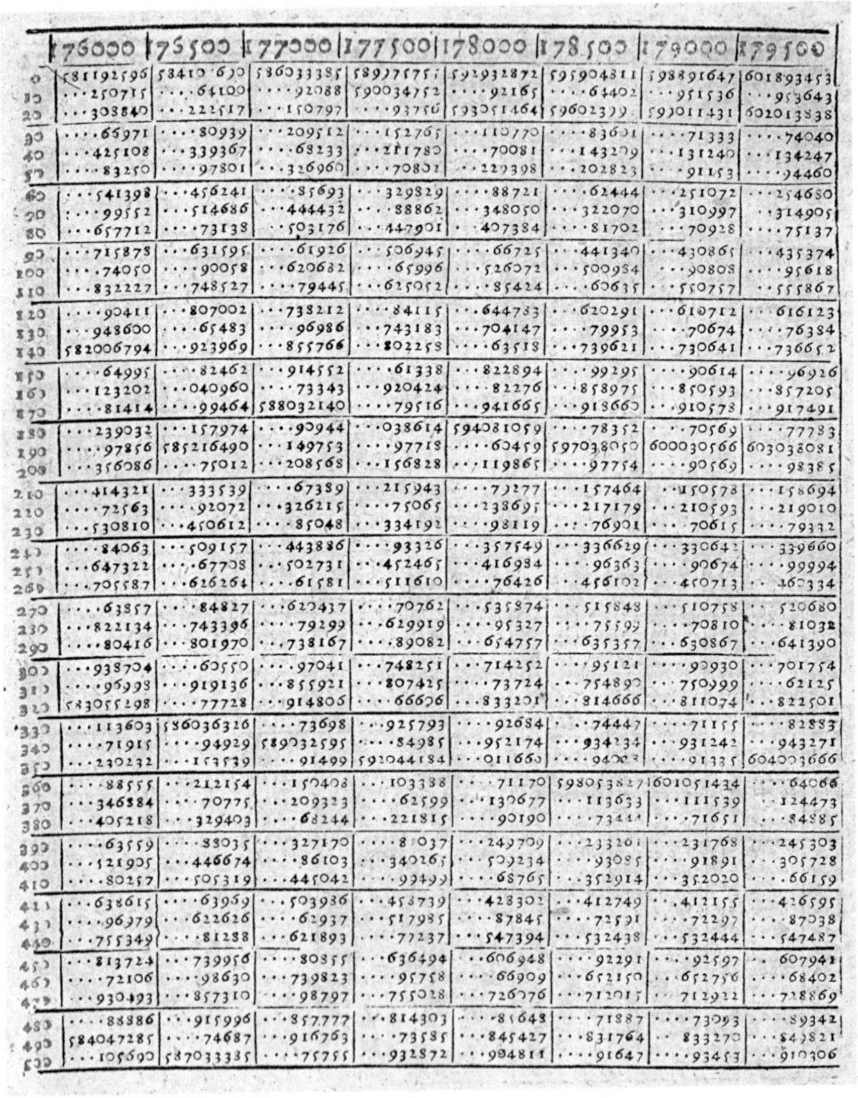

| | 176000 | 176500 | 177000 | 177500 | 178000 | 178500 | 179000 | 179500 |
|---|---|---|---|---|---|---|---|---|
| 0 | 581192596 | 584100630 | 586033385 | 589017575 | 592932872 | 595904811 | 598891647 | 601893453 |
| 10 | ···250715 | ····64100 | ····92088 | 590034752 | ····92165 | ····64402 | ···951536 | ···953643 |
| 20 | ···308840 | ···222517 | ···150797 | ····93750 | 593051464 | 596023?? | 599011431 | 602013838 |
| 30 | ····65971 | ····80939 | ···209512 | ···152765 | ···110770 | ····83601 | ····71333 | ····74040 |
| 40 | ···425108 | ···339367 | ····68233 | ···211780 | ····70081 | ···143209 | ···131240 | ···134247 |
| 50 | ····83250 | ····97801 | ···326960 | ····70801 | ···229398 | ···202823 | ····91153 | ····94460 |
| 60 | ···541398 | ···456241 | ····85693 | ···329829 | ····88721 | ····62444 | ···251072 | ···254680 |
| 70 | ····99552 | ···514686 | ···444432 | ····88862 | ···348050 | ···322070 | ···310997 | ···314905 |
| 80 | ···657712 | ····73138 | ···503176 | ···447901 | ···407384 | ····81702 | ····70928 | ····75137 |
| 90 | ···715878 | ···631595 | ····61926 | ···506945 | ····66725 | ···441340 | ···430865 | ···435374 |
| 100 | ····74050 | ····90058 | ···620682 | ····65996 | ···526072 | ···500984 | ····90808 | ····95618 |
| 110 | ···832227 | ···748527 | ····79445 | ···625052 | ····85424 | ····60635 | ···550757 | ···555867 |
| 120 | ····90411 | ···807002 | ···738212 | ····84115 | ···644783 | ···620291 | ···610712 | ···616123 |
| 130 | ···948600 | ····65483 | ····96986 | ···743183 | ···704147 | ····79953 | ····70674 | ····76384 |
| 140 | 582006794 | ···923969 | ···855766 | ···802258 | ····63518 | ···739621 | ···730641 | ···736652 |
| 150 | ····64995 | ····82462 | ···914552 | ····61338 | ···822894 | ····99295 | ····90614 | ····96926 |
| 160 | ···123202 | ···040960 | ····73343 | ···920424 | ····82276 | ···858975 | ···850593 | ···857205 |
| 170 | ····81414 | ····99464 | 588032140 | ····79516 | ···941665 | ···918660 | ···910578 | ···917491 |
| 180 | ···239032 | ···157974 | ····90944 | ···038614 | 594081059 | ····78352 | ····70569 | ····77783 |
| 190 | ····97856 | 585216490 | ···149753 | ····97718 | ····60459 | 597038050 | 600030566 | 603038081 |
| 200 | ···356086 | ····75012 | ···208568 | ···156828 | ···119865 | ····97754 | ····90569 | ····98385 |
| 210 | ···414321 | ···333539 | ····67389 | ···215943 | ····79277 | ···157464 | ···150578 | ···158694 |
| 220 | ····72563 | ····92072 | ···326215 | ····75065 | ···238695 | ···217179 | ···210593 | ···219010 |
| 230 | ···530810 | ···450612 | ····85048 | ···334192 | ····98119 | ····76901 | ····70615 | ····79332 |
| 240 | ····84063 | ···509157 | ···443886 | ····93326 | ···357549 | ···336629 | ···330642 | ···339660 |
| 250 | ···647322 | ····67708 | ···502731 | ···452465 | ···416984 | ····96363 | ····90674 | ····99994 |
| 260 | ···705587 | ···626264 | ····61581 | ···511610 | ····76426 | ···456102 | ···450713 | ···460334 |
| 270 | ····63857 | ····84827 | ···620437 | ····70762 | ···535874 | ···515848 | ···510758 | ···520680 |
| 280 | ···822134 | ···743396 | ····79299 | ···629919 | ····95327 | ····75599 | ····70810 | ····81032 |
| 290 | ····80416 | ···801970 | ···738167 | ····89082 | ···654757 | ···635357 | ···630867 | ···641390 |
| 300 | ···938704 | ····60550 | ····97041 | ···748251 | ···714252 | ····95121 | ····90930 | ···701754 |
| 310 | ····96998 | ···919136 | ···855921 | ···807425 | ····73724 | ···754890 | ···750999 | ····62125 |
| 320 | 583055298 | ····77728 | ···914806 | ····66606 | ···833201 | ···814666 | ···811074 | ···822501 |
| 330 | ···113603 | 586036326 | ····73698 | ···925793 | ····92684 | ····74447 | ····71155 | ····82883 |
| 340 | ····71915 | ····94929 | 589032595 | ····84985 | ···952174 | ···934234 | ···931242 | ···943271 |
| 350 | ···230232 | ···153539 | ····91499 | 592044184 | ···011660 | ····94003 | ····91335 | 604003666 |
| 360 | ····88555 | ···212154 | ···150408 | ···103388 | ····71170 | 598053827 | 601051434 | ····64066 |
| 370 | ···346884 | ····70775 | ···209323 | ····62599 | ···130677 | ···113633 | ···111539 | ···124473 |
| 380 | ···405218 | ···329403 | ····68244 | ···221815 | ····90190 | ····73444 | ····71651 | ····84885 |
| 390 | ····63559 | ····88035 | ···327170 | ····81037 | ···249709 | ···233201 | ···231768 | ···245303 |
| 400 | ···521905 | ···446674 | ····86103 | ···340265 | ···509234 | ····93085 | ····91891 | ···305728 |
| 410 | ····80257 | ···505319 | ···445042 | ····99499 | ····68765 | ···352914 | ···352020 | ····66159 |
| 420 | ···638615 | ····63959 | ···503936 | ···458739 | ···428302 | ···412749 | ···412155 | ···426595 |
| 430 | ····96979 | ···622626 | ····62937 | ···517985 | ····87845 | ····72591 | ····72297 | ····87038 |
| 440 | ···755349 | ····81288 | ···621893 | ····77237 | ···547394 | ···532438 | ···532444 | ···547487 |
| 450 | ···813724 | ···739956 | ····80855 | ···636494 | ···606948 | ····92291 | ····92597 | ···607941 |
| 460 | ····72106 | ····98630 | ···739823 | ····95758 | ····66909 | ···652150 | ···652756 | ····68402 |
| 470 | ···930493 | ···857310 | ····98797 | ···755028 | ···726076 | ···712015 | ···712922 | ···728869 |
| 480 | ····88886 | ···915996 | ···857777 | ···814303 | ····81648 | ····71887 | ····73093 | ····89342 |
| 490 | 584047285 | ····74687 | ···916763 | ····73585 | ···845427 | ···831764 | ···833270 | ···849821 |
| 500 | ···105690 | 587033385 | ····75755 | ···932872 | ···904811 | ····91647 | ····93453 | ···910306 |

**Figure C.45**

| | 180000 | 180500 | 181000 | 181500 | 182000 | 182500 | 183000 | 183500 |
|---|---|---|---|---|---|---|---|---|
| 0 | 604910300 | 607942279 | 610989450 | 614051894 | 617129687 | 620222908 | 623331032 | 626455931 |
| 10 | ···70797 | 608003073 | 611050549 | ···113299 | ···91400 | ···84930 | ···93565 | ···518584 |
| 20 | 605031294 | ···62874 | ···111664 | ···74710 | ···253120 | ···346955 | ···456305 | ···81236 |
| 30 | ···91797 | ···124680 | ···72765 | ···236128 | ···314845 | ···408993 | ···518650 | ···643894 |
| 40 | ···152306 | ···85493 | ···233882 | ···97551 | ···76576 | ···71034 | ···81002 | ···706558 |
| 50 | ···212821 | ···246311 | ···95006 | ···358981 | ···438314 | ···53308 | ···643369 | ···69229 |
| 60 | ···73343 | ···307136 | ···356135 | ···420417 | ···500058 | ···95135 | ···705725 | ···831906 |
| 70 | ···333870 | ···67966 | ···417271 | ···81859 | ···61808 | ···657194 | ···68095 | ···54589 |
| 80 | ···94403 | ···4[illegible]8803 | ···76413 | ···543307 | ···623564 | ···719260 | ···830472 | ···957279 |
| 90 | ···454943 | ···89646 | ···539560 | ···604762 | ···85326 | ···81332 | ···92855 | 627019974 |
| 100 | ···515488 | ···550495 | ···600714 | ···66222 | ···747095 | ···843410 | ···955244 | ···82676 |
| 110 | ···76040 | ···611350 | ···61874 | ···727689 | ···808870 | ···905494 | 624017640 | ···145385 |
| 120 | ···636597 | ···72211 | ···723041 | ···8[illegible]161 | ···70651 | ···67585 | ···80042 | ···208099 |
| 130 | ···97161 | ···33075 | ···84213 | ···850640 | ···932438 | 621029681 | ···142450 | ···70820 |
| 140 | ···757731 | ···93952 | ···845391 | ···912125 | ···94231 | ···91784 | ···204864 | ···333547 |
| 150 | ···318307 | ···54831 | ···906576 | ···73617 | 618056030 | ···153894 | ···67284 | ···96280 |
| 160 | ···788[illegible]8 | ···15717 | ···67767 | 615035114 | ···117836 | ···216009 | ···329711 | ···459020 |
| 170 | ···939[illegible]6 | ···76608 | 612028963 | ···96618 | ···79648 | ···78131 | ···92144 | ···521766 |
| 180 | 606000070 | 6090375[illegible]6 | ···90166 | ···158127 | ···241466 | ···340258 | ···454583 | ···84518 |
| 190 | ···60670 | ···98410 | ···151375 | ···219643 | ···303290 | ···402392 | ···517029 | ···647277 |
| 200 | ···1[illegible]1276 | ···159[illegible]20 | ···212590 | ···81161 | ···65120 | ···64533 | ···79481 | ···710041 |
| 210 | ···81888 | ···220236 | ···73812 | ···342693 | ···426957 | ···526679 | ···641938 | ···72812 |
| 220 | ···242507 | ···81158 | ···335039 | ···404227 | ···88799 | ···88832 | ···704403 | ···835590 |
| 230 | ···303131 | ···342086 | ···96273 | ···65768 | ···550648 | ···650991 | ···66873 | ···98373 |
| 240 | ···63761 | ···403020 | ···457512 | ···527314 | ···612503 | ···713156 | ···829350 | ···961163 |
| 250 | ···424398 | ···63960 | ···518758 | ···88867 | ···74364 | ···75327 | ···91833 | 628023959 |
| 260 | ···85040 | ···524907 | ···80010 | ···650476 | ···736232 | ···837505 | ···954322 | ···86762 |
| 270 | ···545689 | ···85859 | ···641268 | ···711991 | ···98105 | ···99689 | 625016817 | ···149570 |
| 280 | ···606343 | ···646818 | ···702532 | ···73562 | ···859985 | ···961879 | ···79319 | ···212385 |
| 290 | ···67004 | ···707782 | ···63802 | ···835140 | ···921871 | 622024075 | ···141827 | ···75206 |
| 300 | ···727670 | ···68753 | ···825079 | ···96723 | ···83763 | ···86277 | ···204341 | ···338034 |
| 310 | ···88343 | ···829730 | ···86361 | ···958303 | 619045662 | ···148486 | ···66862 | ···400868 |
| 320 | ···849022 | ···90713 | ···947650 | 616019899 | ···107566 | ···210701 | ···329388 | ···63708 |
| 330 | ···909707 | ···951702 | 613008945 | ···81501 | ···69477 | ···72922 | ···91921 | ···526554 |
| 340 | ···70398 | 610012697 | ···70245 | ···143109 | ···231394 | ···335149 | ···454460 | ···89407 |
| 350 | 607031095 | ···73699 | ···131552 | ···204723 | ···93317 | ···97382 | ···517006 | ···652266 |
| 360 | ···91798 | ···134706 | ···92866 | ···66344 | ···355247 | ···459622 | ···79557 | ···715131 |
| 370 | ···152507 | ···95720 | ···254185 | ···327970 | ···417182 | ···521868 | ···642115 | ···78003 |
| 380 | ···213223 | ···256739 | ···315510 | ···89603 | ···79124 | ···84120 | ···704680 | ···840880 |
| 390 | ···73944 | ···317765 | ···76842 | ···451242 | ···541072 | ···646379 | ···67250 | ···903764 |
| 400 | ···334671 | ···78796 | ···438180 | ···512887 | ···603026 | ···708643 | ···829827 | ···66655 |
| 410 | ···95405 | ···439834 | ···99523 | ···74538 | ···64986 | ···70914 | ···92410 | 629029551 |
| 420 | ···456144 | ···50[illegible]878 | ···560873 | ···636196 | ···726953 | ···833191 | ···954999 | ···97454 |
| 430 | ···516890 | ···61928 | ···622229 | ···97859 | ···88925 | ···95475 | 626017596 | ···155364 |
| 440 | ···77642 | ···622985 | ···83592 | ···759529 | ···850904 | ···9[illegible]7764 | ···80197 | ···218279 |
| 450 | ···638399 | ···84047 | ···744960 | ···821215 | ···912889 | 623020060 | ···142804 | ···81201 |
| 460 | ···99163 | ···745115 | ···806335 | ···82897 | ···74881 | ···82362 | ···205419 | ···344129 |
| 470 | ···759933 | ···806190 | ···67715 | ···944586 | 620036878 | ···144670 | ···68039 | ···407063 |
| 480 | ···820709 | ···67270 | ···929102 | 617006280 | ···98882 | ···206985 | ···330666 | ···70004 |
| 490 | ···81491 | ···928357 | ···90495 | ···67981 | ···160892 | ···69305 | ···93299 | ···532951 |
| 500 | ···942279 | ···89450 | 614051894 | ···129687 | ···222908 | ···331632 | ···455938 | ···95904 |

**Figure C.46**

| | 192000 | 192500 | 193000 | 193500 | 194000 | 194500 | 195000 | 195500 |
|---|---|---|---|---|---|---|---|---|
| 0 | 682030373 | 685448893 | 688884548 | 692337423 | 695807605 | 699295180 | 702800236 | 706322861 |
| 10 | ···98576 | ···51743 | ···953437 | ···406657 | ···77186 | ···365110 | ···70516 | ···93493 |
| 20 | ···166786 | ···85990 | 689022332 | ···75898 | ···946773 | ···435046 | ···940803 | ···464132 |
| 30 | ···235003 | ···654549 | ···91234 | ···545145 | 696016368 | ···504990 | 703011098 | ···534779 |
| 40 | ···303226 | ···723114 | ···160143 | ···614400 | ···85970 | ···74940 | ···81399 | ···605432 |
| 50 | ···71457 | ···91686 | ···2290[illegible]9 | ···83651 | ···155578 | ···644898 | ···151707 | ···76093 |
| 60 | ···439694 | ···860266 | ···97982 | ···52909 | ···225194 | ···714862 | ···222022 | ···746760 |
| 70 | ···507938 | ···928852 | ···366912 | ···822185 | ···94816 | ···84834 | ···92344 | ···817435 |
| 80 | ···76189 | ···97444 | ···435849 | ···91467 | ···364446 | ···854812 | ···362673 | ···88117 |
| 90 | ···644446 | 686066044 | ···504792 | ···960456 | ···434082 | ···924798 | ···433010 | ···958805 |
| 100 | ···712711 | ···134651 | ···73743 | 693030052 | ···503726 | ···94790 | ···503353 | 707029501 |
| 110 | ···80982 | ···203264 | ···642700 | ···99355 | ···73376 | 700064790 | ···73703 | ···100204 |
| 120 | ···849260 | ···71885 | ···711664 | ···168665 | ···643034 | ···134796 | ···644061 | ···70914 |
| 130 | ···917545 | ···340512 | ···80636 | ···237982 | ···712698 | ···204810 | ···714425 | ···241631 |
| 140 | ···85837 | ···409146 | ···849614 | ···307306 | ···82362 | ···74839 | ···84796 | ···312356 |
| 150 | 683054135 | ···77787 | ···918599 | ···76657 | ···852047 | ···344858 | ···855175 | ···83087 |
| 160 | ···122441 | ···546435 | ···87591 | ···445994 | ···921733 | ···414882 | ···925560 | ···453825 |
| 170 | ···90713 | ···615089 | 690506589 | ···515339 | ···91425 | ···84924 | ···95953 | ···524571 |
| 180 | ···259072 | ···83751 | ···125595 | ···84690 | 697061124 | ···554972 | 704066353 | ···95323 |
| 190 | ···327398 | ···752419 | ···94608 | ···654049 | ···130839 | ···625038 | ···136759 | ···666083 |
| 200 | ···95731 | ···821094 | ···263627 | ···723414 | ···200543 | ···95100 | ···207173 | ···736849 |
| 210 | ···464070 | ···89776 | ···332653 | ···92787 | ···70263 | ···765170 | ···77594 | ···807623 |
| 220 | ···532417 | ···958465 | ···401687 | ···862166 | ···339990 | ···835246 | ···348021 | ···878404 |
| 230 | ···600770 | 687027161 | ···70727 | ···931552 | ···409724 | ···905330 | ···418456 | ···49191 |
| 240 | ···69130 | ···95864 | ···53977 | 694000945 | ···79465 | ···75420 | ···88898 | 708019988 |
| 250 | ···737497 | ···164574 | ···608828 | ···70345 | ···549283 | 701045518 | ···559348 | ···90788 |
| 260 | ···805870 | ···233290 | ···77889 | ···139752 | ···618968 | ···115622 | ···629804 | ···161597 |
| 270 | ···74251 | ···302013 | ···746957 | ···209166 | ···88730 | ···85734 | ···700267 | ···232414 |
| 280 | ···942638 | ···70744 | ···816031 | ···78587 | ···758499 | ···255852 | ···70737 | ···303237 |
| 290 | 684011033 | ···439481 | ···85113 | ···348015 | ···828275 | ···325978 | ···841214 | ···74067 |
| 300 | ···79434 | ···508225 | ···954201 | ···417450 | ···98057 | ···96111 | ···911697 | ···444905 |
| 310 | ···147842 | ···76975 | 691023297 | ···86892 | ···967847 | ···466250 | ···82138 | ···515749 |
| 320 | ···216257 | ···645733 | ···92399 | ···556340 | 698037644 | ···536397 | 705052686 | ···86661 |
| 330 | ···84678 | ···714498 | ···161509 | ···625796 | ···107448 | ···606550 | ···123192 | ···657459 |
| 340 | ···353197 | ···83269 | ···230624 | ···95259 | ···77258 | ···76711 | ···93704 | ···728325 |
| 350 | ···421542 | ···852047 | ···99747 | ···764728 | ···247076 | ···746879 | ···264223 | ···99198 |
| 360 | ···89984 | ···920833 | ···368877 | ···834205 | ···316901 | ···817054 | ···334750 | ···879978 |
| 370 | ···558433 | ···89625 | ···438014 | ···903688 | ···86733 | ···87235 | ···485283 | ···940865 |
| 380 | ···626889 | 688058424 | ···507158 | ···73178 | ···456571 | ···957424 | ···75824 | 709011859 |
| 399 | ···95352 | ···127229 | ···76309 | 695042676 | ···526417 | 702027620 | ···546371 | ···82760 |
| 400 | ···763821 | ···96042 | ···645466 | ···112180 | ···596270 | ···97822 | ···616926 | ···153668 |
| 410 | ···832298 | ···264862 | ···714631 | ···81691 | ···666129 | ···168032 | ···87488 | ···224584 |
| 420 | ···900781 | ···333688 | ···83802 | ···251209 | ···735996 | ···238249 | ···758056 | ···95506 |
| 430 | ···69271 | ···402522 | ···852981 | ···320734 | ···805869 | ···308473 | ···828632 | ···366436 |
| 440 | 685037768 | ···71362 | ···922166 | ···90267 | ···75750 | ···78704 | ···99215 | ···437372 |
| 450 | ···106272 | ···540209 | ···91358 | ···459806 | ···945638 | ···448941 | ···969805 | ···508316 |
| 460 | ···74790 | ···609063 | 692060557 | ···529352 | 699015532 | ···519186 | 706040402 | ···79267 |
| 470 | ···243308 | ···77924 | ···129763 | ···98904 | ···85434 | ···89438 | ···111006 | ···650225 |
| 490 | ···311832 | ···746792 | ···98976 | ···668464 | ···155342 | ···659697 | ···81617 | ···721190 |
| 480 | ···80365 | ···815667 | ···262196 | ···738031 | ···225258 | ···729963 | ···252235 | ···92162 |
| 500 | ···448893 | ···84548 | ···337423 | ···807605 | ···95180 | ···800236 | ···322861 | ···863141 |

**Figure C.47**

| | 188000 | 188500 | 189000 | 189500 | 190000 | 190500 | 191000 | 191500 |
|---|---|---|---|---|---|---|---|---|
| 0 | 655288890 | 658573375 | 661874322 | 665191815 | 668525936 | 671876768 | 675244955 | 678628902 |
| 10 | ...354419 | ...639232 | ...940510 | ...258334 | ....92788 | ...943955 | ...311920 | ....56765 |
| 20 | ...419955 | ...705096 | 662006704 | ...324860 | ...659647 | 672011150 | ....79451 | ...764635 |
| 30 | ....85497 | ....70967 | ....72904 | ....91392 | ...726513 | ....78351 | ...446989 | ...832511 |
| 40 | ...551045 | ...836844 | ...139112 | ...457931 | ....93386 | ...145559 | ...514533 | ...900395 |
| 50 | ...616600 | ...902727 | ...205326 | ...524477 | ...860265 | ...212773 | ....82085 | ....68285 |
| 60 | ....82162 | ....68618 | ....71546 | ....91030 | ...927151 | ....79995 | ...649643 | 679036181 |
| 70 | ...747730 | 659034515 | ...337774 | ...657589 | ....94044 | ...347223 | ...717208 | ...104085 |
| 80 | ...813305 | ...100418 | ...404008 | ...724154 | 669060943 | ...414457 | ....84780 | ....71995 |
| 90 | ....78886 | ....66328 | ....70248 | ....90727 | ...127850 | ....81699 | ...852358 | ...239913 |
| 100 | ...944474 | ...232245 | ...536495 | ...857306 | ....94762 | ...548946 | ...919944 | ...307837 |
| 110 | 656010069 | ....98168 | ...602749 | ...923892 | ...261682 | ...616202 | ....87536 | ....75767 |
| 120 | ....75670 | ...364098 | ....69009 | ....90484 | ...328608 | ....83463 | 676055134 | ...443705 |
| 130 | ...141277 | ...430034 | ...735276 | 666057083 | ....95541 | ...750732 | ...122740 | ...511649 |
| 140 | ...206891 | ....95977 | ...801549 | ...123689 | ...462481 | ...818007 | ....90352 | ....79601 |
| 150 | ....72512 | ...561927 | ....67829 | ....90301 | ...529427 | ....85289 | ...257971 | ...647559 |
| 160 | ...338139 | ...627883 | ...934116 | ...256920 | ....96380 | ....952577 | ...325597 | ...715523 |
| 170 | ...403773 | ....93846 | 663000409 | ...323546 | ...663339 | 673019872 | ....93230 | ....83495 |
| 180 | ....69413 | ...759815 | ....66709 | ....90178 | ...730306 | ....87174 | ...460869 | ...851473 |
| 190 | ...535060 | ...825791 | ...133016 | ...456817 | ....97279 | ...154483 | ...528515 | ...919458 |
| 200 | ...600714 | ....91774 | ....99329 | ...523463 | ...864258 | ...221799 | ....96168 | ....87450 |
| 210 | ....66374 | ...557763 | ...265649 | ....90115 | ...931245 | ....89121 | ...663828 | 680055449 |
| 220 | ...732041 | 660023759 | ...331976 | ...656774 | ....98238 | ...356450 | ...731494 | ...123454 |
| 230 | ....97714 | ....89761 | ....98309 | ...723440 | 670065238 | ...423785 | ....99167 | ....91467 |
| 240 | ...863394 | ...155770 | ...464649 | ....90112 | ...132244 | ....98128 | ...866847 | ...259486 |
| 250 | ...929080 | ...221786 | ...530999 | ...856791 | ....99257 | ...558477 | ...9345[illegible]4 | ...327512 |
| 260 | ....94773 | ....87808 | ....97348 | ...923477 | ...266277 | ...625833 | 677002227 | ....95545 |
| 270 | 657060472 | ...353837 | ...663798 | ....90169 | ...333304 | ....93195 | ....69927 | ...463584 |
| 280 | ...126178 | ...419872 | ...730074 | 667056868 | ...400337 | ...760564 | ...137634 | ...531631 |
| 290 | ....918[illegible]1 | ....85914 | ....96447 | ...123574 | ....67377 | ...887941 | ...205348 | ....99684 |
| 300 | ...257610 | ...551963 | ...862827 | ....90287 | ...534424 | ....95323 | ....73069 | ...667744 |
| 310 | ...323336 | ...618018 | ...929213 | ...257006 | ...601477 | ...962713 | ...340796 | ...735811 |
| 320 | ....89068 | ....84080 | ....95606 | ...323731 | ....68537 | 674030109 | ...408530 | ...803884 |
| 330 | ...454807 | ...750148 | 664062006 | ....90464 | ...735604 | ....97512 | ....76271 | ....71965 |
| 340 | ...520553 | ...816223 | ...128412 | ...457203 | ...802678 | ...164922 | ...544018 | ...940052 |
| 350 | ....86305 | ....82305 | ....94825 | ...523948 | ....69758 | ...232339 | ...611773 | 681008146 |
| 360 | ...652063 | ...948393 | ...261244 | ....90701 | ...936845 | ....99762 | ....70534 | ....76246 |
| 370 | ...717829 | 661014488 | ...327670 | ...657460 | 671003939 | ...367192 | ...797302 | ...144354 |
| 380 | ....82600 | ....80589 | ....94103 | ...724226 | ....71040 | ...434629 | ...815[illegible]77 | ...212468 |
| 390 | ...849379 | ...146697 | ...460543 | ....90998 | ...138147 | ...502072 | ....82858 | ....80590 |
| 400 | ...915164 | ...212812 | ...526989 | ...857777 | ...205260 | ....69522 | ...950646 | ...348718 |
| 410 | ....80955 | ....78933 | ....93441 | ...924563 | ....72381 | ...636979 | 678018442 | ...416853 |
| 420 | 658046753 | ...345061 | ...659901 | ....91355 | ...339508 | ...714443 | ....86243 | ....84994 |
| 430 | ...112558 | ...411195 | ...726367 | 668058155 | ...406642 | ....71913 | ...154052 | ...553143 |
| 440 | ....78269 | ....77337 | ....92839 | ...124960 | ....73783 | ...839391 | ...221867 | ...621298 |
| 450 | ...244187 | ...543484 | ...859319 | ....91773 | ...540930 | ...906874 | ....89690 | ....89460 |
| 460 | ...310011 | ...609639 | ...925805 | ...258592 | ...608084 | ....74365 | ...357519 | ...757629 |
| 470 | ....75342 | ....75800 | ....92297 | ...325418 | ....75245 | 675041862 | ...425354 | ...825805 |
| 480 | ...441680 | ...741967 | 665058796 | ....92250 | ...742413 | ...109367 | ....93197 | ....93988 |
| 490 | ...507524 | ...808141 | ...125302 | ...459090 | ...809587 | ....76878 | ...561046 | ...962177 |
| 500 | ....73375 | ....74322 | ....91815 | ...525936 | ....76768 | ...244395 | ...628902 | 682030373 |

G

**Figure C.48**

| | 184000 | 184500 | 185000 | 185500 | 186000 | 186500 | 187000 | 187500 |
|---|---|---|---|---|---|---|---|---|
| 0 | 629595904 | 632751609 | 635923131 | 639110547 | 642313943 | 6455333[illegible] | 64[illegible]768981 | 652020786 |
| 01 | ...658864 | ...814884 | ...86723 | ...7416[illegible] | ...73175 | ...9794 | ...833858 | ...84988 |
| 20 | ...721830 | ...78166 | 646050322 | ...235[illegible]7 | ...[illegible]12412 | ...66250 | ...98742 | ...151197 |
| 30 | ...84832 | ...941453 | ...113927 | ...312301 | ...500657 | ...727073 | ...963631 | ...216[illegible]12 |
| 40 | ...847781 | 633004743 | ...77535 | ...66231 | ...70907 | ...91646 | 649028528 | ...81634 |
| 50 | ...910765 | ...68043 | ...241156 | ...430168 | ...635164 | ...856225 | ...93431 | ...346862 |
| 60 | ...73757 | ...131355 | ...304780 | ...94111 | ...99428 | ...930811 | ...158340 | ...412097 |
| 70 | 630036754 | ...94668 | ...68410 | ...553060 | ...63698 | ...85403 | ...223256 | ...77338 |
| 80 | ...99758 | ...257987 | ...432047 | ...622016 | ...827974 | 646050001 | ...88178 | ...542526 |
| 90 | ...162768 | ...321313 | ...95690 | ...85978 | ...92257 | ...114606 | ...353107 | ...607840 |
| 100 | ...225784 | ...84645 | ...559340 | ...749947 | ...956546 | ...79218 | ...418042 | ...73101 |
| 110 | ...88807 | ...447984 | ...622996 | ...813922 | 643020842 | ...243836 | ...82984 | ...738368 |
| 120 | ...351835 | ...511329 | ...86658 | ...77903 | ...85144 | ...308460 | ...547932 | ...803642 |
| 130 | ...414871 | ...74630 | ...750327 | ...941891 | ...149453 | ...73091 | ...612883 | ...68922 |
| 140 | ...77912 | ...638037 | ...814002 | 640005885 | ...213768 | ...437728 | ...77848 | ...934209 |
| 150 | ...540960 | ...701401 | ...77683 | ...69886 | ...78088 | ...502372 | ...742816 | ...99502 |
| 160 | ...604014 | ...64771 | ...941371 | ...133893 | ...342417 | ...67022 | ...807790 | 653064802 |
| 170 | ...67074 | ...828148 | 637005065 | ...97906 | ...406751 | ...631679 | ...72771 | ...130109 |
| 180 | ...730141 | ...91530 | ...68766 | ...261926 | ...71092 | ...96372 | ...937759 | ...95422 |
| 190 | ...93214 | ...954920 | ...132473 | ...325952 | ...535439 | ...761012 | 650002752 | ...260741 |
| 200 | ...856293 | 634018315 | ...96186 | ...89985 | ...99792 | ...825688 | ...67753 | ...326068 |
| 210 | ...919379 | ...81717 | ...259905 | ...454024 | ...664152 | ...90370 | ...132759 | ...91400 |
| 220 | ...82471 | ...145125 | ...323631 | ...518069 | ...728519 | ...955059 | ...97773 | ...456739 |
| 230 | 631045569 | ...208540 | ...87364 | ...82121 | ...92891 | 647019755 | ...262793 | ...522085 |
| 240 | ...108674 | ...71961 | ...451102 | ...646179 | ...857271 | ...84457 | ...327819 | ...87447 |
| 250 | ...71784 | ...335388 | ...514848 | ...710244 | ...921656 | ...149165 | ...92852 | ...652806 |
| 260 | ...234902 | ...98821 | ...78599 | ...74315 | ...86049 | ...213880 | ...457891 | ...718161 |
| 270 | ...98025 | ...462261 | ...642357 | ...838392 | 644050447 | ...78602 | ...522937 | ...83533 |
| 280 | ...361155 | ...525707 | ...706121 | ...902476 | ...114852 | ...343330 | ...87989 | ...848911 |
| 290 | ...4[illegible]4291 | ...89160 | ...69892 | ...66566 | ...79264 | ...408054 | ...653048 | ...914296 |
| 300 | ...87433 | ...652619 | ...833669 | 641030663 | ...243682 | ...72805 | ...718113 | ...79688 |
| 310 | ...550582 | ...716084 | ...97452 | ...94766 | ...308106 | ...537552 | ...83[illegible]35 | 654045086 |
| 320 | ...66[illegible]737 | ...79556 | ...961242 | ...158876 | ...72537 | ...602306 | ...848263 | ...110490 |
| 330 | ...70[illegible]99 | ...843034 | 638025038 | ...222992 | ...436974 | ...67066 | ...913348 | ...75901 |
| 340 | ...740066 | ...906518 | ...88841 | ...87114 | ...501418 | ...731833 | ...78439 | ...241319 |
| 350 | ...803240 | ...70009 | ...152649 | ...351243 | ...65868 | ...96606 | 651043537 | ...306743 |
| 360 | ...65421 | 635033506 | ...216465 | ...415378 | ...630325 | ...861386 | ...108642 | ...72174 |
| 370 | ...929607 | ...97009 | ...80286 | ...79519 | ...94788 | ...926172 | ...73752 | ...437611 |
| 380 | ...92800 | ...160519 | ...344114 | ...543667 | ...759257 | ...90964 | ...238870 | ...503[illegible]55 |
| 390 | 632055990 | ...224035 | ...407949 | ...607822 | ...823733 | 64805576[illegible] | ...303994 | ...68505 |
| 400 | ...119205 | ...87557 | ...71790 | ...71982 | ...88215 | ...12056[illegible] | ...69124 | ...633962 |
| 410 | ...82417 | ...351086 | ...535637 | ...736150 | ...952704 | ...85381 | ...43426[illegible] | ...9[illegible]425 |
| 420 | ...245635 | ...414621 | ...99490 | ...800323 | 645017200 | ...250200 | ...99404 | ...764895 |
| 430 | ...308860 | ...78163 | ...663350 | ...64503 | ...81701 | ...315025 | ...564554 | ...830372 |
| 440 | ...72091 | ...541710 | ...727217 | ...928697 | ...146209 | ...79856 | ...629711 | ...95855 |
| 450 | ...435328 | ...605264 | ...91089 | ...92883 | ...210724 | ...414694 | ...94874 | ...961344 |
| 460 | ...98572 | ...68825 | ...854968 | 642057082 | ...75245 | ...509539 | ...760043 | 655026840 |
| 470 | ...561821 | ...732392 | ...918854 | ...121788 | ...339773 | ...74390 | ...825219 | ...92343 |
| 480 | ...625078 | ...95965 | ...82746 | ...85500 | ...404307 | ...639247 | ...90402 | ...15785 |
| 490 | ...88340 | ...859545 | 639046644 | ...249718 | ...78847 | ...704111 | ...955591 | ...22336 |
| 500 | ...751609 | ...923131 | ...110549 | ...313943 | ...533394 | ...68981 | 652020786 | ...8889[illegible] |

**Figure C.49**

| | 196000 | 196500 | 197000 | 197500 | 198000 | 198500 | 199000 | 199500 |
|---|---|---|---|---|---|---|---|---|
| 0 | 709863141 | 713421167 | 716997026 | 720590809 | 724202604 | 727832502 | 731480597 | 735146973 |
| 10 | ...934127 | ...92509 | 717068726 | ...662868 | ...75024 | ...905286 | ...553745 | ...220488 |
| 20 | 710005121 | ...563858 | ...140433 | ...734934 | ...347452 | ...78076 | ...626901 | ...94010 |
| 30 | ...76121 | ...635214 | ...212147 | ...807008 | ...419886 | 728050874 | ...700063 | ...367539 |
| 40 | ...147129 | ...706578 | ...83868 | ...79088 | ...92328 | ...123679 | ...73233 | ...441076 |
| 50 | ...218144 | ...77949 | ...355596 | ...951176 | ...564777 | ...96491 | ...846409 | ...514620 |
| 60 | ...89166 | ...849326 | ...427332 | 721023271 | ...637234 | ...269311 | ...919593 | ...88172 |
| 70 | ...360194 | ...920711 | ...99074 | ...95373 | ...709698 | ...342138 | ...92785 | ...661730 |
| 80 | ...431230 | ...92103 | ...570824 | ...167483 | ...82169 | ...414972 | 732065985 | ...735297 |
| 90 | ...502274 | 714063503 | ...642581 | ...239599 | ...854647 | ...87814 | ...139191 | ...808870 |
| 100 | ...73324 | ...134909 | ...714346 | ...311723 | ...927132 | ...560663 | ...212405 | ...82451 |
| 110 | ...644381 | ...206322 | ...86117 | ...83855 | ...99625 | ...633519 | ...85626 | ...956039 |
| 120 | ...715445 | ...77743 | ...857896 | ...455993 | 725072125 | ...706382 | ...358855 | 736029635 |
| 130 | ...86517 | ...349171 | ...929681 | ...528139 | ...144632 | ...79253 | ...432091 | ...103238 |
| 140 | ...857596 | ...420606 | 718001474 | ...600291 | ...217147 | ...852131 | ...505334 | ...76848 |
| 150 | ...928681 | ...92048 | ...73275 | ...72451 | ...89668 | ...925016 | ...78584 | ...250466 |
| 160 | ...99774 | ...563497 | ...145082 | ...744619 | ...362197 | ...97908 | ...651842 | ...324091 |
| 170 | 711070874 | ...634953 | ...216896 | ...816793 | ...434734 | 7 9070808 | ...725108 | ...97723 |
| 180 | ...141981 | ...706417 | ...98718 | ...83975 | ...507277 | ...143715 | ...98380 | ...471363 |
| 190 | ...213096 | ...77887 | ...360547 | ...961164 | ...79828 | ...216630 | ...871660 | ...545010 |
| 200 | ...84217 | ...849365 | ...432383 | 722033360 | ...652386 | ...89551 | ...944947 | ...618665 |
| 210 | ...355345 | ...920850 | ...504226 | ...105563 | ...724951 | ...362480 | 733018242 | ...92327 |
| 220 | ...426481 | ...92342 | ...76077 | ...77774 | ...97537 | ...435416 | ...91543 | ...765996 |
| 230 | ...97624 | 715063841 | ...647934 | ...249992 | ...870103 | ...508360 | ...164853 | ...839673 |
| 240 | ...568773 | ...135348 | ...719799 | ...322217 | ...942690 | ...81311 | ...238169 | ...91331[illegible] |
| 250 | ...639930 | ...206861 | ...91671 | ...94449 | 726015285 | ...654269 | ...311493 | ...87048 |
| 260 | ...711094 | ...78382 | ...863550 | ...466688 | ...87886 | ...727234 | ...84824 | 737060747 |
| 270 | ...82265 | ...349910 | ...935437 | ...538935 | ...160495 | ...800207 | ...458162 | ...134453 |
| 280 | ...853444 | ...421445 | 719007330 | ...611189 | ...233111 | ...73187 | ...531508 | ...208166 |
| 290 | ...924629 | ...92987 | ...79231 | ...83450 | ...305734 | ...946174 | ...604861 | ...8188[illegible] |
| 300 | ...95821 | ...564536 | ...151139 | ...755718 | ...78365 | 730019169 | ...78222 | ...355615 |
| 310 | 712067021 | ...636093 | ...223054 | ...827984 | ...451003 | ...92171 | ...751590 | ...42935[illegible] |
| 320 | ...138228 | ...707657 | ...94976 | ...900277 | ...523648 | ...165180 | ...874965 | [illegible] |
| 330 | ...209441 | ...79227 | ...366906 | ...72567 | ...96300 | ...238196 | ...98347 | [illegible] |
| 340 | ...80662 | ...850805 | ...438842 | 723044864 | ...668960 | ...311220 | ...971737 | ...6506[illegible] |
| 350 | ...351891 | ...922390 | ...510786 | ...117168 | ...741627 | ...84251 | 734045134 | ...724367 |
| 360 | ...423126 | ...93983 | ...82737 | ...89480 | ...814301 | ...457200 | ...118539 | ...98139 |
| 370 | ...94368 | 716065582 | ...654696 | ...261799 | ...86982 | ...530335 | ...91951 | ...871919 |
| 380 | ...565617 | ...137189 | ...726661 | ...334125 | ...959671 | ...603388 | ...265370 | ...945706 |
| 390 | ...636874 | ...208802 | ...98634 | ...406459 | 727032367 | ...76449 | ...338796 | 738019501 |
| 400 | ...708138 | ...80423 | ...870614 | ...78799 | ...105070 | ...749519 | ...412230 | ...93303 |
| 410 | ...79409 | ...352051 | ...942601 | ...551147 | ...77781 | ...822594 | ...85672 | ...167112 |
| 420 | ...850666 | ...423686 | 720014595 | ...623502 | ...250498 | ...95676 | ...559120 | ...240979 |
| 430 | ...921972 | ...95329 | ...86597 | ...95865 | ...323223 | ...968766 | ...632576 | ...314753 |
| 440 | ...93264 | ...566978 | ...158605 | ...768234 | ...95956 | 731041863 | ...706039 | ...88584 |
| 450 | 713064563 | ...638635 | ...230621 | ...840611 | ...468695 | ...114967 | ...79510 | ...462423 |
| 460 | ...135870 | ...710299 | ...302644 | ...912995 | ...541442 | ...88078 | ...852988 | ...536269 |
| 470 | ...207133 | ...81970 | ...74675 | ...85386 | ...614196 | ...261197 | ...926473 | ...610123 |
| 480 | ...78504 | ...853648 | ...446712 | 724057785 | ...86958 | ...334323 | ...99966 | ...83984 |
| 490 | ...349832 | ...925333 | ...518757 | ...130191 | ...759726 | ...407457 | 735073466 | ...757852 |
| 500 | ...421167 | ...97026 | ...90809 | ...202604 | ...832502 | ...80597 | ...146973 | ...831728 |



**Figure C.50**

| | 200000 | 200500 | 201000 | 201500 | 202000 | 202500 | 203000 | 203500 |
|---|---|---|---|---|---|---|---|---|
| 0 | 738831728 | 742534952 | 746256737 | 749997477 | 753756305 | 757534395 | 761331362 | 765147360 |
| 10 | ...905611 | ...609205 | ...331363 | 750072177 | ...831741 | ...610149 | ...407495 | ...223875 |
| 20 | ....79502 | ....83466 | ...405996 | ...147184 | ...907124 | ....85910 | ....83636 | ...300397 |
| 30 | 739053400 | ...757735 | ....80637 | ...222199 | ..82515 | ...761678 | ...559784 | ....76927 |
| 40 | ...127305 | ...832010 | ...555285 | ....97221 | 754057913 | ...837455 | ..635940 | ...453465 |
| 50 | ..201218 | ...906294 | ...629940 | ...372251 | ...133319 | ...913238 | ...712104 | ...530010 |
| 60 | ....75138 | ....80584 | ...704603 | ...447288 | ...208732 | ....89030 | ....88275 | ...606563 |
| 70 | ...349065 | 743054582 | ....79275 | ...522332 | ....84153 | 758064828 | ...864454 | ....83124 |
| 80 | ...423000 | ...129188 | ...853952 | ....97385 | ..359581 | ...140635 | ...940640 | ...759692 |
| 90 | ....96943 | ...203501 | ...928637 | ...672444 | ...435017 | ...219449 | 762016834 | ...836268 |
| 100 | ...570892 | ....77821 | 747003330 | ...747512 | ...510461 | ....92271 | ....93036 | ...912852 |
| 110 | ...644849 | ...352149 | ....78030 | ...822587 | ....85912 | ...368100 | ...169245 | ....89443 |
| 120 | ...718814 | ...426484 | ...152738 | ....97669 | ...661371 | ...443937 | ...245462 | 766066042 |
| 130 | ....92786 | ...500827 | ...227453 | ...972759 | ...736837 | ...519781 | ...321687 | ...142649 |
| 140 | ...866765 | ....75177 | ...302176 | 751047856 | ...812310 | ....95633 | ....97919 | ...219263 |
| 150 | ...940752 | ...549534 | ....76906 | ...122961 | ....87792 | ...671493 | ...474159 | ....95885 |
| 160 | 740014746 | ...723899 | ...451644 | ....98073 | ...963280 | ...747360 | ...550406 | ..372514 |
| 170 | ....88747 | ....98272 | ...526389 | ...273193 | 755038777 | ...823235 | ...626661 | ...449152 |
| 180 | ...162756 | ...872651 | ...601142 | ...348320 | ...114281 | ....99117 | ...702924 | ...525797 |
| 190 | ...236773 | ...947039 | ....75902 | ...423455 | ....89792 | ....97500 | ....79194 | ...602449 |
| 200 | ...310796 | 744021433 | ...750669 | ....98597 | ...265311 | 759050904 | ...855472 | ....79109 |
| 210 | ....84827 | ....95835 | ...825445 | ...573747 | ...340837 | ...126809 | ...931758 | ...755777 |
| 220 | ...458866 | ...170245 | ...900227 | ...648905 | ...416372 | ...202722 | 763008051 | ...832453 |
| 230 | ...532912 | ...244662 | ....75017 | ...724070 | ....91913 | ....78642 | ....8435[illegible] | ...909136 |
| 240 | ...606965 | ...319087 | 748049815 | ....99242 | ...567462 | ...354570 | ...160660 | ....85827 |
| 250 | ....81026 | ....93518 | ...124620 | ...874422 | ...643019 | ...430506 | ...236976 | 767062526 |
| 260 | ...755094 | ...467958 | ....99432 | ...949609 | ...718583 | ...506449 | ...313300 | ...139232 |
| 270 | ...829169 | ...542405 | ...274252 | 752024804 | ....94155 | ....82399 | ....89631 | ...215946 |
| 280 | ...903252 | ...616859 | ...349079 | ...100007 | ...869735 | ...658358 | ...465970 | ....92667 |
| 290 | ....77342 | ....91321 | ...423914 | ....75217 | ...945322 | ..734323 | ...542317 | ...369397 |
| 300 | 741051440 | ...765790 | ....98757 | ...250434 | 756020916 | ...810297 | ...618671 | ...446134 |
| 310 | ...125546 | ...840266 | ...573607 | ...325659 | ....96518 | ....86278 | ....95033 | ...522878 |
| 320 | ....99658 | ...914750 | ...648464 | ...400892 | ...172128 | ...962267 | ...771402 | ....99831 |
| 330 | ...273778 | ....89242 | ...723329 | ....76132 | ...247745 | 760038263 | ...847779 | ...676391 |
| 340 | ...347905 | 745063741 | ....98201 | ...551380 | ...323370 | ...114267 | ...924164 | ...753158 |
| 350 | ...422040 | ...138247 | ...873081 | ...626635 | ....99002 | ....90278 | 764000557 | ...829933 |
| 360 | ....96182 | ...212761 | ...947968 | ...701897 | ...474642 | ...266297 | ....76957 | ..906716 |
| 370 | ...570332 | ....87282 | 749022863 | ....77168 | ...550290 | ...342324 | ...153364 | ....83507 |
| 380 | ...644489 | ...361811 | ....97765 | ...852445 | ...625945 | ...418358 | ...229780 | 768060306 |
| 390 | ...718653 | ...436347 | ...172675 | ...927730 | ...701607 | ....94400 | ...306203 | ...137112 |
| 400 | ....92825 | ...510891 | ...247592 | 753003023 | ....77278 | ...570449 | ....82633 | ...213925 |
| 410 | ..867004 | ....85442 | ...822517 | ....78324 | ...852955 | ...646506 | ...459072 | ....90747 |
| 420 | ...941191 | ..660000 | ....97449 | ...153631 | ...928641 | ...722571 | ...535517 | ...367576 |
| 430 | 742015385 | ...734566 | ...472389 | ...228947 | 757004333 | ....98643 | ...611971 | ...444413 |
| 440 | ....89587 | ...809140 | ...547336 | ...304270 | ....80034 | ...874723 | ....88432 | ...521257 |
| 450 | ...163796 | ....83721 | ...622291 | ....79600 | ...155742 | ...950811 | ...764901 | ....98109 |
| 460 | ...238012 | ...958309 | ....97253 | ...454938 | ...231457 | 761026906 | ...841378 | ...674969 |
| 470 | ...312236 | 746032905 | ...772223 | ...530284 | ...307181 | ...103008 | ...917862 | ...751836 |
| 480 | ....86467 | .2.107508 | ...847200 | ...605637 | ....82911 | ....79119 | ....94353 | ...828712 |
| 490 | ...460706 | ....82119 | ...922185 | ....80997 | ...458650 | ...255237 | 765070853 | ...905594 |
| 500 | ...534952 | ...256737 | ....97177 | ...756365 | ...534395 | ...331362 | ...147360 | ....82485 |

**Figure C.51**

| | 204000 | 204500 | 205000 | 205500 | 206000 | 206500 | 207000 | 207500 |
|---|---|---|---|---|---|---|---|---|
| 0 | 768982485 | 772836833 | 776710499 | 780603582 | 784516177 | 788448384 | 792400300 | 796372024 |
| 10 | 769059383 | ···914116 | ···88170 | ···81642 | ···94629 | ···527229 | ···79540 | ···451661 |
| 20 | ···136289 | ···91408 | ···86584 | ···759710 | ···673088 | ···606081 | ···558788 | ···531306 |
| 30 | ···213203 | 773068707 | ···943536 | ···837786 | ···751556 | ···84942 | ···638044 | ···610959 |
| 40 | ···90124 | ···146014 | 777021230 | ···915870 | ···830031 | ···763810 | ···717307 | ···90620 |
| 50 | ···367053 | ···223328 | ···98932 | ···93962 | ···908514 | ···842687 | ···96579 | ···770289 |
| 60 | ···443990 | ···300651 | ···176642 | 781072061 | ···87005 | ···921571 | ···875859 | ···849966 |
| 70 | ···520934 | ···77981 | ···254360 | ···150168 | 785065503 | 789000463 | ···955146 | ···929651 |
| 80 | ···97586 | ···455318 | ···332085 | ···228283 | ···144010 | ···79363 | 793034442 | 797009344 |
| 90 | ···674846 | ···532664 | ···409818 | ···306406 | ···222524 | ···158271 | ···113745 | ···89045 |
| 100 | ···751813 | ···610017 | ···87559 | ···84537 | ···301046 | ···237187 | ···93057 | ···168754 |
| 110 | ···828789 | ···87378 | ···565308 | ···462675 | ···79577 | ···316111 | ···272376 | ···248471 |
| 120 | ···905772 | ···764747 | ···643065 | ···540821 | ···458115 | ···95042 | ···351703 | ···328196 |
| 130 | ···82762 | ···842123 | ···720829 | ···618975 | ···536660 | ···473982 | ···431038 | ···407929 |
| 140 | 8700059760 | ···919508 | ···98601 | ···97137 | ···615214 | ···552929 | ···510381 | ···87669 |
| 150 | ···136766 | ···96900 | ···876381 | ···775307 | ···93776 | ···631884 | ···89732 | ···567418 |
| 160 | ···213780 | 774074299 | ···954168 | ···853484 | ···772345 | ···710848 | ···669091 | ···647175 |
| 170 | ···90802 | ···151707 | 778031964 | ···931670 | ···850922 | ···89819 | ···748458 | ···726940 |
| 180 | ···367831 | ···229122 | ···109767 | 782009863 | ···929507 | ···868798 | ···827833 | ···806712 |
| 190 | ···444867 | ···306545 | ···87578 | ···88064 | 786008100 | ···947785 | ···907216 | ···86493 |
| 200 | ···521912 | ···83976 | ···265397 | ···166273 | ···86701 | 790026779 | ···86607 | ···966282 |
| 210 | ···98964 | ···461414 | ···343223 | ···244489 | ···165310 | ···105782 | 794066005 | 798046078 |
| 220 | ···676024 | ···538860 | ···421058 | ···322714 | ···243926 | ···84793 | ···145412 | ···125883 |
| 230 | ···753092 | ···616314 | ···98900 | ···400946 | ···322551 | ···263811 | ···224826 | ···205695 |
| 240 | ···830167 | ···93776 | ···576750 | ···79186 | ···401183 | ···342838 | ···304249 | ···85516 |
| 250 | ···907750 | ···771245 | ···654607 | ···557434 | ···79823 | ···421872 | ···83679 | ···365345 |
| 260 | ···84341 | ···848722 | ···732473 | ···635690 | ···558471 | ···500914 | ···463118 | ···445181 |
| 270 | 771061439 | ···926207 | ···810346 | ···713953 | ···637127 | ···79964 | ···542564 | ···525026 |
| 280 | ···138545 | 775003699 | ···88227 | ···92225 | ···715790 | ···659022 | ···622018 | ···604878 |
| 290 | ···215659 | ···81200 | ···966116 | ···870504 | ···94462 | ···738088 | ···701481 | ···84739 |
| 300 | ···92781 | ···158708 | 779044012 | ···948791 | ···873141 | ···816162 | ···80951 | ···764607 |
| 310 | ···369910 | ···236224 | ···121917 | 783027086 | ···951829 | ···96244 | ···860429 | ···844484 |
| 320 | ···447047 | ···313747 | ···99829 | ···105389 | 787030524 | ···975333 | ···939915 | ···924368 |
| 330 | ···524192 | ···91279 | ···277749 | ···83699 | ···109227 | 791054431 | 795019409 | 799004260 |
| 340 | ···601344 | ···468818 | ···355677 | ···262018 | ···87938 | ···133536 | ···98911 | ···84161 |
| 350 | ···78504 | ···546365 | ···433612 | ···340344 | ···266657 | ···212650 | ···178421 | ···164069 |
| 360 | ···755672 | ···623919 | ···511556 | ···418678 | ···345383 | ···91771 | ···257938 | ···243986 |
| 370 | ···83284 | ···701482 | ···89507 | ···97020 | ···424118 | ···370900 | ···337464 | ···323910 |
| 380 | ···910031 | ···79052 | ···667466 | ···575369 | ···502860 | ···450037 | ···416998 | ···403842 |
| 390 | ···87222 | ···856630 | ···745433 | ···653727 | ···81011 | ···529182 | ···96540 | ···83783 |
| 400 | 772064421 | ···934216 | ···823407 | ···732092 | ···660369 | ···608335 | ···576089 | ···563731 |
| 410 | ···141627 | 776011809 | ···901389 | ···810465 | ···739135 | ···87496 | ···655647 | ···64[illegible]688 |
| 420 | ···218841 | ···89410 | ···79380 | ···88846 | ···817909 | ···766665 | ···735212 | ···723652 |
| 430 | ···96063 | ···167019 | 780057377 | ···967235 | ···96691 | ···845841 | ···814786 | ···803624 |
| 440 | ···373293 | ···244636 | ···135383 | 784045632 | ···975480 | ···925026 | ···94367 | ···83605 |
| 450 | ···4505[illegible] | ···322260 | ···213397 | ···124037 | 788054278 | 792004212 | ···973957 | ···963593 |
| 460 | ···527775 | ···99893 | ···91418 | ···202449 | ···33083 | ···83419 | 796053554 | 800043590 |
| 470 | ···6050[illegible] | ···4775[illegible] | ···369447 | ···80869 | ···211896 | ···162627 | ···133160 | ···123594 |
| 480 | ···82288 | ···555180 | ···447484 | ···359297 | ···90718 | ···241843 | ···212773 | ···203606 |
| 490 | ···759557 | ···632836 | ···525529 | ···437733 | ···369547 | ···321068 | ···292394 | ···83627 |
| 500 | ···836833 | ···710499 | ···603582 | ···516177 | ···448384 | ···400300 | ···372024 | ···363655 |

Figure C.52

| | 208000 | 208500 | 209000 | 209500 | 210000 | 210500 | 211000 | 211500 |
|---|---|---|---|---|---|---|---|---|
| 0 | 800363655 | 804375293 | 808407039 | 812458993 | 816531257 | 82062393[illegible] | 824737120 | 828870925 |
| 10 | ···443692 | ···455731 | ···87880 | ···540239 | ···612910 | ···705994 | ···819594 | ···953812 |
| 20 | ···523736 | ···536176 | ···568729 | ···621493 | ···94571 | ···88065 | ···902076 | 82903670[illegible] |
| 30 | ···603788 | ···616630 | ···649526 | ···702755 | ···776241 | ···870143 | ···84566 | ···119611 |
| 40 | ···83849 | ···97092 | ···735451 | ···84025 | ···857918 | ···952230 | 825067064 | ···202523 |
| 50 | ···763917 | ···777561 | ···811324 | ···865374 | ···939604 | 821034326 | ···149571 | ···85443 |
| 60 | ···843393 | ···858039 | ···92205 | ···946590 | 817021298 | ···116429 | ···232086 | ···368372 |
| 70 | ···924078 | ···938525 | ···973094 | 813027885 | ···103000 | ···98541 | ···314609 | ···451309 |
| 80 | 801004170 | 805019019 | 809053991 | ···109188 | ···84710 | ···280661 | ···97141 | ···534254 |
| 90 | ···84270 | ···99521 | ···134897 | ···90499 | ···266429 | ···362789 | ···479680 | ···617207 |
| 100 | ···164379 | ···180031 | ···251810 | ···271813 | ···348156 | ···444925 | ···562228 | ···700169 |
| 110 | ···244495 | ···260549 | ···96732 | ···353145 | ···429890 | ···527069 | ···644785 | ···83139 |
| 120 | ···324620 | ···341075 | ···377661 | ···434480 | ···511633 | ···609222 | ···727349 | ···866117 |
| 130 | ···404752 | ···421609 | ···458599 | ···515824 | ···93385 | ···91383 | ···809922 | ···949104 |
| 140 | ···84893 | ···502151 | ···539545 | ···97175 | ···675144 | ···773552 | ···92503 | 830032099 |
| 150 | ···565041 | ···82701 | ···620499 | ···678535 | ···756911 | ···855729 | ···975092 | ···115102 |
| 160 | ···645198 | ···663260 | ···701461 | ···759903 | ···838687 | ···937915 | 826057690 | ···98113 |
| 170 | ···725362 | ···743826 | ···82431 | ···841279 | ···920471 | 822020109 | ···140295 | ···281133 |
| 180 | ···805535 | ···824400 | ···863409 | ···922663 | 818002263 | ···102311 | ···222909 | ···364161 |
| 190 | ···85715 | ···904983 | ···944396 | 814004055 | ···84063 | ···84521 | ···305532 | ···447198 |
| 200 | ···965904 | ···85573 | 810025390 | ···85456 | ···165872 | ···266740 | ···88162 | ···530242 |
| 210 | 802045100 | 806066172 | ···106393 | ···166864 | ···247688 | ···348966 | ···470801 | ···613295 |
| 220 | ···126305 | ···146778 | ···87403 | ···248281 | ···329513 | ···431201 | ···553448 | ···96357 |
| 230 | ···206518 | ···227393 | ···268422 | ···329706 | ···411346 | ···513444 | ···636103 | ···779426 |
| 240 | ···86738 | ···308016 | ···349449 | ···411139 | ···93187 | ···95696 | ···718767 | ···862504 |
| 250 | ···366967 | ···88647 | ···430484 | ···92580 | ···575036 | ···677955 | ···801439 | ···945590 |
| 260 | ···447204 | ···469285 | ···511527 | ···574029 | ···656894 | ···760223 | ···84119 | 83102868[illegible] |
| 270 | ···527448 | ···549932 | ···92578 | ···655487 | ···738760 | ···842499 | ···966807 | ···11178[illegible] |
| 280 | ···607701 | ···630587 | ···673637 | ···736952 | ···820633 | ···924783 | 827049504 | ···94899 |
| 290 | ···87962 | ···711250 | ···754705 | ···818426 | ···902516 | 823007076 | ···132209 | ···278019 |
| 300 | ···768231 | ···91921 | ···835780 | ···99908 | ···84406 | ···89376 | ···214922 | ···361147 |
| 310 | ···848507 | ···872601 | ···916364 | ···981398 | 819066304 | ···171685 | ···97644 | ···444283 |
| 320 | ···928792 | ···953288 | ···97956 | 815062896 | ···148201 | ···254002 | ···380374 | ···527427 |
| 330 | 803009085 | 807033983 | 811079055 | ···144402 | ···230126 | ···336328 | ···463112 | ···610580 |
| 340 | ···89386 | ···114687 | ···160163 | ···225916 | ···312049 | ···418661 | ···545858 | ···93741 |
| 350 | ···169695 | ···95398 | ···241279 | ···307439 | ···93980 | ···501003 | ···628612 | ···776910 |
| 360 | ···250012 | ···276118 | ···322403 | ···88970 | ···475919 | ···83353 | ···711375 | ···860088 |
| 370 | ···330337 | ···356845 | ···403536 | ···470509 | ···557867 | ···665712 | ···94146 | ···943274 |
| 380 | ···410670 | ···437581 | ···84676 | ···552056 | ···639823 | ···748078 | ···876926 | 83202646[illegible] |
| 390 | ···91011 | ···518325 | ···565824 | ···633611 | ···721787 | ···830453 | ···959714 | ···109671 |
| 400 | ···571360 | ···99077 | ···646981 | ···715174 | ···803759 | ···912836 | 828042510 | ···92882 |
| 410 | ···651717 | ···679836 | ···728146 | ···96746 | ···85739 | ···95228 | ···125314 | ···276101 |
| 420 | ···732083 | ···760604 | ···809318 | ···878326 | ···967728 | 824077637 | ···208126 | ···359329 |
| 430 | ···812456 | ···841381 | ···90499 | ···959913 | 820049725 | ···160035 | ···90947 | ···442565 |
| 440 | ···92837 | ···922165 | ···971688 | 816041509 | ···131730 | ···242451 | ···373776 | ···525809 |
| 450 | ···973226 | 808002957 | 812052885 | ···123114 | ···213743 | ···324875 | ···456614 | ···609062 |
| 460 | 804053624 | ···83757 | ···134091 | ···204726 | ···95764 | ···407308 | ···539459 | ···92322 |
| 470 | ···134029 | ···164566 | ···215304 | ···86341 | ···377794 | ···89748 | ···622313 | ···7[illegible]592 |
| 480 | ···214442 | ···245382 | ···96526 | ···367975 | ···459831 | ···572197 | ···705175 | ···858869 |
| 490 | ···94864 | ···326207 | ···377755 | ···449612 | ···541877 | ···654654 | ···88046 | ···942155 |
| 500 | ···375293 | ···407039 | ···458993 | ···531257 | ···623932 | ···737120 | ···870925 | 833025449 |

**Figure C.53**

| | 212000 | 212500 | 213000 | 213500 | 214000 | 214500 | 215000 | 215500 |
|---|---|---|---|---|---|---|---|---|
| 0 | 833025449 | 837200797 | 841397074 | 845614383 | 849852830 | 854112521 | 858393564 | 862696064 |
| 10 | ...108752 | ...84518 | ...481213 | ...98944 | ...937815 | ...97933 | ...479403 | ...782333 |
| 20 | ...92063 | ...368246 | ...565361 | ...783514 | 850022809 | ...283353 | ...565251 | ...868611 |
| 30 | ...275382 | ...451983 | ...649518 | ...868092 | ...107811 | ...368781 | ...651107 | ...954898 |
| 40 | ...358709 | ...535728 | ...733683 | ...952679 | ...92822 | ...454218 | ...736972 | 863041194 |
| 50 | ...442045 | ...619450 | ...817856 | 846037274 | ...277841 | ...539663 | ...822846 | ...127498 |
| 60 | ...525390 | ...703244 | ...902038 | ...121878 | ...362869 | ...625117 | ...908728 | ...213811 |
| 70 | ...608742 | ...87014 | ...86228 | ...206490 | ...447905 | ...710580 | ...94619 | ...300132 |
| 80 | ...92103 | ...870793 | 842070427 | ...91111 | ...532950 | ...96051 | 859080519 | ...86462 |
| 90 | ...775472 | ...954580 | ...154634 | ...375740 | ...618003 | ...881530 | ...166427 | ...472801 |
| 100 | ...858850 | 838038375 | ...238849 | ...460378 | ...703065 | ...967019 | ...252344 | ...559148 |
| 110 | ...942236 | ...122179 | ...323073 | ...545024 | ...88136 | 855052515 | ...338269 | ...645504 |
| 120 | 834025630 | ...205991 | ...407306 | ...629678 | ...873214 | ...138021 | ...424203 | ...731868 |
| 130 | ...109032 | ...89812 | ...91546 | ...714341 | ...958302 | ...223534 | ...510145 | ...818242 |
| 140 | ...92443 | ...373641 | ...575796 | ...99013 | 851043397 | ...309057 | ...96096 | ...904623 |
| 150 | ...275863 | ...457478 | ...660053 | ...883692 | ...128502 | ...94587 | ...682056 | ...91014 |
| 160 | ...359290 | ...541324 | ...744319 | ...968381 | ...213615 | ...480127 | ...768024 | 864077413 |
| 170 | ...442726 | ...625178 | ...828593 | 847053078 | ...98736 | ...565675 | ...854001 | ...163821 |
| 180 | ...526170 | ...709040 | ...912876 | ...137783 | ...383866 | ...651231 | ...939986 | ...250237 |
| 190 | ...609623 | ...92911 | ...97168 | ...222497 | ...469004 | ...736797 | 860025980 | ...336662 |
| 200 | ...93084 | ...876791 | 843081467 | ...307219 | ...554151 | ...822370 | ...111983 | ...423096 |
| 210 | ...776553 | ...960678 | ...165775 | ...91950 | ...639307 | ...907952 | ...97994 | ...509538 |
| 220 | ...860031 | 839044575 | ...250092 | ...476689 | ...724471 | ...93543 | ...284014 | ...95989 |
| 230 | ...943517 | ...128479 | ...334417 | ...561437 | ...809643 | 856079143 | ...370042 | ...682449 |
| 240 | 835027011 | ...212392 | ...418750 | ...646193 | ...94824 | ...164751 | ...456079 | ...768917 |
| 250 | ...110514 | ...96313 | ...503092 | ...730957 | ...980013 | ...250367 | ...542125 | ...855394 |
| 260 | ...94025 | ...380243 | ...87443 | ...815730 | 852065211 | ...335992 | ...628179 | ...941879 |
| 270 | ...277544 | ...464181 | ...671801 | ...900512 | ...150418 | ...421626 | ...714242 | 865028374 |
| 280 | ...361072 | ...548127 | ...756169 | ...85302 | ...235633 | ...507268 | ...800313 | ...114876 |
| 290 | ...444608 | ...632082 | ...840544 | 848070101 | ...320857 | ...92919 | ...86393 | ...201388 |
| 300 | ...528153 | ...716045 | ...924928 | ...154908 | ...406089 | ...678578 | ...972482 | ...87908 |
| 310 | ...611706 | ...800017 | 844009321 | ...239723 | ...91329 | ...764246 | 861058579 | ...374437 |
| 320 | ...95267 | ...83997 | ...93722 | ...324547 | ...576578 | ...849922 | ...144685 | ...460974 |
| 330 | ...778836 | ...967985 | ...178131 | ...409380 | ...661836 | ...935607 | ...230799 | ...547520 |
| 340 | ...862414 | 840051982 | ...262549 | ...94221 | ...747102 | 857021301 | ...316922 | ...634075 |
| 350 | ...946000 | 1...135987 | ...346975 | ...579071 | ...832377 | ...107003 | ...403054 | ...720639 |
| 360 | 836029595 | ...220001 | ...431410 | ...663929 | ...917660 | ...92714 | ...89194 | ...807211 |
| 370 | ...113198 | ...304023 | ...515853 | ...748795 | 853002952 | ...278433 | ...575343 | ...93791 |
| 380 | ...96809 | ...88053 | ...600305 | ...833670 | ...88252 | ...364161 | ...661501 | ...980381 |
| 390 | ...280429 | ...472092 | ...84765 | ...918554 | ...173561 | ...449897 | ...747667 | 866066979 |
| 400 | ...364057 | 0...556139 | ...769233 | 849003444 | ...258878 | ...535642 | ...833842 | ...153585 |
| 410 | ...447693 | ...640195 | ...853710 | ...88345 | ...344204 | ...621396 | ...920025 | ...240201 |
| 420 | ...531338 | ...724259 | ...938195 | ...173253 | ...429539 | ...707158 | 862006217 | ...326825 |
| 430 | ...614991 | ...808331 | 845022689 | ...258171 | ...514882 | ...92928 | ...92418 | ...413458 |
| 440 | ...98653 | ...92412 | ...107192 | ...343096 | ...600233 | ...878708 | ...178627 | ...500099 |
| 450 | ...782323 | ...976501 | ...91702 | ...428034 | ...85593 | ...964496 | ...264345 | ...86749 |
| 460 | ...866001 | 841060599 | ...276221 | ...512974 | ...770962 | 858050292 | ...351071 | ...673408 |
| 470 | ...949687 | ...144705 | ...360749 | ...97925 | ...856339 | ...136097 | ...437307 | ...760075 |
| 480 | 837033387 | ...228819 | ...445285 | ...682885 | ...941725 | ...221911 | ...523550 | ...846751 |
| 490 | ...117086 | ...312942 | ...529830 | ...767853 | 854027119 | ...307733 | ...609803 | ...933436 |
| 500 | ...200797 | ...97074 | ...614383 | ...852830 | ...112521 | ...93564 | ...96064 | 867020129 |

**Figure C.54**

| | 216000 | 216500 | 217000 | 217500 | 218000 | 218500 | 219000 | 219500 |
|---|---|---|---|---|---|---|---|---|
| 0 | 867020129 | 871365868 | 875833388 | 880122800 | 884534213 | 888967737 | 893423483 | 897901562 |
| 10 | ···106831 | ···453004 | ···820962 | ···210812 | ···622666 | 889056634 | ···512825 | ···91352 |
| 20 | ···93542 | ···540150 | ···908544 | ···98833 | ···711129 | ···145539 | ···602176 | 898081151 |
| 30 | ··280261 | ···627304 | ···96135 | ···386863 | ···99600 | ···234454 | ···91537 | ···170960 |
| 40 | ···366989 | ···714466 | 876083734 | ···474902 | ···888080 | ···323377 | ···780906 | ···260777 |
| 50 | ···453726 | ···801638 | ···171343 | ···562950 | ···976568 | ···412310 | ···870284 | ···350603 |
| 60 | ···540471 | ···88818 | ···258960 | ···651006 | 885065066 | ···501251 | ···959671 | ···440438 |
| 70 | ···627225 | ···976007 | ···346586 | ···737071 | ···153573 | ···90201 | 894049067 | ···530282 |
| 80 | ···713988 | 872063204 | ···434220 | ···827145 | ···242088 | ···679160 | ···138472 | ···620135 |
| 90 | ···800759 | ···150411 | ···521864 | ···915228 | ···330612 | ··768128 | ···227886 | ···709997 |
| 100 | ···87539 | ···237626 | ···609516 | 881003319 | ···419145 | ···857105 | ···317308 | ···99868 |
| 110 | ···974328 | ···324849 | ···97177 | ···91419 | ···507687 | ···946090 | ···406740 | ···889748 |
| 120 | 868061126 | ···412082 | ···784847 | ···179528 | ···96238 | 890035085 | ···96181 | ···979637 |
| 130 | 5··147932 | ···99323 | ···872525 | ···267646 | ···684798 | ···124088 | ···585630 | 899069535 |
| 140 | ···234746 | ···586573 | ···960212 | ···355773 | ···753366 | ···213101 | ···675089 | ···159442 |
| 150 | ···321570 | ···673832 | 877047908 | ···443909 | ···861943 | ···302122 | ···764557 | ···249358 |
| 160 | ···408402 | ···761099 | ···135613 | ···532053 | ···950530 | ···91152 | ···854033 | ···339283 |
| 170 | ···95243 | ···848375 | ···223327 | ···620206 | 886039125 | ···480191 | ···943518 | ···429217 |
| 180 | ···582092 | ···935660 | ···311049 | ···708368 | ···127729 | ···569239 | 895033013 | ···519160 |
| 190 | ···668951 | 873022954 | ···98780 | ···896539 | ···216341 | ···658296 | ···122516 | ···609111 |
| 200 | ···755817 | ···110256 | ···486520 | ···884719 | ···304963 | ···747362 | ···212028 | ···99072 |
| 210 | ···842693 | ···97567 | ···574269 | ···972907 | ···93593 | ···836437 | ···301550 | ···789042 |
| 220 | ···929577 | ···284887 | ···662026 | 882061105 | ···482233 | ···925521 | ···91080 | ···879021 |
| 230 | 869016470 | ···372215 | ···749792 | ···149311 | ···570881 | 891014613 | ···480619 | ···969009 |
| 240 | ···103372 | ···459552 | ···837567 | ···237526 | ···659538 | ···103715 | ···570167 | 900059006 |
| 250 | ···90282 | ···546898 | ···925351 | ···325750 | ···748204 | ···92825 | ···659724 | ···149012 |
| 260 | ···277201 | ···634253 | 878013143 | ···413982 | ···836879 | ···281y44 | ···749290 | ···239027 |
| 270 | ···364129 | ···721616 | ···100945 | ···502224 | ···925563 | ···371073 | ···838865 | ···329051 |
| 280 | ···451065 | ···808989 | ···88755 | ···90474 | 887014255 | ···460210 | ···928449 | ···419084 |
| 290 | ···538010 | ···96370 | ···276574 | ···678733 | ···102957 | ···549356 | 896018042 | ···509125 |
| 300 | ···624964 | ···983759 | ···364401 | ···767001 | ···91667 | ··638511 | ···107643 | ···99176 |
| 310 | ···711927 | 874071158 | ···452238 | ···855277 | ···280386 | ···727675 | ···97259 | ···689236 |
| 320 | ···98898 | ···158565 | ···540083 | ···943563 | ···369114 | ···816847 | ···286874 | ···779305 |
| 330 | ···885878 | ···245981 | ···627937 | 883031857 | ···457851 | ···906029 | ···376502 | ···869383 |
| 340 | ···972867 | ···333405 | ···715890 | ···120160 | ···546597 | ···95220 | ···466140 | ···959470 |
| 350 | 870059864 | ···420838 | ···803671 | ···208472 | ···635351 | 892084419 | ···555787 | 901049566 |
| 360 | ···146870 | ···508281 | ···91552 | ···96793 | ···724115 | ···173627 | ···64544 | ···139671 |
| 370 | ···233885 | ···95731 | ···979441 | ···385123 | ···812887 | ···262845 | ···735107 | ···229785 |
| 380 | ···320908 | ···683191 | 879067339 | ···473461 | ···901669 | ···352071 | ···824780 | ···319908 |
| 390 | ···407940 | ···770659 | ···155246 | ···561809 | ···90459 | ···441306 | ···914463 | ···410040 |
| 400 | ···94981 | ···858136 | ···243161 | ···650165 | 888079258 | ···530550 | 897004154 | ···500181 |
| 410 | ···582030 | ···945622 | ···331085 | ···738530 | ···168066 | ···619804 | ···93855 | ···90331 |
| 420 | ···669089 | 875033117 | ···419019 | ···826904 | ···256883 | ···709066 | ···183564 | ···680490 |
| 430 | ···756155 | ···120620 | ···506960 | ···915287 | ···345708 | ···98336 | ···273282 | ···770658 |
| 440 | ···843231 | ···208132 | ···94911 | 884003678 | ···434543 | ···887616 | ···363010 | ···860835 |
| 450 | ···930315 | ···95653 | ···682871 | ···92078 | ···523386 | ···976905 | ···452746 | ···951021 |
| 460 | 871017408 | ···383183 | ···770839 | ···180488 | ···612239 | 893066103 | ···542491 | 902041216 |
| 470 | ···104510 | ···470721 | ···858816 | ···268906 | ···701100 | ···155509 | ···632246 | ···131420 |
| 480 | ···91621 | ···558268 | ···946802 | ···357333 | ···89970 | ···244825 | ···722009 | ···221633 |
| 490 | ···278740 | ···645824 | 880034797 | ···445768 | ···878849 | ···334149 | ···811781 | ···311856 |
| 500 | ···365868 | ···733388 | ···122800 | ···534213 | ···967737 | ···423483 | ···901562 | ···42087 |

**Figure C.55**

| | 220000 | 220500 | 221000 | 221500 | 222000 | 222500 | 223000 | 223500 |
|---|---|---|---|---|---|---|---|---|
| 0 | 902402087 | 906925169 | 911470923 | 916039461 | 920630898 | 925245348 | 929882927 | 934543751 |
| 10 | ···92327 | 907015862 | ···562070 | ···131065 | ···722961 | ···337873 | ···975915 | ···637206 |
| 20 | ···582576 | ···106564 | ···653226 | ···222678 | ···815033 | ···430406 | 930068913 | ···730669 |
| 30 | ···672835 | ···97274 | ···744392 | ···314300 | ···907115 | ···522949 | ···161920 | ···824142 |
| 40 | ···763102 | ···287994 | ···835566 | ···405932 | ···99205 | ···615502 | ···254936 | ···917625 |
| 50 | ···853378 | ···378723 | ···926750 | ···97572 | 921091305 | ···708063 | ···347962 | 935011117 |
| 60 | ···943664 | ···469461 | 912017942 | ···589222 | ···183414 | ···800634 | ···440996 | ···104618 |
| 70 | 903033958 | ···560208 | ···109144 | ···680881 | ···275533 | ···93214 | ···534041 | ···98128 |
| 80 | ···124261 | ···650964 | ···200355 | ···772549 | ···367660 | ···985803 | ···627094 | ···291648 |
| 90 | ···214574 | ···741729 | ···91575 | ···864226 | ···459797 | 926078402 | ···720157 | ···385177 |
| 100 | ···304895 | ···832503 | ···382804 | ···955913 | ···551943 | ···171010 | ···813229 | ···478716 |
| 110 | ···95226 | ···923286 | ···474042 | 917047608 | ···644098 | ···263627 | ···906310 | ···572264 |
| 120 | ···485565 | 908014078 | ···565290 | ···139313 | ···736263 | ···356253 | ···99401 | ···665821 |
| 130 | ···575914 | ···104880 | ···656546 | ···231027 | ···828436 | ···448859 | 931092500 | ···759387 |
| 140 | ···666271 | ···95690 | ···747812 | ···322750 | ···920619 | ···541534 | ···185610 | ···852963 |
| 150 | ···756638 | ···286510 | ···839087 | ···414483 | 922012811 | ···634188 | ···278728 | ···946548 |
| 160 | ···847014 | ···377339 | ···930371 | ···506224 | ···105012 | ···726851 | ···371856 | 936040143 |
| 170 | ···937398 | ···458176 | 913021664 | ···97975 | ···97223 | ···819524 | ···464993 | ···133747 |
| 180 | 904027792 | ···559023 | ···112966 | ···689734 | ···289443 | ···912206 | ···558140 | ···227361 |
| 190 | ···118195 | ···649879 | ···204277 | ···781503 | ···381672 | 927004897 | ···651296 | ···320983 |
| 200 | ···208607 | ···740744 | ···95598 | ···873281 | ···473910 | ···97598 | ···744461 | ···414615 |
| 210 | ···99028 | ···831618 | ···386927 | ···965069 | ···566157 | ···190308 | ···837635 | ···508257 |
| 220 | ···389457 | ···922501 | ···478266 | 918056865 | ···658414 | ···283027 | ···930819 | ···601908 |
| 230 | ···479896 | 909013394 | ···569614 | ···148671 | ···750680 | ···375755 | 932024012 | ···95568 |
| 240 | ···570344 | ···104295 | ···660971 | ···240486 | ···842955 | ···468493 | ···117215 | ···789237 |
| 250 | ···660801 | ···95205 | ···752337 | ···332310 | ···935239 | ···561239 | ···210426 | ···882916 |
| 260 | ···751267 | ···286125 | ···843712 | ···424143 | 923027533 | ···653995 | ···303647 | ···976605 |
| 270 | ···841743 | ···377053 | ···935096 | ···515985 | ···119835 | ···746761 | ···96878 | 937070302 |
| 280 | ···932227 | ···467991 | 914026490 | ···607837 | ···212147 | ···839535 | ···490117 | ···164009 |
| 290 | 905022720 | ···558938 | ···117892 | ···99698 | ···304468 | ···932319 | ···583366 | ···257726 |
| 300 | ···113222 | ···649894 | ···209304 | ···791568 | ···96799 | 928025113 | ···676625 | ···351451 |
| 310 | ···203734 | ···740859 | ···300725 | ···883447 | ···489139 | ···117915 | ···769892 | ···445187 |
| 320 | ···94254 | ···831833 | ···92155 | ···975335 | ···581488 | ···210727 | ···863169 | ···538931 |
| 330 | ···384783 | ···922816 | ···483594 | 919067233 | ···673846 | ···303548 | ···956456 | ···632685 |
| 340 | ···475322 | 910013808 | ···575043 | ···159140 | ···766213 | ···96378 | 933049751 | ···726448 |
| 350 | ···565869 | ···104810 | ···666500 | ···251056 | ···858590 | ···489218 | ···143056 | ···820221 |
| 360 | ···656426 | ···95820 | ···757967 | ···342981 | ···950976 | ···582067 | ···236371 | ···914003 |
| 370 | ···746992 | ···286840 | ···849443 | ···434915 | 924043371 | ···674925 | ···329694 | 938007794 |
| 380 | ···837566 | ···377869 | ···940928 | ···526858 | ···135775 | ···767793 | ···423027 | ···101595 |
| 390 | ···928150 | ···468906 | 915032422 | ···618811 | ···228189 | ···860669 | ···516370 | ···95405 |
| 400 | 906018743 | ···559953 | ···123925 | ···710773 | ···320611 | ···953555 | ···609721 | ···289225 |
| 410 | ···109345 | ···651009 | ···215438 | ···802744 | ···413043 | 929046451 | ···703082 | ···383054 |
| 420 | ···99956 | ···742074 | ···306959 | ···94724 | ···505485 | ···139355 | ···96452 | ···476892 |
| 430 | ···290576 | ···833149 | ···98490 | ···986714 | ···97935 | ···232?69 | ···889832 | ···570740 |
| 440 | ···381205 | ···924232 | ···490030 | 920078713 | ···690395 | ···325193 | ···983221 | ···664597 |
| 450 | ···471843 | 911015324 | ···581579 | ···170721 | ···782864 | ···418125 | 934076619 | ···758463 |
| 460 | ···562490 | ···106426 | ···673137 | ···262738 | ···875342 | ···511067 | ···170027 | ···852339 |
| 470 | ···653146 | ···97536 | ···764704 | ···354764 | ···967830 | ···604018 | ···263444 | ···946224 |
| 480 | ···743812 | ···288656 | ···856281 | ···446799 | 625060327 | ···96978 | ···356870 | 939040119 |
| 490 | ···834486 | ···379785 | ···947866 | ···538844 | ···152833 | ···78994 | ···450306 | ···134023 |
| 500 | ···925169 | ···470923 | 916039461 | ···630898 | ···245348 | ···882927 | ···543751 | ···227936 |

H

Figure C.56

| | 224000 | 224500 | 225000 | 225500 | 226000 | 226500 | 227000 | 227500 |
|---|---|---|---|---|---|---|---|---|
| 0 | 939227936 | 943935600 | 948666800 | 953421834 | 958200641 | 963003401 | 967830234 | 972681260 |
| 10 | ···321859 | 944029993 | ···761726 | ···517176 | ···96461 | ···99701 | ···927017 | ···778528 |
| 20 | ···415791 | ···124396 | ···856603 | ···612528 | ···392291 | ···196011 | 968023809 | ···875806 |
| 30 | ···509733 | ···218809 | ···951488 | ···707889 | ···488130 | ···292331 | ···120612 | ···973093 |
| 40 | ···603684 | ···313231 | 949046383 | ···803260 | ···583979 | ···388660 | ···217424 | 973070391 |
| 50 | ···97644 | ···407662 | ···141288 | ···98640 | ···679837 | ···484999 | ···314246 | ···167698 |
| 60 | ···791614 | ···502103 | ···236202 | ···994030 | ···775705 | ···581348 | ···411077 | ···265015 |
| 70 | ···885593 | ···96553 | ···331126 | 954089430 | ···871583 | ···677706 | ···507918 | ···362341 |
| 80 | ···979582 | ···691013 | ···426059 | ···184838 | ···967470 | ···774074 | ···604769 | ···459677 |
| 90 | 940073580 | ···785482 | ···521001 | ···280257 | 959063367 | ···870451 | ···701629 | ···557023 |
| 100 | ···167587 | ···879961 | ···615954 | ···375685 | ···159273 | ···966838 | ···98500 | ···654379 |
| 110 | ···261604 | ···974449 | ···710915 | ···471122 | ···255189 | 964063235 | ···895379 | ···751744 |
| 120 | ···355630 | 945068946 | ···805886 | ···566570 | ···351114 | ···159641 | ···992269 | ···849120 |
| 130 | ···449666 | ···163453 | ···900867 | ···662026 | ···447050 | ···256057 | 969089168 | ···946505 |
| 140 | ···543711 | ···257969 | ···95857 | ···757492 | ···542994 | ···352483 | ···186077 | 974043899 |
| 150 | ···637765 | ···352495 | 950090857 | ···852968 | ···638949 | ···448918 | ···282995 | ···141304 |
| 160 | ···731829 | ···446930 | ···185866 | ···948453 | ···734913 | ···545363 | ···379924 | ···238718 |
| 170 | ···825902 | ···541575 | ···280884 | 955043948 | ···830886 | ···641817 | ···476862 | ···336142 |
| 180 | ···919985 | ···636029 | ···375912 | ···139453 | ···926869 | ···738281 | ···573810 | ···433575 |
| 190 | 941014077 | ···730693 | ···470950 | ···234967 | 960022862 | ···834755 | ···670767 | ···531019 |
| 200 | ···108178 | ···825266 | ···565997 | ···330490 | ···118864 | ···931239 | ···767734 | ···628472 |
| 210 | ···202289 | ···919848 | ···661054 | ···426023 | ···214876 | 965027732 | ···864711 | ···725934 |
| 220 | ···96409 | 946014440 | ···756120 | ···521566 | ···310898 | ···124235 | ···961697 | ···823407 |
| 230 | ···390539 | ···109042 | ···851195 | ···617118 | ···406929 | ···220747 | 970058694 | ···920889 |
| 240 | ···484678 | ···203653 | ···946280 | ···712680 | ···502969 | ···317269 | ···155699 | 975018381 |
| 250 | ···578826 | ···98273 | 951041375 | ···808251 | ···99020 | ···413801 | ···252715 | ···115883 |
| 260 | ···672984 | ···392903 | ···136479 | ···903832 | ···695080 | ···510342 | ···349740 | ···213395 |
| 270 | ···767151 | ···487542 | ···231593 | ···99422 | ···791149 | ···606893 | ···446775 | ···310916 |
| 280 | ···861328 | ···582191 | ···326716 | 956095022 | ···887228 | ···703454 | ···543820 | ···408447 |
| 290 | ···955514 | ···676849 | ···421849 | ···190632 | ···983317 | ···800024 | ···640875 | ···505988 |
| 300 | 942049710 | ···771517 | ···516991 | ···286251 | 961079415 | ···96604 | ···737939 | ···603539 |
| 310 | ···143915 | ···866194 | ···612143 | ···381879 | ···175523 | ···993194 | ···835012 | ···701099 |
| 320 | ···238129 | ···960881 | ···707304 | ···477517 | ···271641 | 966089793 | ···932096 | ···98669 |
| 330 | ···332353 | 947055577 | ···802475 | ···573165 | ···367768 | ···186402 | 971029189 | ···896249 |
| 340 | ···426586 | ···150282 | ···97655 | ···668823 | ···463905 | ···283021 | ···126292 | ···993839 |
| 350 | ···520829 | ···244997 | ···992835 | ···764489 | ···560051 | ···379649 | ···223405 | 976091438 |
| 360 | ···615081 | ···339722 | 952088034 | ···860166 | ···656207 | ···476287 | ···320527 | ···189047 |
| 370 | ···709342 | ···434456 | ···183243 | ···955852 | ···752373 | ···572935 | ···417559 | ···286666 |
| 380 | ···803613 | ···529199 | ···278461 | 957051547 | ···848548 | ···669592 | ···514801 | ···384295 |
| 390 | ···97894 | ···623952 | ···373689 | ···147253 | ···944733 | ···766259 | ···611952 | ···481933 |
| 400 | ···992183 | ···718714 | ···468926 | ···242967 | 962040927 | ···862936 | ···709113 | ···579581 |
| 410 | 943086482 | ···813486 | ···564173 | ···338692 | ···137131 | ···959622 | ···806284 | ···677239 |
| 420 | ···180791 | ···908268 | ···659429 | ···434426 | ···233345 | 967056318 | ···903465 | ···774907 |
| 430 | ···275109 | 948003058 | ···754695 | ···530169 | ···329568 | ···153024 | 972000655 | ···872685 |
| 440 | ···369437 | ···97859 | ···849976 | ···625922 | ···425801 | ···249739 | ···97855 | ···970272 |
| 450 | ···463774 | ···192669 | ···945266 | ···721685 | ···522044 | ···346464 | ···195065 | 977067969 |
| 460 | ···558120 | ···287488 | 953040560 | ···817457 | ···618296 | ···443198 | ···292285 | ···165676 |
| 470 | ···652476 | ···382317 | ···135865 | ···913238 | ···714558 | ···539943 | ···389514 | ···263392 |
| 480 | ···746841 | ···477155 | ···231178 | 958009030 | ···810829 | ···636697 | ···486753 | ···361119 |
| 490 | ···841216 | ···572002 | ···326501 | ···104831 | ···907110 | ···733460 | ···584001 | ···458855 |
| 500 | ···935600 | ···666860 | ···421834 | ···200641 | 963003401 | ···830234 | ···681260 | ···556601 |

**Figure C.57**

|  | 228000 | 228500 | 229000 | 229500 | 230000 | 230000 |
|---|---|---|---|---|---|---|
| 0 | 977556601 | 982456378 | 987380714 | 992329732 | ···0000 | 997303557 |
| 10 | ···654356 | ···554623 | ···479452 | ···428965 | ···10000 | ···403287 |
| 20 | ···752122 | ···652879 | ···578200 | ···528208 | ···20000 | ···503027 |
| 30 | ···849897 | ···751144 | ···676958 | ···627461 | ···30000 | ···602778 |
| 40 | ···947682 | ···849419 | ···775726 | ···726724 | ···40000 | ···702538 |
| 50 | 978045477 | ···947704 | ···874503 | ···825996 | ···50000 | ···802308 |
| 60 | ···143281 | 983045999 | ···973291 | ···925279 | ···60000 | ···902088 |
| 70 | ···241096 | ···144304 | 988072088 | 993024572 | ···70000 | 998001879 |
| 80 | ···338920 | ···242618 | ···170895 | ···123874 | ···80000 | ···101679 |
| 90 | ···436754 | ···340942 | ···269712 | ···223187 | ···90000 | ···201489 |
| 100 | ···534597 | ···439276 | ···368539 | ···322509 | 230100000 | ···301309 |
| 110 | ···632451 | ···537620 | ···467376 | ···421841 | ···10000 | ···401139 |
| 120 | ···730314 | ···635974 | ···566223 | ···521183 | ···20000 | ···500979 |
| 130 | ···828187 | ···734338 | ···665080 | ···620535 | ···30000 | ···600829 |
| 140 | ···926070 | ···832711 | ···763946 | ···719898 | ···40000 | ···700690 |
| 150 | 979023962 | ···931094 | ···862822 | ···819269 | ···50000 | ···800560 |
| 160 | ···121865 | 984029488 | ···961709 | ···918651 | ···60000 | ···900440 |
| 170 | ···219777 | ···127890 | 989060605 | 994018043 | ···70000 | 999000330 |
| 180 | ···317699 | ···226303 | ···159511 | ···117445 | ···80000 | ···100230 |
| 190 | ···415631 | ···324726 | ···258427 | ···216857 | ···90000 | ···200140 |
| 200 | ···513572 | ···423158 | ···357353 | ···316278 | 230200000 | ···300060 |
| 210 | ···611524 | ···521601 | ···456288 | ···415710 | ···10000 | ···399990 |
| 220 | ···709485 | ···620053 | ···555234 | ···515152 | ···20000 | ···499930 |
| 230 | ···807456 | ···718515 | ···654190 | ···614603 | ···30000 | ···599880 |
| 240 | ···905437 | ···816987 | ···753155 | ···714065 | ···40000 | ···699840 |
| 250 | 980003427 | ···915468 | ···852130 | ···813536 | ···50000 | ···799810 |
| 260 | ···101427 | 985013960 | ···951115 | ···913017 | ···60000 | ···899790 |
| 270 | ···99438 | ···112461 | 990050111 | 995012509 | 230270000 | 999999779 |
| 280 | ···297457 | ···210973 | ···149116 | ···112010 | ······10 | ······879 |
| 290 | ···395487 | ···309494 | ···248130 | ···211521 | ······20 | ······979 |
| 300 | ···493527 | ···408025 | ···347155 | ···311042 | ·······1 | ·······89 |
| 310 | ···591576 | ···506565 | ···446190 | ···410573 | 230270022 | 999999999 |
| 320 | ···689635 | ···605116 | ···545235 | ···510115 |  |  |
| 330 | ···787704 | ···703677 | ···64428y | ···609666 |  |  |
| 340 | ···885783 | ···802247 | ···743353 | ···709227 |  |  |
| 350 | ···983872 | ···900827 | ···842428 | ···808797 |  |  |
| 360 | 981081970 | ···99417 | ···941512 | ···908378 |  |  |
| 370 | ···180078 | 986098017 | 991040606 | 996007969 |  |  |
| 380 | ···278196 | ···196627 | ···139711 | ···107570 |  |  |
| 390 | ···376324 | ···295247 | ···238825 | ···207181 |  |  |
| 400 | ···474462 | ···393876 | ···337948 | ···306801 |  |  |
| 410 | ···572609 | ···492516 | ···437082 | ···406432 |  |  |
| 420 | ···670766 | ···591165 | ···536226 | ···506073 |  |  |
| 430 | ···768934 | ···689824 | ···635380 | ···605723 |  |  |
| 440 | ···867110 | ···788493 | ···734543 | ···705384 |  |  |
| 450 | ···965297 | ···887172 | ···833717 | ···805054 |  |  |
| 460 | 982063494 | ···985861 | ···932900 | ···904735 |  |  |
| 470 | ···161700 | 987084559 | 992032093 | 997004425 |  |  |
| 480 | ···259916 | ···183268 | ···131296 | ···104126 |  |  |
| 490 | ···358142 | ···281986 | ···230510 | ···203836 |  |  |
| 500 | ···456378 | ···380714 | ···329732 | ···303557 |  |  |

ALso enden sich die zwo Summen Zalen in 9. Zyphern/ vñ ist die Rote

230270022 —
230270023 +

Die Schwartze aber ist gantz mit 9. nollen als 1000000000 vnd so dieselben gantzen Zalen/nicht gnug geben mögen / so mag man dieselben 2.3.4.5.6.7.8.9. zusammen addieren.

**Figure C.58**

# References


Boyer CB (1991) A history of mathematics, 2nd edn. Wiley, New York

Bramer B (1648) Bericht zu M. Jobsten Burgi seligen Geometrischen Triangular Instruments mit schönen Kupfferstücken hierzu geschnitten. Schütz, Cassel

Buchheim A (1877) Correspondence to the editor of the City of London School Magazine. The City of London School Magazine 1(3):69–70, March 1877

Bürgi J (1620) Aritmet*ische und* Geometri*sche* Progreß *Tabulen/Sambt gründlichem unterricht/ wie solche nützlich in allerley Rechnungen zu gebrauchen/und verstanden werden sol*. Paul Sessen, Prague

Cajori F (1915) Algebra in Napier's day and alleged prior inventions of logarithms. In: Knott CG (ed) Napier tercentenary memorial volume. Longmans, Green and Company, London, pp 93–109

Calinger RS (1999) A contextual history of mathematics. Prentice Hall, Englewood Cliffs

Cantor M (1900) *Vorlesungen über Geschichte der Mathematik, Zweiter Band, von 1200-1668, Zweite Auflage*. BG Teubner, Leipzig

Clark KM (2011) 'In these numbers we use no fractions': a classroom module on Stevin's decimal fractions *Loci*. doi:10.4169/loci003333

Faustmann G (1997) Jost Bürgis Progress Tabulen und die Entwicklung der Logarithmen. Acta Historiae Rerum Naturalium, New Series vol I. Prague

Fauvel J (1995) Revisiting the history of logarithms. In: Swetz F, Fauvel J, Johansson B, Katz V, Bekken O (eds) Learn from the masters. The Mathematical Association of America, Washington, DC, pp 39–48

Folkerts M (2015) Eine bisher unbekannte Schrift von Jost Bürgi zur Trigonometrie. In: Gebhardt R (ed) Arithmetik, Geometrie und Algebra der frühen Neuzeit. Adam-Ries-Bund, Annaberg-Buchholz, pp 107–114

Folkerts M, Launert D, Thom A (2015) Jost Bürgi's method for calculating sines. arXiv 1510.03180v1 [math.HO]

Folta J, Nový L (1968) Zu Bürgi's Anleitung zu den Logarithmentafeln. Acta Hist Rerum Nat Tech 4:97–126

Gaulke K (2015) Perfect in every sense: scientific iconography on an equation clock by Jost Bürgi and the self-understanding of the astronomers at the Kassel Court in the late 1580s. Nuncius 30:37–74

Gieswald HR (1856) Justus Byrg als Mathematiker und dessen Einleitung in seine Logarithmen. In: Bericht über die St. Johannis-Schule, vol 35, pp 1–36. Danzig

González-Velasco EA (2011) Journey through mathematics: creative episodes in its history. Springer, New York


© Springer Science+Business Media New York 2015

K. Clark, *Jost Bürgi's Aritmetische und Geometrische Progreß Tabulen (1620)*, Science Networks. Historical Studies 53, DOI 10.1007/978-1-4939-3161-3

Grattan-Guinness I (1997) The rainbow of mathematics: a history of the mathematical sciences. W. W. Norton and Company, New York

Gronau D (1996) The logarithms—from calculation to functional equations. Not South Afr Math Soc 28:60–66

Havil J (2014) John Napier: life, logarithms, and legacy. Princeton University Press, Princeton

Hutton C (1811) Mathematical tables; containing the common, hyperbolic, and logistic logarithms, also sines, tangents, secants, & versed sines, etc. FCJ Rivington, London

Jacob S (1565) Ein New und Wohlgegründt Rechenbuch. Feyerabend, Rab und Hütter, Nürnberg

Katz VJ (1998) A history of mathematics: an introduction. Addison-Wesley, Reading

Katz VJ (2009) A history of mathematic: an introduction. Pearson Education, Boston

Krabbe J (1609) Newes Astrolabium, sampt dessen Nutz und Gebrauch: nicht allein den Astronomis und Medicis, sondern allen Kriegs Officirern, Bawmeistern, Seefahrenden Schiffern und Bergleuten, item Schantz und Büchsenmeistern. Matthiam Beckern, Frankfurt

List M, Bialas V (1973) Nova Kepleriana: Die Coss von Jost Bürgi in der Redaktion von Johannes Kepler: Ein Beitrag zur frühen Algebra. Verlag der Bayerischen Akademie der Wissenschaften, München

Lutstorf HT (2005) Die Logarithmentafeln Jost Bürgis. Bemerkungen zur Stellenwert- und Basisfrage. Mit Kommentar zu Bürgis "Gründlichem Unterricht". ETH-Bibliothek, Zürich

Lutstorf HT, Walter M (1992) Jost Bürgi's "Progress Tabulen" (Logarithmen). ETH-Bibliothek, Zürich

Montucla JE (1758) Histoire des Mathématiques, Tome Second. Henri Agasse, Paris

Naux C (1966) Histoire des logarithmes de Neper a Euler (Tome 1). Librairie Scientifique et Technique A. Blanchard, Paris

Nový L (1970) Jost Bürgi. In: Gillispie CC (ed) The dictionary of scientific biography, vol 2. Charles Scribner and Sons, New York, pp 602–603

Oechslin L (2001) Jost Bürgi. Verlag Ineichen, Zürich

Pesic P (2010) Hearing the irrational: music and the development of the modern concept of number. Isis 101(3):501–530

Roegel D (2010a) Bürgi's "Progress Tabulen" (1620): logarithmic tables without logarithms. http://locomat.loria.fr/buergi1620/buergi1620doc.pdf. Accessed 15 Mar 2013

Roegel D (2010b) Napier's ideal construction of the logarithms [research report]. https://hal.inria.fr/inria-00543934. Accessed 7 Jan 2015

Sampson RA (1915) The discovery of logarithms by Jost Bürgi. In: Knott CG (ed) Napier tercentenary memorial volume. Longmans, Green and Company, London, pp 208–212

Sesiano J (1986) Dasypodius, Konrad. Nouveau dictionnaire de biographie alsacienne, 7

Smith DE (1958) History of mathematics. Dover, New York

Staudacher F (2014) Jost Bürgi, Johannes Kepler und der Kaiser. Verlag NZZ Libro, Zürich

Stedall J (2008) Mathematics emerging: a sourcebook 1540-1900. Oxford University Press, New York

The City of London School Magazine (1877) vol 1. E. Matthews & Sons, London, 1(3):69–70

Thoren VE (1988) Prosthaphaeresis revisited. Hist Math 15:32–39

Volk W (2009) Zeugnisse zu Mathematikern (Monuments on mathematicians). http://www.w--volk.de/museum/grave30.htm. Accessed 8 Aug 2013

Waldvogel J (2012) Jost Bürgi and the discovery of the logarithms. Seminar für Angewandte Mathematik. ETH-Zürich, Zürich

Waldvogel J (2014) Jost Bürgi and the discovery of the logarithms. Elem Math 69(3):89–117

Wallis J (1685) A treatise of algebra, both historical and practical: shewing the original, progress, and advancement thereof, from time to time, and by what steps it hath attained to the height at which now it is; with some additional treatises, I. Of the cono-cuneus; being a body representing in part a conus, in part a cuneus. II. Of angular sections; and other things relating thereunto, and to trigonometry. III. Of the angle of contact; with other things appertaining to the composition of magnitudes, the inceptives of magnitudes, and the composition of motions, with the results thereof. IV. Of combinations, alternations, and aliquot parts. John Playford for Richard Davis, London

Whiteside D (2014) And John Napier created logarithms…'. Br Soc Hist Math Bull 29(3):154–166

Wolf R (1858) Jost Bürgi von Lichtensteig. In: Biografien zur Kulturgeschichte der Schweiz. Orell Füssli, Zürich, pp 57–80

Wolf R (1872) Johannes Keppler und Jost Bürgi. Vortrag gehalten den 4. Januar 1872 auf dem Rathaus in Zürich. Friedrich Schulthess, Zürich

Zons M (1616) Ein new wolgegründetes Kunst- und artigs Rechenbuch. Schmitz, Köln

# Index of Names



© Springer Science+Business Media New York 2015
K. Clark, *Jost Bürgi's Aritmetische und Geometrische Progreß Tabulen (1620)*, Science Networks. Historical Studies 53, DOI 10.1007/978-1-4939-3161-3

# Index of Places



© Springer Science+Business Media New York 2015
K. Clark, *Jost Bürgi's Aritmetische und Geometrische Progreß Tabulen (1620)*,
Science Networks. Historical Studies 53, DOI 10.1007/978-1-4939-3161-3

# Subject Index



© Springer Science+Business Media New York 2015
K. Clark, *Jost Bürgi's Aritmetische und Geometrische Progreß Tabulen (1620)*,
Science Networks. Historical Studies 53, DOI 10.1007/978-1-4939-3161-3